Aptamers

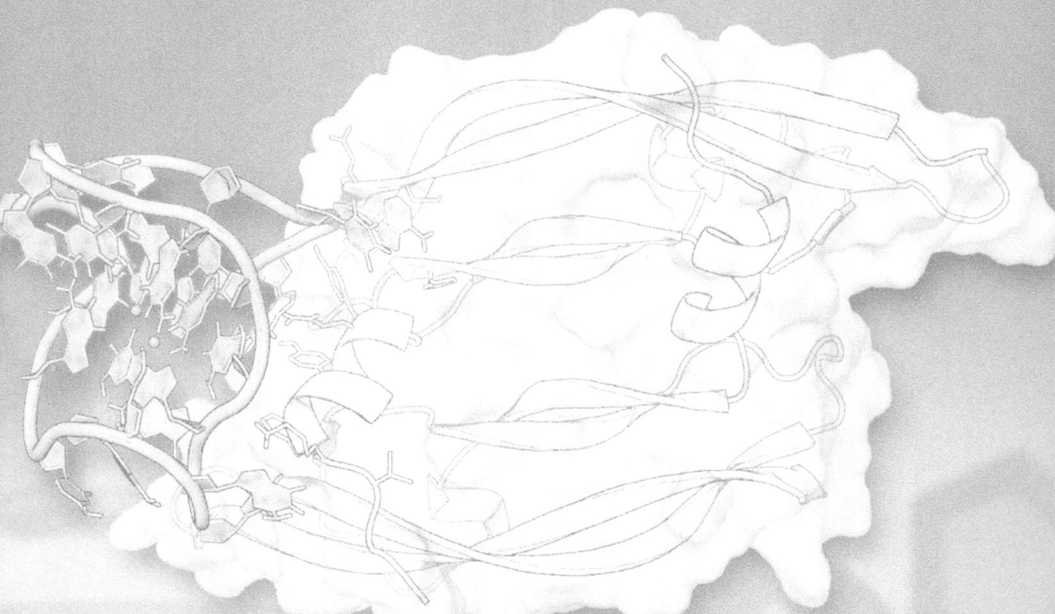

Aptamers

Tools for Nanotherapy and Molecular Imaging

edited by

Rakesh N. Veedu

CRC Press
Taylor & Francis Group
Boca Raton London New York

CRC Press is an imprint of the
Taylor & Francis Group, an **informa** business

Aptamers: Tools for Nanotherapy and Molecular Imaging

First published 2016 by Pan Stanford Publishing Pte. Ltd.

Published 2022 by CRC Press
Taylor & Francis Group
6000 Broken Sound Parkway NW, Suite 300
Boca Raton, FL 33487-2742

ISBN-13: 978-981-4669-83-2 (hbk)
ISBN-13: 978-981-4669-84-9 (ebk)

Visit the Taylor & Francis Web site at
http://www.taylorandfrancis.com

and the CRC Press Web site at
http://www.crcpress.com

British Library Cataloguing-in-Publication Data
A catalogue record for this book is available from the British Library.

Contents

Preface

Target-specific drug delivery and therapy still remains one of the holy grails of the drug development community. Of the many approaches used to date, antibody-based techniques are the most widely used, and this technology has matured over several decades. In the early 1990s a new class of targeting compounds emerged with very high target specificity and affinity, as an alternative to antibodies for targeted drug delivery and therapy. These compounds, aptamers (often referred to as "chemical antibodies"), are short, single-stranded functional nucleic acids that can fold into complex 3D shapes in solution for high-affinity recognition of defined molecular targets ranging from small molecules to large proteins and even whole cells. Typically, aptamers are developed through an in vitro reiterative combinatorial selection process called **s**ystematic **e**volution of **l**igands by **ex**ponential (SELEX) enrichment starting with a large pool of oligonucleotide sequences.

Since their discovery, nucleic acid aptamers have attracted considerable attention across various fields of medicine as a platform technology for targeted therapeutic development for a broad range of disease conditions as well as for molecular imaging and diagnosis. Aptamers possess certain qualities that potentially give them an edge over antibody-based approaches. In particular, they do not require live animals for production and can be synthesised on a large scale. Moreover, they possess high stability with long shelf lives, they can be chemically modified, and their functions can be neutralized using an antidote sequence. In 2004, the first aptamer drug targeting the vascular endothelial growth factor protein, Mucagen (Pegaptanib), was approved for clinical use by the US Food and Drug Administration (FDA) for the treatment of age-related macular degeneration (AMD). These developments have been instrumental in reinforcing the potential of aptamers as a clinical tool and have led to a plethora of publications and patents exploring the use of aptamer technology in various fields. The primary purpose of this book is to review the history, development, and applications of aptamers and discuss their potential as a transformational technology for target-specific therapeutics and diagnostics.

This book covers all aspects of aptamers across 13 chapters. Chapters 1–4 provide a general introduction, while Chapters 5–7 describe the development of therapeutic aptamer candidates for cancer, neurological diseases, and infectious and parasitic diseases. Chapters 8 and 9 provide an overview of aptamers as tools for targeted drug delivery and nanotherapy. The application of aptamers as biosensors for medical diagnosis is described in Chapter 10, and Chapter 11 highlights the use of aptamers as a platform technology for target-specific molecular imaging. Finally, Chapters 12 and 13 are devoted to an in-depth analysis of the clinical development of aptamers and their scope, limitations, and future prospects. As the editor of this book, with many years of experience in the aptamer field, I remain convinced that aptamers have tremendous potential as a tool for delivering targeted nanotherapies and in molecular imaging. I feel immensely proud to have had the opportunity to edit this book and wish to thank the book's many distinguished contributors. I hope that the book will encourage scientists and clinicians from a variety of fields to consider aptamer-based technologies as a tool in their own research.

Rakesh N. Veedu

Chapter 1

History of Aptamer Development

Nasa Savory, Koichi Abe, Taiki Saito, and Kazunori Ikebukuro
Department of Biotechnology and Life Science, Tokyo University of Agriculture and Technology, 2-24-16 Naka-cho, Koganei, Tokyo 184-8588, Japan
ikebu@cc.tuat.ac.jp

1.1 Introduction

Aptamers are ribo- and deoxyribo-oligonucleotide ligands, first described in 1990 by Tuerk and Gold [1] and Ellington and Szostak [2]. The name "aptamer" comes from the Greek *aptus*, meaning "to fit," and *merus*, meaning "particle." Aptamers adopt complex three-dimensional structures capable of recognizing target molecules with high affinity and specificity comparable with those of antibodies. Over the past two decades, aptamers have been developed for a variety of target molecules such as metal ions, small molecules, peptides, proteins, microorganisms, cells, and tissues. Because aptamers are chemically produced, there is little or no batch-to-batch variation and aptamers can be easily modified with functional groups and molecular probes. Furthermore, complex aptamer structures can be reversibly denatured at high temperatures. Their structures and structural changes occurring upon target binding can

Aptamers: Tools for Nanotherapy and Molecular Imaging
Edited by Rakesh N. Veedu

ISBN 978-981-4669-83-2 (Hardcover), 978-981-4669-84-9 (eBook)
www.panstanford.com

be designed on the basis of Watson–Crick base pairing. Aptamers also elicit little or no immunogenicity in therapeutic applications. Owing to those characteristics, aptamers have rapidly become alternatives to antibodies [3] in various applications including but not limited to therapeutics, drug delivery, molecular imaging, clinical diagnostics, biomarker discovery, food safety, and environmental monitoring. As the first therapeutic aptamer, pegaptanib (Macugen) was approved in 2004 by the US Food and Drug Administration for the treatment of neovascular age-related macular degeneration. Some aptamers are commercially used in analytical and medical applications, and more than 10 aptamers are in clinical trials [4].

Systematic evolution of ligands by exponential enrichment (SELEX) has been the most common method for identifying aptamers since the first reports in 1990 [1, 2]. The SELEX procedure comprises repetitive cycles of sequential steps. The process is initiated by binding target molecules with a synthesized random RNA/DNA library containing more than 10^{14} different sequences. Oligonucleotides bound to target molecules are separated to yield a population displaying high affinity toward target molecules. The target-bound oligonucleotides are then amplified by polymerase chain reaction (PCR) to prepare a single-stranded oligonucleotide library for the next cycle of sequential steps. This selection process is usually continued for 5–20 rounds with increased selection stringency. SELEX with many methodological improvements has been used for the development of aptamers for various target molecules.

Although almost all aptamers are selected through SELEX, there are some drawbacks in the SELEX process, and aptamers selected by this method often exhibit a lack of affinity and/or specificity. For example, by-product formation in the PCR amplification step of SELEX is a typical problem in the process [5]. Separation of a binding population from the oligonucleotide library is also a critical point for obtaining aptamers. To overcome these problems and to improve the efficiency of aptamer development, many methodological improvements have been made in the SELEX process.

Post-SELEX strategies to improve and fine-tune aptamer functions have also been established. In vitro recombination and reselection from partially randomized aptamers have been shown to achieve improvement of binding properties for specific targets. Post-SELEX modification and engineering of aptamers have been attempted to improve aptamer functions. Artificial nucleotides and unnatural bases have been widely introduced to increase resistance to nucleases and pharmacokinetics of aptamers and to improve target-binding ability. In addition, in silico selection and sequence optimization have been used to optimize aptamer sequences and structures to achieve desired improvements in aptamer functions. These various post-SELEX strategies for improving aptamer function have led to the development of promising aptamers and are opening new opportunities for the application of aptamers.

In this chapter, we describe SELEX strategies with methodological improvements, well established since the first application in aptamer development. We further present a brief description of post-SELEX improvements of aptamers that enable the development of aptamers with desired binding properties and functions.

1.2 In vitro Selection

Since the first aptamer selections were reported, most aptamers have been selected by SELEX. Many researchers have also engaged in improvement of the SELEX methodology to efficiently select aptamers, as comprehensively reviewed by Darmostuk et al. [6]. SELEX consists of four steps: incubation of an oligonucleotide library with target proteins, partitioning of oligonucleotides bound to target molecules, amplification of eluted oligonucleotides by PCR, and preparation of single-stranded DNA/RNA for subsequent cycles (Fig. 1.1). The simplicity of these cycles has allowed their automation, potentially reducing the time and cost of aptamer development [7]. Among the SELEX processes, separation of bound from unbound oligonucleotides is one of the most important steps, given that efficient fractionation allows the reduction of the SELEX cycle number, shortening the development time of aptamers and saving costs. Moreover, efficient fractionation enables the reduction of PCR amplification steps, which sometimes leads to the loss

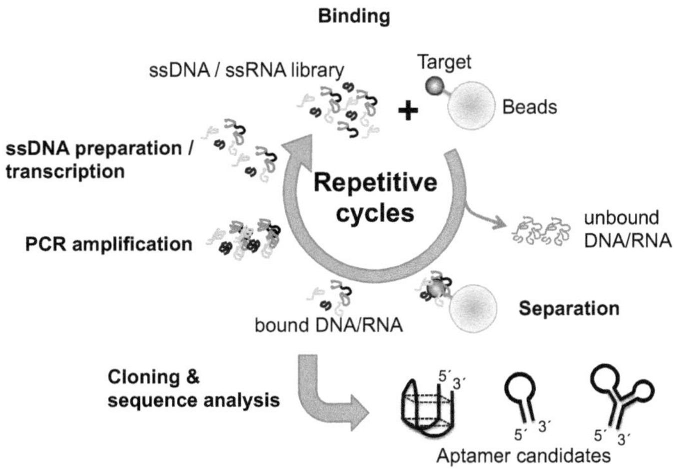

Figure 1.1 The basic method of conventional SELEX for aptamer selection. The selection step starts with binding of an initial single-stranded (ss) DNA or RNA library to the target, which is usually immobilized on a solid matrix such as magnetic beads. After removal of unbound or weakly bound oligonucleotides by washing steps, the target-bound oligonucleotides are collected. These oligonucleotides are then amplified by PCR (reverse transcription into DNA is required when selection is of RNA aptamers) and ssDNAs/ssRNAs are prepared for the next selection round. After several rounds of selection, the oligonucleotide pool is sequenced to identify enriched aptamer candidates for evaluation of binding affinity and specificity to the target.

of high-affinity aptamers and the amplification of unexpected oligonucleotides [8, 9].

Membranes and beads are commonly used to immobilize target proteins for fractionations. They can readily separate oligonucleotides bound to target molecules from other oligonucleotides. However, we need to consider the possibility of the enrichment of oligonucleotides binding to these supports. Capillary electrophoresis (CE) is a useful strategy for removing unbound oligonucleotides without supports [10]. Because CE-SELEX can efficiently partition oligonucleotides binding to target molecules, Berezovski proposed selection of aptamers without amplification steps, a process named non-SELEX [11]. Other researchers have also reported efficient fractionations using a microfluidic system combined with magnetic beads (reviewed in Ref. [12]) or flow cytometry combined with emulsion PCR [13].

1.2.1 *Library for SELEX*

Diversity of the oligonucleotide library is another important factor for identification of aptamers [14]. Because in vitro selection of aptamers is analogous to random search in a sequence–function space, the diversity of the library is representative of the search space. We can typically use 1–10 nmol of oligonucleotides as the initial library equivalent of 10^{14}–10^{15} molecules, which theoretically covers the diversity of 24–26-mer randomized oligonucleotides. Although the whole sequence space cannot be searched if we use longer oligonucleotides, longer sequences are sometimes favored to be used for selection of aptamers, because a library consisting of longer sequences can contain more complex functional modules and several simple functional motifs in a sequence. However, we need to handle longer libraries carefully because they may easily form multimers via intermolecular hybridization.

The sequences flanking constant regions of a library, which hybridize with primers during PCR process in SELEX, are also important for successful aptamer selection. We need to select constant regions to amplify selected DNA efficiently with limited nonspecific amplification. In addition, we need to consider the interference of hybridization between constant and random regions, resulting in decreased diversity of functional sequence motifs. Hybridization of antisense oligonucleotides with single-stranded DNA in a library helps in reducing interference, and it is very useful for constructing truncated mutants of aptamers [15–17]. Primerless SELEX has also been reported [18]. However, many researchers still use a library harboring a primer hybridization region. Hybridization of antisense DNA to the defined region helps minimize the effect of a defined flanking region on the folding of a random region and also in truncating the defined sequence effectively [16].

Parallel-type sequencers, as represented by Illumina instruments, have caused a revolution in biology. We can sequence not only whole genomes in a day but also whole transcriptomes, including unidentified transcripts, which is impossible by microarray analysis.

They have also caused a revolution in aptamer selection. In conventional SELEX, we analyze some 100–200 sequences by Sanger sequencing after cloning into an appropriate plasmid vector. To discriminate enriched sequences from others, we should perform many SELEX cycles to enrich sequences to more than 10% of the population. However, because parallel-type sequencers permit the analysis of several billion sequences in a single run, we can easily identify enriched sequences accounting for less than 1% of the population after a few SELEX cycles [19]. In addition, we can determine the enrichment of individual sequences by sequencing and comparing the library sequences obtained at each round of SELEX in a single run by addition of barcode sequences. In this manner, we can identify promising aptamers on the basis of not only read counts but also transitions of read counts through SELEX cycles. Various bioinformatics analyses help in the identification of aptamers in large sequence datasets [20, 21].

1.2.2 *Target Applicability for SELEX*

The applicability of SELEX has expanded to various targets. Given that SELEX does not involve living organisms, SELEX can be employed for aptamer selection for any kind of target, including toxic molecules and proteins forming special conformations. In cell-SELEX, we can directly use cells as a target for selection of aptamers against membrane proteins with native structures under physiological conditions. In addition, we can identify biomarkers by selection of aptamers following biomarker enrichment by identified aptamers and analysis of biomarkers by mass spectrometry [22, 23]. Cell-SELEX can also achieve selection of aptamers internalized into cells by receptor-dependent endocytosis upon binding to their target receptors [24, 25].

While cell-SELEX can target only membrane-associated proteins, aptamers against any unpurified proteins in a crude sample or tissue samples can be obtained by SELEX combined with polyacrylamide gel electrophoresis (PAGE) and electroblotting. Aptamer selection against protein targets separated by native PAGE, sodium dodecyl sulfate (SDS)-PAGE, or two-dimensional electrophoresis from crude samples followed by electroblotting onto a nitrocellulose or polyvinylidene difluoride (PVDF) membrane has been demonstrated; these methods offer potential as powerful tools for novel biomarker discovery [26–28]. Oligonucleotides binding to target proteins on a membrane can be extracted by excision of the membrane region, immobilizing the target protein. By comparing PAGE patterns between samples from normal and abnormal cells or tissues, we can simultaneously identify biomarker candidates and aptamers against the candidate.

Aptamer selection can also target a slight difference between target and nontarget proteins, such as methylation of a protein [29]. We have also reported aptamers recognizing special conformations of the target protein. α-Synuclein and amyloid-β are considered to be causative agents for neurodegenerative diseases and can form soluble amyloidogenic oligomer and amyloid fibrils. Because soluble oligomers show higher cytotoxicity than monomeric protein and amyloid fibrils, soluble oligomers in blood or cerebrospinal fluid are expected to be biomarkers for neurodegenerative disease diagnosis [30]. We have reported aptamers specifically recognizing soluble oligomers [16]. Although we performed SELEX against a soluble oligomer of α-synuclein, aptamers also recognized soluble oligomers of amyloid-β. This characteristic is similar to that of an antisoluble oligomer antibody A11 [31]. A11 is a very attractive antibody for specific detection of soluble oligomers, but our reported aptamers showed higher specificity than A11.

1.2.3 *Expansion of Functional Space in SELEX*

Introduction of unnatural nucleotides is another option for further improving aptamer performance. One advantage of unnatural nucleotides is better resistance to degradation in cells. DNA and RNA are very unstable in blood and in cells, with half-lives measured in minutes. To use aptamers in therapeutic medicine, we need to modify aptamers to avoid degradation. Pegaptanib is modified at the 2′ OH of all nucleotides except for two adenines by fluorination for pyrimidines and modification with a methoxy group for purines [32]. Given that the 2′ OH can act as a hydrogen bond donor and acceptor, its modification may result in loss of function by changing the conformation of the aptamer. Ruckman et al. used a 2′ F-pyrimidine in a SELEX procedure, whereas the 2′ OH of the purines was modified after SELEX [32]. To select aptamers with modified nucleotides, we need to establish a method of introducing unnatural nucleotides into SELEX processes. Several modified nucleotides are recognized and replicated by DNA polymerases [33–35], RNA polymerases [36], and a reverse transcriptase [37] with various degrees of efficiency and fidelity. Gold et al. used uracil modified by amino acids as a functional group to introduce hydrophobic groups into aptamers [38]. Engineering of polymerase has also been reported [39, 40]. Pinheiro et al. succeeded in engineering DNA polymerase to couple six xeno-nucleic acids (XNAs) and in the screening of aptamers using one of the XNAs or hexitol nucleic acids (HNAs) [41].

Unnatural nucleic acids are very useful for addition of novel functions to aptamers consisting of natural DNA or RNA. However, these modifications are limited to the sugar or phosphate moieties, which are not associated with sequence diversity or replacement of a natural with an unnatural nucleotide. Hirao and coworkers and Sefa et al. have recently reported expansion of the genetic code by introduction of a new base pair that is orthogonal to others and can be replicated by DNA polymerase (Fig. 1.2) [42, 43]. This technology

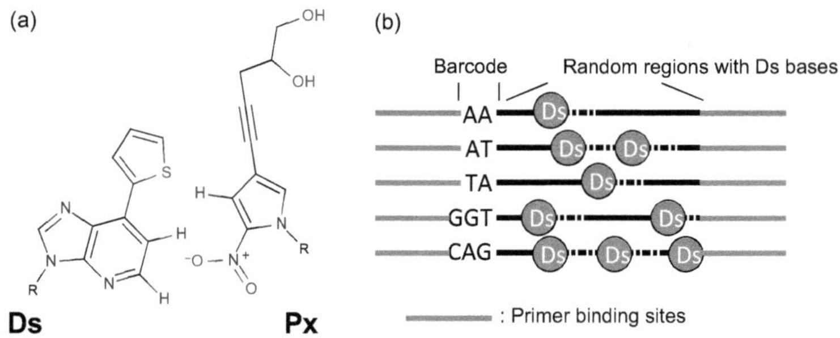

Figure 1.2 Hydrophobic bases that potentially expand the diversity of aptamers (a) Chemical structure of a 7-(2-thienyl)imidazo[4,5-b]pyridine (Ds) base, which enhances the hydrophobicity of aptamers and a diol-modified 2-nitro-4-propynylpyrrole (Px), which pairs exclusively with Ds during DNA replication [42]. (b) Aptamers containing Ds bases are selected by the SELEX procedure using a DNA library containing oligonucleotides with a central random region with Ds bases at predetermined positions, flanked by constant primer-binding regions and a 2–3-natural-base barcode sequence. Modified according to Kimoto et al. (2013) [42].

can potentially expand the diversity of aptamers from 4^n to 6^n, where n is the length of the aptamers.

1.3 Aptamer Improvement

1.3.1 *Artificial Nucleotides*

As described before, aptamers have very attractive advantages over antibodies; however, there are few examples of their commercial application. In addition to an obstacle posed by a strong patent filed in the Untied States (US5270163), which expired in 2010, the target limitation of aptamers is a serious problem. Gold et al. stated that "there were also examples of difficult protein targets for which standard RNA and DNA SELEX did not yield high affinity aptamers" [38]. This limitation may be because of the limited chemical diversity compared to antibodies consisting of 20 amino acids, including cationic, anionic, hydrophilic, hydrophobic, and aromatic amino acids. In particular, hydrophobic interaction is limited in natural nucleic acids. One of the solutions is introduction of unnatural nucleotides into aptamers. Gold et al. used uracil modified by a hydrophobic amino acid as a functional group replacing thymine that could be efficiently replicated by DNA polymerase [38] (Fig. 1.3). They compared selections with modified and unmodified nucleotides targeting 13 "difficult" human proteins that had repeatedly failed SELEX with unmodified DNA. Their overall success rate (pool K_d <30 nM) was 84% (1204 of 1428) for high-quality aptamers to a wide range of human proteins. They called the selected aptamers slow off-rate-modified aptamers because these showed lower dissociation rates than conventional aptamers.

Other approaches to improving aptamer affinity have also been attempted. Because aptamers have many negative charges derived from their phosphate backbone, the conformation of some aptamers would be unstable in the free state. Upon binding to target molecules, their structure is stabilized. This feature is an advantage for constructing novel biosensors, because conformational changes of aptamers upon binding to target molecules can be readily transduced to measurable signals (as reviewed in Ref.

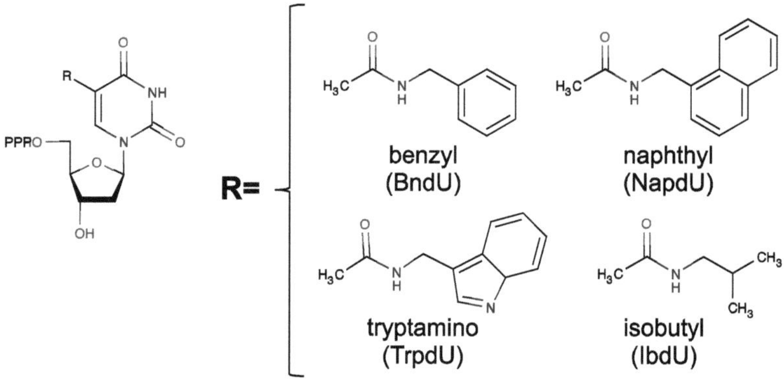

Figure 1.3 Chemical structures of uridine (dUTP) modified with a hydrophobic amino acid as a functional group in place of thymine that can be efficiently replicated by DNA polymerase. Modified according to Gold et al. (2010) [38].

[44]). However, conformational flexibility results in decrease of entropy upon binding to the target molecules, indicating that aptamers would show lower affinity than aptamers with high-stability conformations. Modification of the sugar moiety has commonly been introduced to enhance RNA stability; it may also be introduced to enhance affinity. Locked nucleic acids (LNAs), which have 2′-O,4′-C methylene-linked bicyclic ribonucleosides, reduce the conformational flexibility of the sugar moiety, resulting in enhancing hybridization with DNA or RNA because of the entropic constraint. LNA modification of nucleotides in stem regions of selected aptamers by SELEX increased their affinity and also in vivo stability by stabilization of the stem in the aptamer [45, 46].

Aptamers forming guanine quadruplex (G4) structures are frequently identified against various target proteins [47]. However, some guanine-rich sequences can fold into various types of structures. Because one specific structure of an aptamer may show binding to a target protein, the conformation of the aptamer should be regulated. Nuclear magnetic resonance (NMR) analysis revealed that our reported vascular endothelial growth factor (VEGF)-binding aptamer would fold several types of G4 structures, and the VEGF aptamer was engineered to fold into a specific structure by introduction of LNAs or unlocked nucleic acids [48]. LNAs can also be introduced via SELEX processes (reviewed in Ref. [49]).

1.3.2 *Multimerization Strategies*

The strength of interactions in which multiple binding sites simultaneously interact with the target is known as avidity, particularly for antibody–antigen interactions. In aptamers, multimerization of single or multiple aptamers often offers dramatically higher affinity and specificity than monovalent binding. We have shown that the dimerization of two thrombin-binding aptamers (TBAs) that interact with different target sites of thrombin increased affinity for thrombin [50]. The dimerized thrombin aptamer was designed by tandem attachment of a 15-nt TBA (TBA15) [51] and a 29-nt TBA (TBA29) [52] through 0-, 5-, 10-, or 20-nt thymine linkers. The dimerization led to decreased dissociation constant (k_{off}) values; thus, it led to improvement of the dissociation constant (K_D) for thrombin binding. In addition, the dimerization study of TBAs indicated that the lengths of linker oligonucleotides are important for the dimerized aptamer's inhibitory activity toward fibrinogen degradation by thrombin. A dimerized VEGF-binding aptamer was also developed, resulting in improvement of affinity for VEGF [50].

We also developed a trivalent aptamer for VEGF that improved affinity. In a three-way junction-type scaffold, we trimerized a specific stem-loop structure that is a binding site for VEGF, a homodimeric protein [91]. Surface plasmon resonance (SPR) analysis showed that the trivalent aptamer had a markedly improved k_{on}. Two regions of binding sites in the trivalent aptamer were linked through a stem-loop region, an observation consistent with the length of the target subunits in VEGF and suggesting that the linker length is critical for design of multivalent aptamers. In addition to the distance between binding sites of the aptamer, flexibility is another critical factor in designing a multivalent aptamer for the enhancement of avidity [53].

As a linker for aptamer dimerization, Kim et al. investigated the influence of the length of a phosphoramidite linker on its inhibitory activity [54]. They showed that a linker as long as 16 nm (eight spacers) exhibited the highest inhibitory activity for the thrombin, which is 3–4 nm in diameter. These findings suggested that some extra linker length is required for aptamers to wrap around target proteins and assume the optimum three-dimensional position and orientation for aptamer–target interactions. Polyethylene glycol (PEG) is also commonly used for aptamer multimerization as a linker. Tian et al. showed a marked improvement of K_D by dimerization of TBA15 and TBA29 through 12 and 24 nm PEG linkers [55].

For optimization of linkers for dimerized aptamer design, SELEX approaches instead of rational design were used to determine the scaffold of the dimerized aptamer with appropriate positions and orientations of binding sites. Ahmad et al. performed five rounds of selection from random 35-nt DNA sequences, which connected TBA15 and TBA29 with primer sites [56]. After five rounds of a microfluidic SELEX process, a bivalent aptamer with high affinity for thrombin with an apparent K_D of <10 pM was isolated, representing a 200-fold improvement in binding affinity over monomeric aptamers. It also exhibited a 15-fold improvement over the best rationally designed bivalent construct. Wilson et al. also demonstrated one-round selection of a linker sequence for a dimerized TBA15–TBA29 aptamer from a 30-nt random region, resulting in identification of a consensus sequence and structural motifs for thrombin binding [57].

Although these dimerization or multimerization strategies have proven to be a powerful approach to achieving high-affinity aptamers, the discovery of aptamer pairs that can recognize distinct epitopes on target proteins is often challenging. Although we require only single aptamers to design multivalent aptamers for targeting a homodimer or multimer protein, such as VEGF, we need a pair of aptamers that bind to distinct epitopes on a target protein when targeting a monomer or heteromultimer protein. In addition, SELEX preferentially yields aptamers that interact with a dominant epitope such as the positively charged surface of a target protein. To address these issues, Cho et al. recently reported an array-based discovery platform for multivalent aptamers (AD-MAPs) [58]. In this approach, a custom DNA array is made from a pool of aptamer candidates that are pre-enriched by microfluidic selection and high-throughput sequencing. This array is used to identify the highest-affinity aptamer and then the aptamer–target complex is subjected to identify potential bidentate pairs that simultaneously bind to the target with the highest-affinity aptamer. Aptamer pairs for fluorescently labelled Ang2 protein were obtained using the AD-MAP approach. Such strategies to identify aptamer pairs that recognize distinct epitopes on target proteins are promising approaches for the development of high-affinity multivalent aptamers.

1.3.3 *Genetic Algorithms for Aptamer Improvement*

Sequence refinement and optimization are usually required to achieve desired binding properties of aptamers after SELEX selection. A comprehensive mutagenesis study of IgE-binding aptamers using a microarray technology revealed that point mutations readily decrease the binding ability of aptamers [59]. In addition, sequence–fitness landscape modeling revealed that the sequence–function space of oligonucleotides is a rugged landscape of local optima and minima [60]. To identify optimized aptamer sequences in such a rugged sequence space, genetic algorithms (GAs) have been used for aptamer development. A GA represents an intelligent exploitation of a random search within a defined search space to solve a complex problem.

GAs have been employed for development of biomolecules, including peptides [61, 62] and aptamers [63]. GA-based methods for aptamer generation are called in silico maturation (ISM) [63, 64] and closed-loop aptameric directed evolution (CLADE) [65] (Fig. 1.4). These methods are based on iterative rounds of oligonucleotide synthesis, in vitro functional screening, and in silico sequence recombination and mutation. Random sequences or selected aptamers can be used as a parent generation. These sequences are subjected to in vitro functional screening for analysis of target functions, including but not limiting to binding affinity, specificity, and inhibitory activity. Sequences with high function values are selected as the next parent sequences and subjected to in silico sequence generation. Recombination as well as point mutations are introduced to produce offspring sequences. Recombination has the purpose of preventing the population from

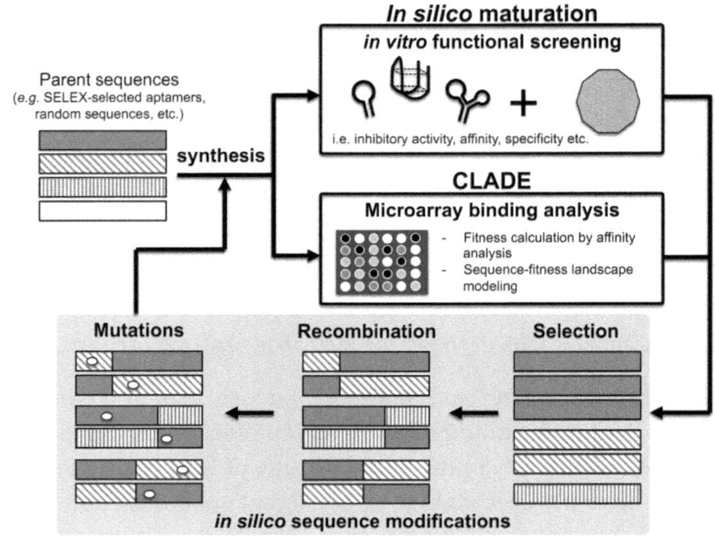

Figure 1.4 Current approaches for improvement of aptamers on the basis of genetic algorithms (GAs) known as in silico maturation (ISM) [63, 64] and closed-loop aptameric directed evolution (CLADE) [65]. GA-based approaches start from parent sequence selection based on prior knowledge. For ISM approaches, known aptamer sequences are usually used as parents because it increases maturation speed. In contrast, CLADE approaches use random sequences as parents, given that they use high-throughput binding analysis. These parent sequences are synthesized for in vitro evaluation. In CLADE, affinity is usually defined as fitness, which can be evaluated with a high-throughput microarray. Sequences showing higher function are then selected for producing offspring sequences through in silico sequence recombination and/or mutation. These steps are repeatedly performed to obtain aptamers showing improved function.

becoming stuck in local optima and enabling it to reach global optima. Offspring sequences are then synthesized for the next round of in vitro functional screening. Iterative cycles of these processes optimize aptamer sequences to achieve improvement of target function(s).

Bittker et al. demonstrated homogeneous recombination to improve aptamer affinity [66], but in vitro recombination is known to offer limited efficiency for recombination of short oligonucleotides. Because in silico sequence modification can be applied to generate any sequence with in vitro chemical synthesis, it would be a powerful approach for optimizing any type of aptamer sequence. The ISM strategy was first applied to optimize aptamer sequences into G4 structures to inhibit thrombin and *Taq* DNA polymerase activities [63, 67–69]. In addition, ISM was employed for refining loop sequences of a three-way junction-type aptamer against VEGF. The improved aptamer displayed high affinity for VEGF with a K_D of 52 nM and a subsequent multimerization strategy achieved a K_D of 0.37 nM [91]. These evolutionary strategies would be a powerful approach for optimizing aptamer sequences to fit a specific known structure motif required for aptamer function.

GA-based aptamer development is also capable of finding a particular sequence or structure for aptamer binding from diverse parent sequences. The sequence convergence of prostate-specific antigen (PSA)-binding DNA aptamers into a particular stem-loop structure from nonconserved parent sequences by ISM suggested that it could be used to find a structure motif required for aptamers with high affinity for PSA [70]. ISM was also employed to improve the specificity of aptamers against *Proteus mirabilis* [71] and *Streptococcus mutans* [72]. The specificity was evaluated as relative binding to the target and nontarget bacterial strains in vitro. Through 11 generations producing a total of more than 300 sequences of aptamers for *S. mutans*, identification of the core sequence that was required for aptamer binding facilitated the improvement of the aptamers. Such a functional sequence region is known as a schema, and identification of a schema is important for the success of a GA process. ISM showed that schema search is also critical for aptamer improvement. It is noteworthy that ISM can target any function of aptamers, including specificity and inhibitory activity for improvement with in vitro or even in vivo evaluation, whereas conventional SELEX approaches are limited to affinity-based selection.

Aptamer development using GAs was also demonstrated using high-throughput DNA microarrays [65, 73]. The CLADE approach that employed a GA process starting with an array of random sequences explored a sequence–fitness landscape of aptamers and developed aptamers with affinity for fluorescent [65] and biotinylated proteins [74, 75]. These high-throughput approaches using DNA microarrays and next-generation sequences would also be a powerful strategy for aptamer sequence optimization.

1.4 In silico Selection

In vitro selection and improvement have been extensively explored for aptamer development. In addition to in vitro strategies, in silico approaches have been developed for rational design of aptamers. For example, Chushak and Stone developed an in silico approach for designing a starting pool of RNA sequences to screen RNA aptamers for small-molecule ligands [76]. In their approach, a set of criteria was developed to design a starting pool with limited numbers of sequences with abundant simple structural motifs and maximum numbers of stable low-energy structures. This secondary structure and thermodynamic characterization were aimed at increasing the potential for selecting high-affinity aptamers. The Rosetta package [77] was then used to predict 3D structures of the selected sequences. Computational docking using the DOVIS package with AutoDock4 software was then performed to screen aptamers with high affinity to the target ligand. A 27-mer RNA library with 21-mer randomized sequences was used for designing a starting pool, and the RNA sequence search space could be reduced by 4 to 5 orders of magnitude, markedly accelerating the selection of aptamers. More recently, Hu et al. demonstrated a computational simulation of aptamer–protein interaction by the ZDOCK algorithm [78].

Computational approaches have also been developed to select aptamers from SELEX-derived sequences. Hoinka et al. developed Aptamotif, which uses an ensemble-based approach for identifying binding motifs in large SELEX datasets [79]. As a meta-motif-based statistical framework, MPBind was also developed to predict binding motifs of SELEX-derived aptamers [80]. MPBind was shown to be robust to both PCR amplification bias and incomplete sequencing of aptamer pools. For analyzing a large sequence set obtained by high-throughput sequencing, the open-source FASTAptamer toolkit was published [20]. The FASTAptamer toolkit compares populations for sequence distribution and generates clusters of sequence families from high-throughput sequence data to select potential aptamer candidates from a large sequence dataset and predict binding sequences of aptamers. These approaches hold potential not only to extract the binding motifs of aptamers but also to elucidate the evolution of aptamer sequences during SELEX.

We applied an in silico approach to the selection of aptamers from genomic information. For a variety of target molecules, aptamers folding into G4 structures have been selected by SELEX [16, 81–84]. The G4-forming sequence found in the promoter of the insulin gene was identified as an insulin-binding aptamer [85–87]. Moreover, one or more potential G4-forming oligonucleotide sequences have been found in over 40% of promoter regions of the genome by bioinformatics studies [88]. In vitro studies using a DNA micro array with a G4-specific fluorescent ligand revealed that a majority of these potential G4-forming sequences in the genome actually fold into G4 structures [89]. Accordingly, we developed a method for identifying aptamers from potential G4 sequences in the human genome [90]. The process is named G4 promoter-derived aptamer selection (G4PAS). We identified G4-forming aptamers for VEGF165, platelet-derived growth factor-AA, and RB1 by G4PAS. Bioinformatics aptamer selection from genome information is a simple and rapid method, and G4PAS could be applied to identify aptamers against target proteins that contain G4 DNAs in their promoters. ISM of those aptamers selected by G4PAS is currently underway in our laboratory. The combination of in silico selection and in silico improvement of aptamers would ultimately be a powerful strategy for aptamer development without in vitro screening.

1.5 Conclusion

Since the first report of aptamer and SELEX technology in 1990, several improved techniques for aptamer development have been reported. In addition to aptamers identified by SELEX approaches with technical improvements, aptamers have been engineered in vitro and in silico through both rational and screening strategies to achieve desirable functions. Artificial nucleotides and chemical modifications have been used as a successful approach to achieve high-affinity aptamers with suitable pharmacological properties. One of the examples of aptamers with commercial success is Macugen, an anti-angiogenic drug marketed by Pfizer. Aptamers are also recognized as promising ligands for diagnostics and analytical tools as well as therapeutic applications.

In silico approaches have enabled the successful development of aptamers. High-throughput sequencing in SELEX with computational data analysis has revealed the progress of in vitro evolution of aptamers. GAs have been successfully applied to aptamer development via ISM or CLADE. ISM has been shown to result in successful development of aptamers with high affinity or specificity compared with conventional SELEX selection, and CLADE has revealed the sequence–fitness landscape of aptamer sequences. Recently, successful examples of structure-based in silico aptamer selection have been reported. Many of these in silico approaches have great potential to facilitate aptamer development and in particular to identify optimized sequences with artificial nucleotides.

These improvements in aptamer development strategies have increased interest in aptamers for commercial applications. Currently more than 10 aptamer clinical studies are underway, and some aptamer-based research tools are already in the market. Further modification of aptamer development technologies would provide powerful aptamers for many applications in a variety of fields.

Acknowledgments

We would like to dedicate this chapter to celebrate the 20th anniversary of the Department of Biotechnology and Life Science at the Tokyo University of Agriculture and Technology.

References

1. Tuerk, C., and Gold, L. (1990). Systematic evolution of ligands by exponential enrichment: RNA ligands to bacteriophage T4 DNA polymerase, *Science,* **249**, pp. 505–510.

2. Ellington, A. D., and Szostak, J. W. (1990). In vitro selection of RNA molecules that bind specific ligands, *Nature,* **346**, pp. 818–822.
3. Jayasena, S. D. (1999). Aptamers: an emerging class of molecules that rival antibodies in diagnostics, *Clin. Chem.,* **45**, pp. 1628–1650.
4. Sundaram, P., Kurniawan, H., Byrne, M. E., and Wower, J. (2013). Therapeutic RNA aptamers in clinical trials, *Eur. J. Pharm. Sci.,* **48**, pp. 259–271.
5. Tolle, F., Wilke, J., Wengel, J., and Mayer, G. (2014). By-product formation in repetitive PCR amplification of DNA libraries during SELEX, *PLoS One,* **9**, p. e114693.
6. Darmostuk, M., Rimpelova, S., Gbelcova, H., and Ruml, T. (2015). Current approaches in SELEX: an update to aptamer selection technology, *Biotechnol. Adv.,* **33**, pp. 1141–1161.
7. Cox, J. C., Hayhurst, A., Hesselberth, J., Bayer, T. S., Georgiou, G., and Ellington, A. D. (2002). Automated selection of aptamers against protein targets translated in vitro: from gene to aptamer, *Nucleic Acids Res.,* **30**, p. e108.
8. Crameri, A., and Stemmer, W. P. (1993). 10(20)-fold aptamer library amplification without gel purification, *Nucleic Acids Res.,* **21**, p. 4410.
9. Kanagawa, T. (2003). Bias and artifacts in multitemplate polymerase chain reactions (PCR), *J. Biosci. Bioeng.,* **96**, pp. 317–323.
10. Mendonsa, S. D., and Bowser, M. T. (2004). In vitro evolution of functional DNA using capillary electrophoresis, *J. Am. Chem. Soc.,* **126**, pp. 20–21.
11. Berezovski, M., Musheev, M., Drabovich, A., and Krylov, S. N. (2006). Non-SELEX selection of aptamers, *J. Am. Chem. Soc.,* **128**, pp. 1410–1411.
12. Weng, C. H., Huang, C. J., and Lee, G. B. (2012). Screening of aptamers on microfluidic systems for clinical applications, *Sensors,* **12**, pp. 9514–9529.
13. Wang, J., Gong, Q., Maheshwari, N., Eisenstein, M., Arcila, M. L., Kosik, K. S., and Soh, H. T. (2014). Particle display: a quantitative screening method for generating high-affinity aptamers, *Angew. Chem.,* **53**, pp. 4796–4801.
14. Klug, S. J., and Famulok, M. (1994). All you wanted to know about SELEX, *Mol. Biol. Rep.,* **20**, pp. 97–107.
15. Shtatland, T., Gill, S. C., Javornik, B. E., Johansson, H. E., Singer, B. S., Uhlenbeck, O. C., Zichi, D. A., and Gold, L. (2000). Interactions of

Escherichia coli RNA with bacteriophage MS2 coat protein: genomic SELEX, *Nucleic Acids Res.,* **28**, p. E93.

16. Tsukakoshi, K., Abe, K., Sode, K., and Ikebukuro, K. (2012). Selection of DNA aptamers that recognize alpha-synuclein oligomers using a competitive screening method, *Anal. Chem.,* **84**, pp. 5542–5547.
17. Ouellet, E., Lagally, E. T., Cheung, K. C., and Haynes, C. A. (2014). A simple method for eliminating fixed-region interference of aptamer binding during SELEX, *Biotechnol. Bioeng.,* **111**, pp. 2265–2279.
18. Jarosch, F., Buchner, K., and Klussmann, S. (2006). In vitro selection using a dual RNA library that allows primerless selection, *Nucleic Acids Res.,* **34**, p. e86.
19. Cho, M., Soo Oh, S., Nie, J., Stewart, R., Eisenstein, M., Chambers, J., Marth, J. D., Walker, F., Thomson, J. A., and Soh, H. T. (2013). Quantitative selection and parallel characterization of aptamers, *Proc. Natl. Acad. Sci. U S A,* **110**, pp. 18460–18465.
20. Alam, K. K., Chang, J. L., and Burke, D. H. (2015). FASTAptamer: a bioinformatic toolkit for high-throughput sequence analysis of combinatorial selections, *Mol. Ther. Nucleic Acids*, **4**, p. e230.
21. Hoinka, J., Berezhnoy, A., Sauna, Z. E., Gilboa, E., and Przytycka, T. M. (2014). AptaCluster: a method to cluster HT-SELEX aptamer pools and lessons from its application, *Research in Computational Molecular Biology: Annual International Conference, RECOMB: Proceedings,* **8394**, pp. 115–128.
22. Berezovski, M. V., Lechmann, M., Musheev, M. U., Mak, T. W., and Krylov, S. N. (2008). Aptamer-facilitated biomarker discovery (AptaBiD), *J. Am. Chem. Soc.,* **130**, pp. 9137–9143.
23. Shangguan, D., Cao, Z., Meng, L., Mallikaratchy, P., Sefah, K., Wang, H., Li, Y., and Tan, W. (2008). Cell-specific aptamer probes for membrane protein elucidation in cancer cells, *J. Proteome. Res.,* **7**, pp. 2133–2139.
24. Yan, A., and Levy, M. (2014). Cell internalization SELEX: in vitro selection for molecules that internalize into cells, *Methods Mol. Biol.,* **1103**, pp. 241–265.
25. Thiel, W. H., Thiel, K. W., Flenker, K. S., Bair, T., Dupuy, A. J., McNamara, J. O., 2nd, Miller, F. J., and Giangrande, P. H. (2015). Cell-internalization SELEX: method for identifying cell-internalizing RNA aptamers for delivering siRNAs to target cells, *Methods Mol. Biol.,* **1218**, pp. 187–199.
26. Noma, T., Ikebukuro, K., Sode, K., Ohkubo, T., Sakasegawa, Y., Hachiya, N., and Kaneko, K. (2006). A screening method for DNA aptamers that bind

to a specific, unidentified protein in tissue samples, *Biotechnol. Lett.*, **28**, pp. 1377–1381.

27. Liu, Y., Wang, C., Li, F., Shen, S., Tyrrell, D. L., Le, X. C., and Li, X. F. (2012). DNase-mediated single-cycle selection of aptamers for proteins blotted on a membrane, *Anal. Chem.*, **84**, pp. 7603–7606.
28. Savory, N., Goto, S., Yoshida, W., Unuma, Y., Nakamura, M., Abe, K., Ferri, S., and Ikebukuro, K. (2013). Two-dimensional electrophoresis-based selection of aptamers against an unidentified protein in a tissue sample, *Anal. Lett.*, **46**, pp. 2954–2963.
29. Williams, B. A., Lin, L., Lindsay, S. M., and Chaput, J. C. (2009). Evolution of a histone H4-K16 acetyl-specific DNA aptamer, *J. Am. Chem. Soc.*, **131**, pp. 6330–6331.
30. Tokuda, T., Qureshi, M. M., Ardah, M. T., Varghese, S., Shehab, S. A., Kasai, T., Ishigami, N., Tamaoka, A., Nakagawa, M., and El-Agnaf, O. M. (2010). Detection of elevated levels of alpha-synuclein oligomers in CSF from patients with Parkinson disease, *Neurology*, **75**, pp. 1766–1772.
31. Kayed, R., Head, E., Thompson, J. L., McIntire, T. M., Milton, S. C., Cotman, C. W., and Glabe, C. G. (2003). Common structure of soluble amyloid oligomers implies common mechanism of pathogenesis, *Science*, **300**, pp. 486–489.
32. Ruckman, J., Green, L. S., Beeson, J., Waugh, S., Gillette, W. L., Henninger, D. D., Claesson-Welsh, L., and Janjic, N. (1998). 2'-Fluoropyrimidine RNA-based aptamers to the 165-amino acid form of vascular endothelial growth factor (VEGF165). Inhibition of receptor binding and VEGF-induced vascular permeability through interactions requiring the exon 7-encoded domain, *J. Biol. Chem.*, **273**, pp. 20556–20567.
33. Hutter, D., and Benner, S. A. (2003). Expanding the genetic alphabet: non-epimerizing nucleoside with the pyDDA hydrogen-bonding pattern, *J. Org. Chem.*, **68**, pp. 9839–9842.
34. Mitsui, T., Kimoto, M., Sato, A., Yokoyama, S., and Hirao, I. (2003). An unnatural hydrophobic base, 4-propynylpyrrole-2-carbaldehyde, as an efficient pairing partner of 9-methylimidazo[(4,5)-b]pyridine, *Bioorg. Med. Chem. Lett.*, **13**, pp. 4515–4518.
35. Roychowdhury, A., Illangkoon, H., Hendrickson, C. L., and Benner, S. A. (2004). 2'-deoxycytidines carrying amino and thiol functionality: synthesis and incorporation by Vent (exo-) polymerase, *Org. Lett.*, **6**, pp. 489–492.
36. Switzer, C. Y., Moroney, S. E., and Benner, S. A. (1993). Enzymatic recognition of the base pair between isocytidine and isoguanosine, *Biochemistry*, **32**, pp. 10489–10496.

37. Sismour, A. M., Lutz, S., Park, J. H., Lutz, M. J., Boyer, P. L., Hughes, S. H., and Benner, S. A. (2004). PCR amplification of DNA containing non-standard base pairs by variants of reverse transcriptase from human immunodeficiency virus-1, *Nucleic Acids Res.*, **32**, pp. 728–735.
38. Gold, L., Ayers, D., Bertino, J., Bock, C., Bock, A., Brody, E. N., Carter, J., Dalby, A. B., Eaton, B. E., Fitzwater, T., Flather, D., Forbes, A., Foreman, T., Fowler, C., Gawande, B., Goss, M., Gunn, M., Gupta, S., Halladay, D., Heil, J., Heilig, J., Hicke, B., Husar, G., Janjic, N., Jarvis, T., Jennings, S., Katilius, E., Keeney, T. R., Kim, N., Koch, T. H., Kraemer, S., Kroiss, L., Le, N., Levine, D., Lindsey, W., Lollo, B., Mayfield, W., Mehan, M., Mehler, R., Nelson, S. K., Nelson, M., Nieuwlandt, D., Nikrad, M., Ochsner, U., Ostroff, R. M., Otis, M., Parker, T., Pietrasiewicz, S., Resnicow, D. I., Rohloff, J., Sanders, G., Sattin, S., Schneider, D., Singer, B., Stanton, M., Sterkel, A., Stewart, A., Stratford, S., Vaught, J. D., Vrkljan, M., Walker, J. J., Watrobka, M., Waugh, S., Weiss, A., Wilcox, S. K., Wolfson, A., Wolk, S. K., Zhang, C., and Zichi, D. (2010). Aptamer-based multiplexed proteomic technology for biomarker discovery, *PLoS One,* **5**, p. e15004.
39. Burmeister, P. E., Lewis, S. D., Silva, R. F., Preiss, J. R., Horwitz, L. R., Pendergrast, P. S., McCauley, T. G., Kurz, J. C., Epstein, D. M., Wilson, C., and Keefe, A. D. (2005). Direct in vitro selection of a 2'-O-methyl aptamer to VEGF, *Chem. Biol.,* **12**, pp. 25–33.
40. Chelliserrykattil, J., and Ellington, A. D. (2004). Evolution of a T7 RNA polymerase variant that transcribes 2'-O-methyl RNA, *Nat. Biotechnol.,* **22**, pp. 1155–1160.
41. Pinheiro, V. B., Taylor, A. I., Cozens, C., Abramov, M., Renders, M., Zhang, S., Chaput, J. C., Wengel, J., Peak-Chew, S. Y., McLaughlin, S. H., Herdewijn, P., and Holliger, P. (2012). Synthetic genetic polymers capable of heredity and evolution, *Science,* **336**, pp. 341–344.
42. Kimoto, M., Yamashige, R., Matsunaga, K., Yokoyama, S., and Hirao, I. (2013). Generation of high-affinity DNA aptamers using an expanded genetic alphabet, *Nat. Biotechnol.,* **31**, pp. 453–457.
43. Sefah, K., Yang, Z., Bradley, K. M., Hoshika, S., Jimenez, E., Zhang, L., Zhu, G., Shanker, S., Yu, F., Turek, D., Tan, W., and Benner, S. A. (2014). In vitro selection with artificial expanded genetic information systems, *Proc. Natl. Acad. Sci. U S A,* **111**, pp. 1449–1454.
44. Li, D., Song, S., and Fan, C. (2010). Target-responsive structural switching for nucleic acid-based sensors, *Acc. Chem. Res.,* **43**, pp. 631–641.
45. Forster, C., Zydek, M., Rothkegel, M., Wu, Z., Gallin, C., Gessner, R., Lisdat, F., and Furste, J. P. (2012). Properties of an LNA-modified ricin RNA aptamer, *Biochem. Biophys. Res. Commun.,* **419**, pp. 60–65.

46. Schmidt, K. S., Borkowski, S., Kurreck, J., Stephens, A. W., Bald, R., Hecht, M., Friebe, M., Dinkelborg, L., and Erdmann, V. A. (2004). Application of locked nucleic acids to improve aptamer in vivo stability and targeting function, *Nucleic Acids Res.*, **32**, pp. 5757–5765.
47. Tucker, W. O., Shum, K. T., and Tanner, J. A. (2012). G-quadruplex DNA aptamers and their ligands: structure, function and application, *Curr. Pharmaceut. Design*, **18**, pp. 2014–2026.
48. Marusic, M., Veedu, R. N., Wengel, J., and Plavec, J. (2013). G-rich VEGF aptamer with locked and unlocked nucleic acid modifications exhibits a unique G-quadruplex fold, *Nucleic Acids Res.*, **41**, pp. 9524–9536.
49. Kuwahara, M., and Obika, S. (2013). In vitro selection of BNA (LNA) aptamers, *Artif. DNA PNA XNA*, **4**, pp. 39–48.
50. Hasegawa, H., Taira, K. I., Sode, K., and Ikebukuro, K. (2008). Improvement of aptamer affinity by dimerization, *Sensors*, **8**, pp. 1090–1098.
51. Bock, L. C., Griffin, L. C., Latham, J. A., Vermaas, E. H., and Toole, J. J. (1992). Selection of single-stranded-DNA molecules that bind and inhibit human thrombin, *Nature*, **355**, pp. 564–566.
52. Tasset, D. M., Kubik, M. F., and Steiner, W. (1997). Oligonucleotide inhibitors of human thrombin that bind distinct epitopes, *J. Mol. Biol.*, **272**, pp. 688–698.
53. Zhao, X., Lis, J. T., and Shi, H. (2013). A systematic study of the features critical for designing a high avidity multivalent aptamer, *Nucleic Acid Ther.*, **23**, pp. 238–242.
54. Kim, Y., Cao, Z., and Tan, W. (2008). Molecular assembly for high-performance bivalent nucleic acid inhibitor, *Proc. Natl. Acad. Sci. U S A*, **105**, pp. 5664–5669.
55. Tian, L., and Heyduk, T. (2009). Bivalent ligands with long nanometer-scale flexible linkers, *Biochemistry*, **48**, pp. 264–275.
56. Ahmad, K. M., Xiao, Y., and Soh, H. T. (2012). Selection is more intelligent than design: improving the affinity of a bivalent ligand through directed evolution, *Nucleic Acids Res.*, **40**, pp. 11777–11783.
57. Wilson, R., Bourne, C., Chaudhuri, R. R., Gregory, R., Kenny, J., and Cossins, A. (2014). Single-step selection of bivalent aptamers validated by comparison with SELEX using high-throughput sequencing, *PLoS One*, **9**, p. e100572.
58. Cho, M., Oh, S. S., Nie, J., Stewart, R., Radeke, M. J., Eisenstein, M., Coffey, P. J., Thomson, J. A., and Soh, H. T. (2015). Array-based discovery of aptamer pairs, *Anal. Chem.*, **87**, pp. 821–828.

59. Katilius, E., Flores, C., and Woodbury, N. W. (2007). Exploring the sequence space of a DNA aptamer using microarrays, *Nucleic Acids Res.*, **35**, pp. 7626–7635.
60. Rowe, W., Platt, M., Wedge, D. C., Day, P. J., Kell, D. B., and Knowles, J. (2010). Analysis of a complete DNA-protein affinity landscape, *J. R. Soc. Interface*, **7**, pp. 397–408.
61. Singh, J., Ator, M. A., Jaeger, E. P., Allen, M. P., Whipple, D. A., Soloweij, J. E., Chowdhary, S., and Treasurywala, A. M. (1996). Application of genetic algorithms to combinatorial synthesis: a computational approach to lead identification and lead optimization, *J. Am. Chem. Soc.*, **118**, pp. 1669–1676.
62. Yokobayashi, Y., Ikebukuro, K., McNiven, S., and Karube, I. (1996). Directed evolution of trypsin inhibiting peptides using a genetic algorithm, *J. Chem. Soc. Perkin Trans. 1*, pp. 2435–2437.
63. Ikebukuro, K., Okumura, Y., Sumikura, K., and Karube, I. (2005). A novel method of screening thrombin-inhibiting DNA aptamers using an evolution-mimicking algorithm, *Nucleic Acids Res.*, **33**, p. e108.
64. Savory, N., Abe, K., Yoshida, W., and Ikebukuro, K. (2014). In silico maturation: processing sequences to improve biopolymer functions based on genetic algorithms, in *Applications of Metaheuristics in Process Engineering* (Valadi, J., and Siarry, P., eds.), pp. 271–288, Springer International.
65. Knight, C. G., Platt, M., Rowe, W., Wedge, D. C., Khan, F., Day, P. J., McShea, A., Knowles, J., and Kell, D. B. (2009). Array-based evolution of DNA aptamers allows modelling of an explicit sequence-fitness landscape, *Nucleic Acids Res.*, **37**, p. e6.
66. Bittker, J. A., Le, B. V., and Liu, D. R. (2002). Nucleic acid evolution and minimization by nonhomologous random recombination, *Nat. Biotechnol.*, **20**, pp. 1024–1029.
67. Ikebukuro, K., Yoshida, W., Noma, T., and Sode, K. (2006). Analysis of the evolution of the thrombin-inhibiting DNA aptamers using a genetic algorithm, *Biotechnol. Lett.*, **28**, pp. 1933–1937.
68. Noma, T., and Ikebukuro, K. (2006). Aptamer selection based on inhibitory activity using an evolution-mimicking algorithm, *Biochem. Biophys. Res. Commun.*, **347**, pp. 226–231.
69. Noma, T., Sode, K., and Ikebukuro, K. (2006). Characterization and application of aptamers for Taq DNA polymerase selected using an evolution-mimicking algorithm, *Biotechnol. Lett.*, **28**, pp. 1939–1944.

70. Savory, N., Abe, K., Sode, K., and Ikebukuro, K. (2010). Selection of DNA aptamer against prostate specific antigen using a genetic algorithm and application to sensing, *Biosens. Bioelectron.*, **26**, pp. 1386–1391.

71. Savory, N., Lednor, D., Tsukakoshi, K., Abe, K., Yoshida, W., Ferri, S., Jones, B. V., and Ikebukuro, K. (2013). In silico maturation of binding-specificity of DNA aptamers against Proteus mirabilis, *Biotechnol. Bioeng.*, **110**, pp. 2573–2580.

72. Savory, N., Takahashi, Y., Tsukakoshi, K., Hasegawa, H., Takase, M., Abe, K., Yoshida, W., Ferri, S., Kumazawa, S., Sode, K., and Ikebukuro, K. (2014). Simultaneous improvement of specificity and affinity of aptamers against Streptococcus mutans by in silico maturation for biosensor development, *Biotechnol. Bioeng.*, **111**, pp. 454–461.

73. Asai, R., Nishimura, S. I., Aita, T., and Takahashi, K. (2005). In vitro selection of DNA aptamers on chips using a method for generating point mutations, *Anal. Lett.*, **37**, pp. 645–656.

74. Rowe, W., Platt, M., Wedge, D. C., Day, P. J., Kell, D. B., and Knowles, J. D. (2010). Convergent evolution to an aptamer observed in small populations on DNA microarrays, *Phys. Biol.*, **7**, p. 036007.

75. Platt, M., Rowe, W., Wedge, D. C., Kell, D. B., Knowles, J., and Day, P. J. (2009). Aptamer evolution for array-based diagnostics, *Anal. Biochem.*, **390**, pp. 203–205.

76. Chushak, Y., and Stone, M. O. (2009). In silico selection of RNA aptamers, *Nucleic Acids Res.*, **37**, p. e87.

77. Das, R., and Baker, D. (2007). Automated de novo prediction of native-like RNA tertiary structures, *Proc. Natl. Acad. Sci. U S A*, **104**, pp. 14664–14669.

78. Hu, W.-P., Kumar, J. V., Huang, C.-J., and Chen, W.-Y. (2015). Computational selection of RNA aptamer against angiopoietin-2 and experimental evaluation, *Biomed Res. Int.*, **2015**, p. 658712.

79. Hoinka, J., Zotenko, E., Friedman, A., Sauna, Z. E., and Przytycka, T. M. (2012). Identification of sequence-structure RNA binding motifs for SELEX-derived aptamers, *Bioinformatics*, **28**, pp. i215–i223.

80. Jiang, P., Meyer, S., Hou, Z., Propson, N. E., Soh, H. T., Thomson, J. A., and Stewart, R. (2014). MPBind: a meta-motif-based statistical framework and pipeline to predict binding potential of SELEX-derived aptamers, *Bioinformatics*, **30**, pp. 2665–2667.

81. Morita, Y., Yoshida, W., Savory, N., Han, S. W., Tera, M., Nagasawa, K., Nakamura, C., Sode, K., and Ikebukuro, K. (2011). Development of a

novel biosensing system based on the structural change of a polymerized guanine-quadruplex DNA nanostructure, *Biosens. Bioelectron.*, **26**, pp. 4837–4841.

82. Nonaka, Y., Sode, K., and Ikebukuro, K. (2010). Screening and improvement of an anti-VEGF DNA aptamer, *Molecules*, **15**, pp. 215–225.
83. Yoshida, W., Mochizuki, E., Takase, M., Hasegawa, H., Morita, Y., Yamazaki, H., Sode, K., and Ikebukuro, K. (2009). Selection of DNA aptamers against insulin and construction of an aptameric enzyme subunit for insulin sensing, *Biosens. Bioelectron.*, **24**, pp. 1116–1120.
84. Ogihara, K., Savory, N., Abe, K., Yoshida, W., Asahi, M., Kamohara, S., and Ikebukuro, K. (2015). DNA aptamers against the Cry j 2 allergen of Japanese cedar pollen for biosensing applications, *Biosens. Bioelectron.*, **63**, pp. 159–165.
85. Connor, A. C., Frederick, K. A., Morgan, E. J., and McGown, L. B. (2006). Insulin capture by an insulin-linked polymorphic region G-quadruplex DNA oligonucleotide, *J. Am. Chem. Soc.*, **128**, pp. 4986–4991.
86. Xiao, J., Carter, J. A., Frederick, K. A., and McGown, L. B. (2009). A genome-inspired DNA ligand for the affinity capture of insulin and insulin-like growth factor-2, *J. Sep. Sci.*, **32**, pp. 1654–1664.
87. Xiao, J., and McGown, L. B. (2009). Mass spectrometric determination of ILPR G-quadruplex binding sites in insulin and IGF-2, *J. Am. Soc. Mass Spectrom.*, **20**, pp. 1974–1982.
88. Huppert, J. L., and Balasubramanian, S. (2007). G-quadruplexes in promoters throughout the human genome, *Nucleic Acids Res.*, **35**, pp. 406–413.
89. Iida, K., Nakamura, T., Yoshida, W., Tera, M., Nakabayashi, K., Hata, K., Ikebukuro, K., and Nagasawa, K. (2013). Fluorescent-ligand-mediated screening of G-quadruplex structures using a DNA microarray, *Angew. Chem.*, **52**, pp. 12052–12055.
90. Yoshida, W., Saito, T., Yokoyama, T., Ferri, S., and Ikebukuro, K. (2013). Aptamer selection based on G4-forming promoter region, *PLoS One*, **8**, p. e65497.
91. Fukaya, T., Abe, K., Savory, N., Tsukakoshi, K., Yoshida, W., Ferri, S., Sode, K., and Ikebukuro, K. (2015). Improvement of the VEGF binding ability of DNA aptamers through in silico maturation and multimerization strategy, *J. Biotechnol.*, **212**, pp. 99–105.

Chapter 2

Thioredoxin-Based Peptide Aptamers: Development and Applications

David S. Burz, Sergey Reverdatto, and Alexander Shekhtman

Department of Chemistry, University at Albany, State University of New York, 1400 Washington Ave., Albany, NY 12222, USA

dsburz@albany.edu

2.1 Introduction

The ability to control the dynamic network of complex cellular signaling and macromolecular interactions that govern living processes is the Holy Grail of biomedicine [1]. Designing small molecules to regulate these cellular processes through selective high-affinity binding to discrete sites on biological target molecules remains one of the greatest challenges of modern biological science. DNA and RNA aptamers were among the first choices for such regulatory molecules because of the ease with which high-affinity binders can be selected [2]. However, the emergence of peptide aptamers (PAs) [3] provided a distinct advantage over these molecules in general by virtue of a smaller binding footprint, which allows for a more thorough and exacting interrogation of the target than

Aptamers: Tools for Nanotherapy and Molecular Imaging
Edited by Rakesh N. Veedu

ISBN 978-981-4669-83-2 (Hardcover), 978-981-4669-84-9 (eBook)
www.panstanford.com

that afforded by nucleic acid–based probes [4]. The ability to produce large combinatorial libraries [5–8] of candidate molecules, combined with more robust selection technology, has advanced the field of PA technology considerably in the past decade, and today PAs are widely used in therapy and diagnostics [1, 9–11], targeted drug delivery [12–14], the area of molecular imaging [15, 16], and biosensors [17–20].

PAs consist of a short, 5–20-residue, peptide loop embedded within a small and very stable protein backbone (**scaffold**). This **loop-on-a-frame** design provides the source of variability for selecting high-affinity binders to a target protein or small molecule from combinatorial libraries. PAs exhibit high stability, high solubility, and rapid folding and are available in large quantities through chemical synthesis or bacterial expression [21]. Embedding is important to stabilize the peptide loop, making it more likely to fold and recognize cognate surfaces. The binding affinity of constrained aptamers can be up to 1000 times greater than that of the free peptide, a fact attributed to the lower conformational entropy of the restricted peptide loop [22, 23]. PAs also exhibit a high degree of variability through functional groups (i.e., different pH dependence on hydrogen and electrostatic bond formation) and conformational flexibility [21], which translates to a plethora of configurations capable of binding to various molecular surfaces, thus improving the chances of finding an ideal interactor.

Similar to antibodies in binding affinity with nanomolar to picomolar binding constants, PAs are less immunogenic and smaller (10–50 kDa) and can be used in a variety of environmental conditions. But the greatest advantage is the speed of generating, selecting, and evolving PAs [24]. Aptamers can be engineered and generated in vitro and are easily modified during and after chemical synthesis to increase the stability and variability of the library [25]. And whereas inefficient cellular uptake and intracellular processing of endosome-targeted RNAs [26] can impede selection of DNA and RNA aptamers prepared in vitro, PAs can be produced and selected in vivo via yeast two hybrid (Y2H) and other similar techniques, making PAs ideal candidates for interrogating intracellular targets in a physiological environment [27] and exceptional research tools [28, 29].

2.2 Thioredoxin Scaffolds

The choice of scaffold protein is critical to creating a combinatorial PA library. The ideal scaffold is a compact, rigid, monomeric, stable protein core capable of displaying variable target interaction surfaces in a manner analogous to the immunoglobulin complementarity-determining region. The scaffold must be soluble, nontoxic, and capable of high-level expression in prokaryotes. The cloned scaffold gene should be engineered to accommodate functional protein elements such as localization signals and epitope and purification tags [30]. It is critical that permutations introduced into the variable regions not adversely affect solubility, folding, and aggregating properties of the resulting combinatorial product.

Bacterial thioredoxin A (TrxA) is among the most widely used scaffolds for in vivo selection. The design was introduced in 1996 to display conformationally constrained peptides [3]. TrxA is a small (12 kDa), structurally rigid, highly soluble, rapidly folding oxidoreductase enzyme involved in maintaining cytosolic thiol/disulfide equilibrium in *Escherichia coli*. It is nontoxic even when expressed at high levels in bacteria. Its single domain consists of four α-helices flanking a five-stranded β-sheet core. The thioredoxin fold is a stable structural feature common to enzymes that catalyze the formation of cystines (Fig. 2.1A).

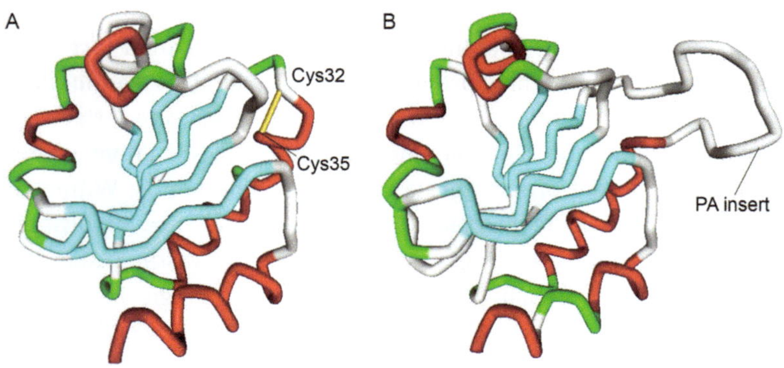

Figure 2.1 Thioredoxin scaffold protein. (A) The active site loop lies between Cys32 and Cys35, and the corresponding disulfide bridge is in yellow. (B) Peptide aptamer sequences are cloned into the active site loop.

The Cys-Gly-Pro-Cys active site of TrxA includes one disulfide group that protrudes from the protein surface [31] (Fig. 2.1A) and can accommodate long peptide insertions, albeit with a concomitant loss of enzymatic function [32]. The surface solvent-accessible loop where PA inserts are presented is essential for recognizing and binding to target compounds (Fig. 2.1B). Comparative studies demonstrated that TrxA possesses superior stability over other scaffolds [33], making it an ideal choice to use in constructing PA libraries.

2.3 Selection and Characterization of PAs

It is an imposing task to identify and isolate high-affinity ligands from combinatorial libraries. A large and continually expanding arsenal of powerful screening methods and techniques has been developed to assist researchers in this endeavor [34]. Selection technologies can be divided into two classes. The first and most widely used approach is selection through extracellular display, such as on the surface of cells, filamentous phages, or in vitro. The second is in vivo screening represented by a plethora of in-cell protein complementation assays, including two-hybrid systems, in which the target and ligand are both produced and interact inside a living cell.

Although time consuming [35], in vivo selection is preferable to in vitro screening because the selection occurs under near-physiological conditions and does not require purified targets. In particular, Y2H screening provides several advantages over methods based on extracellular interactions. Being eukaryotes, the yeast cytosol allows for N-linked glycosylation and oxidative protein folding, ensuring proper formation of disulfide bonds. Within the intracellular setting proteins have a higher chance of properly assembling and interacting. A critical feature that distinguishes Y2H screening is that each PA expressed within a single cell has no competition for binding to a given target during selection [36]. This makes possible isolation of PAs with different affinities that bind to distinct or even overlapping sites on the same target. The resulting collection of PAs allows for a comprehensive characterization of interaction surfaces on target molecules in vitro and in vivo.

The discovery of interactors after extensive library screening is just the first step on a long road. Comprehensive biophysical and biochemical characterization, confirmation of target recognition, and detailed atomic resolution of the binding interface are required [37]. Molecules with moderate target binding affinity or with unfavorable characteristics such as low solubility, aggregation, or toxicity are subjected to several rounds of affinity maturation under conditions of increased stringency with special emphasis on selecting high stability compounds. This is usually carried out through random mutagenesis of variable regions, sometimes involving the scaffold, or by more focused approaches guided by rational computer-aided design based on structural data and the results of comprehensive mutational scanning to generate heat maps of antigen-binding enrichment ratios [37–44].

Quite often maturation rounds are performed in different library screening settings. Display methods used during primary selection can be complemented by Y2H for affinity evolution. Besides random mutagenesis, nonnaturally occurring amino acids and various chemical modifications can be used to increase the interaction space and library complexity [25, 38, 45]. Although requiring sufficient resources and time investment, affinity maturation generally appears to provide desired improvements in protein characteristics, sometimes producing extraordinary results. Only then are the evolved high-affinity products ready for testing in biomedical or industrial settings.

2.4 Applications

Owing to its remarkable characteristics, TrxA was the first widely used PA scaffold (Table 1 in Ref. [30]). Seminal studies explored the utility and versatility of the TrxA scaffold in presenting combinatorial libraries for identifying and isolating PAs. Ensuing work examined the interaction of PAs with key viral or cellular regulatory proteins in prokaryotic and eukaryotic cells and the ability of PAs to regulate mammalian cellular processes and specifically target cancer cells in order to demonstrate that PAs can be utilized to manipulate specific biological processes.

2.4.1 *Combinatorial Libraries*

The earliest study employed Y2H selection to isolate PAs against human cyclin-dependent kinase 2 (Cdk2) from a combinatorial library of 2.9×10^9 molecules containing 20-amino-acid-residue peptide inserts constrained within the TrxA active site loop [3]. The 20-repeat sequence, NNG/T, contained a G or a T at the third position of the codon to reduce the frequency of stop codons, while maintaining the maximum diversity of amino acids. Fourteen PAs were isolated, and six were characterized in vitro and found to have nanomolar binding affinity for Cdk2 and to inhibit phosphorylation of histone H1 by Cdk2/cyclin E kinase, most likely by competing with its H1 substrate.

The same library was used with Y2H selection to isolate PAs against the E2F transcription factor that binds to the DP dimerization and DNA-binding domains of E2Fs. PAs Apt4 and Apt5 prevented E2F/DP heterodimers from binding to an E2F DNA site, and in one case, Apt5 inhibited the formation of heterodimers [46]. The PAs, when transfected or microinjected into fibroblasts, impeded the cells' ability to enter the S phase, consistent with the observation that E2F activity is required for the G1/S transition [47] and demonstrating the utility of PAs in regulating mammalian cellular processes.

To select PAs directly in bacterial cells, a method dubbed ABBIS, for **aptamer-based bacterial inhibition system**, was developed [48]. A 16-amino-acid-residue library was fused to TrxA using the same methodology as in Ref. [3]. Expression of the library vector was regulated by the *E. coli* araBAD promoter [49], which is induced by L-arabinose. The library was used to isolate PAs that disrupt the activity of a specific target, thymidylate synthase (ThyA). In the absence of ThyA activity cells become thymidine auxotrophs and are resistant to trimethoprim, a dihydrofolate reductase inhibitor. ABBIS was also used nonspecifically to identify five PAs that prevent colony growth; of these four were bacteriostatic and one was bactericidal. Target molecules were not identified. The method successfully demonstrated direct in vivo selection of PAs that affect intracellular processes in bacteria.

Using a 20-amino-acid-insert TrxA PA library of $\sim 2 \times 10^8$ members and Y2H selection [50], a PA that targets the oncoprotein

HPV16 E6 was identified. Expression of the E6 and E7 genes is associated with the oncogenic activity of human papilloma virus (HPV). The specificity of the PAs for E6 was shown in vitro. Expressing the aptamers E61-1 and E61-17 in HPV16-positive SiHa and CaSki cervical carcinoma cells resulted in strong inhibition of growth. In addition, E61-1 gave rise to apoptotic cell death in HPV16-positive cells and increased levels of p53 protein, the latter considered an important mediator of apoptotic response of cells to viral oncoproteins [51, 52]. The work demonstrated that PAs directed against antiapoptotic viral protein selectively eliminate virus-positive cancer cells.

Using the same library [50] and Y2H screening, PAs that bind to the intracellular kinase domain of the epidermal growth factor (EGF) receptor, EGFR, were identified [53]. The PAs were shown to interact specifically with EGFR in experiments performed in bacteria and in mouse fibroblast NIH3T3 cells containing human EGFR (Herc). Transfection of PA KDI1 into Herc cells to generate stable cell lines reduced the growth rate of cells. Introducing KDI1 into human epidermoid carcinoma, A431, cells by fusing the PAs to a protein transduction domain, PTD [54], consisting of nine arginine residues, resulted in dose-dependent growth inhibition and at 2 μM completely blocked cell division. KDI1 was also shown to interfere with activation of the Shc- and Stat3-signaling pathways. KDI1 reduced the extent of EGF-induced tyrosine phosphorylation on EGFR and suppressed c-Src-mediated phosphorylation of Tyr-845 and phosphorylation of Tyr-316 on p46 Shc.

The same group identified PAs DD3 and DBD1, which target the dimerization domain and the DNA-binding domain of the transcription factor Stat3, respectively [55]. Stat3 is not essential for the survival of many normal cell types but is crucial for the survival of different tumor lines. Both PAs inhibit transactivation and induce apoptosis in tumor cells. Treatment of B16, mouse melanoma, and U266, human myeloma, cells with the DBD1 peptide sequence fused to a nine-arginine residue PTD led to a dose-dependent drop in viability. These studies revealed the therapeutic potential of PAs and their ability to selectively interfere with signal transduction pathways and regulatory process in cancer cells.

A host of ensuing studies used the TrxA-based combinatorial library to identify and isolate PAs against well-folded, structurally stable proteins, including the Rho protein guanine nucleotide exchange factor (Rho-GEF), Trio [56]; the tomato golden mosaic viral protein AL1 [57]; cell surface receptors ErbB2 [58] and RAGE [28]; prion proteins [59]; and material surfaces such as TiO_2 [60]. And in 2014, Cobbert et al. [61] used an eight-amino-acid-insert PA library of $\sim 3 \times 10^{10}$ members presented on a modified TrxA scaffold [28] and Y2H selection to identify PAs directed against the intrinsically disordered protein (IDP) ubiquitin-like bacterial Pup.

Pup binding to *Mycobacterium tuberculosis* (*Mtb*) Mpa directs pupylated substrate into the *Mtb* proteasome for degradation. The binding specificity of the PAs was characterized in vitro and by in-cell nuclear magnetic resonance (NMR) spectroscopy. The PAs bind to distinct sites on Pup with nanomolar affinity and elicit different effects on rescuing known proteasome substrates and on the survival of the Bacille–Calmette–Guèrin strain of *M. bovis*. By selecting and characterizing PAs in-cell, physiological relevant conformations of Pup were targeted. This study demonstrated that PAs can be selected against even transiently structured targets.

2.4.2 *Rational Design of PAs*

Previous work to isolate and identify PAs utilized combinatorial libraries containing $>10^8$ members. Another methodology entails the rational design of PAs on the basis of binding determinants discerned from physical measurements. There is no selection process since the PA inserts are specifically designed.

A major target for treating cancer is the tumor suppressor protein p53. Activation of p53 is considered a normal cellular response to DNA damage. Accumulation of p53 enhances p53-responsive gene expression that can induce apoptosis and cell cycle arrest. The concentration of p53 is regulated by binding to Mdm2 [62, 63], which directs p53 for ubiquitin-mediated degradation. Originally identified by using phage selection [64], a 12-amino-acid-insert, peptide 12/1, based on the p53-binding motif, FXXXWXXL, binds tightly to p53 [65] and acquires an α-helical structure when bound to Mdm2. Peptide 12/1 was cloned into TrxA and dubbed SuperTIP for thioredoxin insert protein [66]. The SuperTIP PA inhibited p53 binding to Mdm2 in vitro, facilitated the accumulation of p53 protein in mammalian cells, and activated p53-responsive genes, without the need to incur DNA damage.

The rational design of a PA insert to the translation elongation factor 4E (eIF4E) was based on the design of the SuperTIP aptamer. The α-helix-forming sequence of the elongation initiation factor 4GI peptide was aligned with that of the 12/1 peptide and cloned into TrxA [67] to yield PA 4GI-SG. The binding of SuperTIP and 4GI-SG to eIF4E and Mdm2 was determined by isothermal titration calorimetry (ITC) and found to be $\sim$1 μM and 90 nM, respectively. In both previous cases, a peptide sequence identified from earlier PA screening was optimized and inserted into the TrxA scaffold for successful functional studies.

Cui et al. [68] employed rational design to introduce PAs derived from the interaction motifs of Smad-binding proteins CBP (49 amino acids), FoxH1 (24 amino acids), and Lef1 (tandem duplication of 30 amino acids) into the TrxA scaffold to inhibit transforming growth factor beta (TGF-β)-induced gene expression. TGF-β alters gene expression by activating type I serine kinase receptors, which, in turn, phosphorylate Smad2 and Smad 3. Smad2 and Smad 3 form heterooligomeric complexes with Smad4 that accumulate in the nucleus and bind >20 nuclear transcription factors, including CBP, FoxH1, and Lef1. These specific interactions give rise to the cell-type-specific effects of stimulating TGF-β gene expression, such as proliferation, apoptosis, and differentiation. The three PAs examined bind specifically to Smads and differentially inhibit TGF-β-induced gene expression in HepG2 cells.

In a subsequent work a 56-amino-acid PA insert derived from the Smad anchor for receptor activation (SARA) protein, Trx-SARA, was developed that binds to Smad2 and Smad3. Trx-SARA inhibited TGF-β-induced gene expression and the epithelial to mesenchymal transition in murine mammary epithelial (NMuMG) cells [69]. The PA also resulted in lower levels of Smad2–Smad4 and Smad3–Smad4 complexes following TGF-β stimulation but had no effect on the extent of phosphorylation of those species. These studies showed that the rational design of PAs based on known interaction motifs can selectively disrupt the formation of specific signaling complexes and can be used to study gene expression.

Moretto et al. [70] constructed TrxA-based PAs consisting of tandemly repeated inserts derived from the human amyloid-β (Aβ) peptide for use as antigens in an approach called thioredoxin-displayed multipeptide immunogens (TDMI). One, four, and eight copies of the immunodominant B-cell epitope Aβ15 were cloned into the active site loop of TrxA and injected into mice with adjuvant to elicit an immunogenic response. All three constructs gave rise to antibodies with distinct binding specificities. Anti-Trx(Aβ15)$_4$ recognizes oligomeric and fibrillar forms of Aβ, but not monomeric Aβ, binds to human Alzheimer disease (AD) plaques, and depresses the pathology of Aβ protein in transgenic AD mice. The conformational selectivity of anti-Trx(Aβ15)$_4$ was evidenced by cross-reactivity with fibrillary and oligomeric aggregates of another amyloidogenic protein, TTP, suggesting that nonmonomer amyloids may share a common structural motif.

More recently, TDMI was applied to the human papillomavirus (HPV) minor capsid protein L2. This protein contains polypeptide sequences that are conserved among carcinogenic HPVs but is poorly immunogenic. Rubio et al. [71] identified a 19-amino-acid L2 peptide (segments 20–38) that upon tandem insertion into TrxA displayed strong immunogenicity and elicited the production of polyclonal and monoclonal antibodies capable of cross-neutralizing different types of high-risk HPVs. These studies reveal the immunotherapeutic potential of TrxA-PAs for treating disease states and expand the versatility of TrxA-PA technology.

2.5 Constrained vs. Unconstrained PA Inserts

The original concept of constrained PA inserts is based on the variable loop-binding determinant present on immunoglobulins. Unlike singly constrained PAs that are fused to proteins and other macromolecules, inserts doubly constrained within the Trx scaffold are anchored at two sites at the base of the active site loop. The TrxA scaffold is assumed to maintain a rigid, compact, stable core that doesn't undergo structural changes when peptide inserts are introduced into the loop. Constrained PAs have less conformational diversity than unconstrained. The reduced conformation diversity lowers the entropic cost of binding when the target-bound PA adopts a single conformation and contributes to the tight binding affinities exhibited by doubly constrained PAs. Embedding the insert in a scaffold is believed to confer the following advantages: (1) The constraint can help expose amino acid residues that are often buried, (2) constrained inserts frequently display tighter binding affinity than the free peptide, and (3) constrained PAs are generally more stable than unstructured peptides in-cell. Comparative studies using constrained and free peptides have produced mixed results.

Cohen et al. [22] observed that a 20-amino-acid peptide required ~1000-fold greater concentration than the constrained insert to inhibit H1 kinase activity. They concluded that the conformational constraint imposed by the TrxA scaffold facilitates a high-affinity interaction by lowering the entropic cost of binding.

Fabbrizio et al. [46] compared the activities of PA Apt5 and the free peptide Pep5, a 20-amino-acid synthetic peptide with a sequence corresponding to the Apt5 insert. The PA blocked binding of E2F1/DP1, E2F4/DP1, and E2F5/DP1 heterodimers to DNA equally well in gel shift assays, but the free peptide required a tenfold higher concentration to similarly affect E2F4/DP1 and E2F5/DP1 binding to relative to E2F1. Similarly, a stronger effect was elicited by Apt5 relative to Pep5 in functional assays. The authors concluded that the Apt5 insert is more highly structured within the TrxA scaffold than the free peptide and that tighter aptamer target binding is reflected by the smaller loss of entropy associated with the constrained PA compared to the free peptide.

A synthetic 20-amino-acid construct, KDI1-4R, containing three flanking TrxA residues, CGP, was translocated via a 4-arginine PTD into Herc cells [53]. The peptide functioned similarly to the constrained PA, KDI1, indicating that the Trx scaffold was not critical for the inhibitory function of the PA. It was speculated that the cys residues may form a disulfide bond that constrains the peptide sequence in a manner similar to that of the TrxA scaffold.

Klevenz et al. [72] examined the role of the scaffold protein on the ability of PAs, identified using Y2H selection [50, 73], or rational design [74], to function when expressed from other scaffolds: green fluorescent protein (GFP) and catalytically inactive staphylococcal nuclease (SNase). Three PAs were examined, and two of them failed to bind to the target protein when displayed from GFP or SNase. The work showed that TrxA-based PAs can display binding motifs found in naturally interacting partners. The PA that did bind to its target when displayed from the alternative platforms also functioned when expressed as a free peptide, suggesting that particular peptide contains a "relatively high intrinsic conformational stability."

Biophysical characterization of the Mdm2 aptamer, SuperTIP, showed that it was bound 2x more tightly than the free peptide 12/1, but the eIF4E aptamer, 4GI-SG, was bound ~5x weaker than the free peptide [67]. In the case of 4GI-SG, interactions between the TrxA scaffold and the target, Mdm2, were found to contribute to the binding enthalpy. The same group found, by ITC and molecular dynamic modeling, that inserting a peptide into TrxA destabilized the scaffold. They concluded that high-affinity binding occurs when the magnitude of the enthalpy change due to specific binding interactions between the PA and its target overcomes the change in entropy associated with peptide insertion.

The binding free energy is governed by the balance struck between the binding enthalpy and the change in entropy arising from the conformation flexibility of the peptide insert. Longer inserts may provide more determinants for specific binding than shorter ones, resulting in a large enthalpy change upon bond formation, but in general are expected to exact a greater entropic cost due to the wider range of accessible conformations. Individual peptide inserts will doubtless possess more or less intrinsic flexibility, some by virtue of the amino acid sequence. Others, especially larger inserts, may partially fold, limiting the range of conformational flexibility and facilitating tighter binding. Finally, TrxA itself can play a key role in this process as individual insertions may destabilize the scaffold, requiring increased favorable binding enthalpy to overcome the deficit in binding free energy.

2.6 Modifications and Alternatives to the Trx Scaffold

Despite the widespread use of bacterial TrxA as a scaffold for PAs, problems arose that limited the potential therapeutic usefulness of TrxA: The active site cysteines, while stable in the reducing environment of the cell, can oxidize during purification; TrxA-PAs may aggregate making them difficult to purify as monomers; and, although cellular testing is generally accomplished by transforming/transfecting PAs and overexpressing the constructs, administrating recombinant proteins directly has proven to be a better method of controlling dosages in cellular assays and is likely to be a preferred method of therapeutic administration. Several studies have addressed these concerns.

Borghouts et al. [75] investigated methods of improving the thioredoxin scaffold. They discerned that the TrxA active site cysteines can form intermolecular disulfide bridges, giving rise to oligomeric protein. Mutating the cysteines to serines greatly reduced aggregate formation during purification, resulted in efficient cellular uptake, and, for the TrxA-AII-7 aptamer, increased binding to the target protein.

The first generations of combinatorial libraries for in vivo screening utilized nondirectional cloning at a single *Rsr*II restriction site, which limited the library size to $\sim 10^8$ unique molecules. In addition, random peptide sequence insertions frequently destabilize the TrxA scaffold and can create molecules that are prone to aggregation [67, 75]. In general, polar or negatively charged amino acids in the insert loop tend to result in proteins that are more soluble than those containing hydrophobic or positively charged residues. Reverdatto et al. [28] addressed these problems while developing a method for constructing a Combinatorial Library of Improved Peptide aptamers (CLIPs).

The TrxA reaction center was modified by introducing a polylinker site (*Bam*HI, *Eco*RI, *Apa*I), which simplified directional cloning of peptide sequences and reduced the likelihood of PA aggregation by eliminating the cysteine residues that give rise to intra- and intermolecular disulfide bonds [35]. To stabilize the TrxA scaffold residues D26 and K57 were mutated (Fig. 2.2). D26A removes the negative charge at the base of the hydrophobic cleft, improving solubility, solution stability, and providing better dispersion in the 1D NMR spectra. K57 forms a salt bridge with D26. The mutations K57Q and K57E were designed to compensate for the loss of the negative charge in D26A. The D26A-K57E and, to a lesser extent, the D26A-K57Q double mutation markedly increased the in-

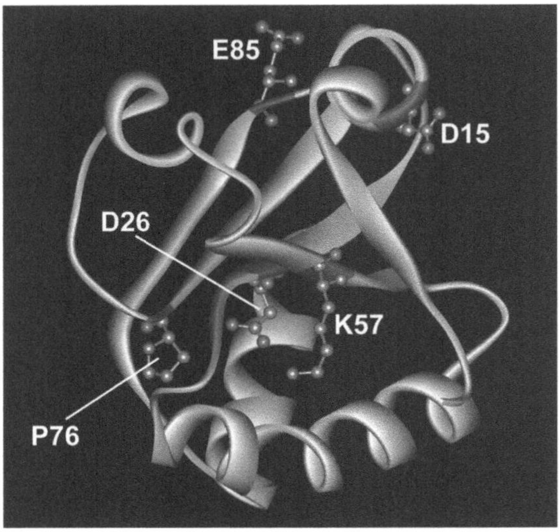

Figure 2.2 Mutations confer stability and improved solubility to TrxA peptide aptamers.

cell solubility of both PAs #97 and #103, two particularly insoluble anti-RAGE C2 domain PAs.

Alternatives to the TrxA scaffold have also been investigated. Although structurally similar to human thioredoxin, hTrx, TrxA exhibits only 30% sequence homology and may be immunogenic in humans [76]. The active site cysteines in hTrx were mutated to glycines, and three additional cysteines were mutated to serines. A second restriction site was introduced to facilitate unidirectional cloning into the active site, and the coding sequence included additional glycine and proline residues. The glycines provide more flexibility for the aptamer structure, and the prolines create protrusions away from the active site to enhance the presentation of the PA. The utility of the modified hTrx scaffold, hTrx-Dcys5, was demonstrated by inserting peptide sequences, previously selected using TrxA, into hTrx-Dcys5 [75]. Inserts presented on the hTrx-Dcys5 construct were highly overexpressed, yielded monomeric protein, displayed enhanced binding to their targets relative to TrxA-PAs, and were more easily transferred via PTD fusion into cells than the wild-type hTrx.

TrxA-PAs have been used to confer immunogenicity to peptide epitopes. To produce more effective peptide immunogens with high thermal stability, no cross-reactivity, and low production cost, Canali et al. [77] identified *Pyrococcus furiosus* thioredoxin (PfTrx) as a superior protein scaffold to be used in constructing PA construct recombinant PA immunogens. PfTrx is a highly thermostable and protease-resistant scaffold with a greater capacity for solubilization than TrxA. PfTrx is nine amino acids shorter than TrxA at the amino terminus, lacks one N-terminal β-strand that is present in all other thioredoxins, and has a CPPC active site. Studies employing PfTrx containing a previously identified HPV16-L2 peptide epitope [71] obtained $>90\%$ pure protein by using a one-step thermal purification protocol. The purified protein was heat stable and retained immunogenic activity when frozen and thawed or reconstituted from a lyophilized state and was as effective as the TrxA-based PA in eliciting neutralizing anti-HPV antibodies. The authors concluded that PfTrx is a superior, general-purpose scaffold for the construction of safe, stable, and low-cost peptide immunogens.

2.7 Conclusions

The original concept of Roger Brent proposed a loop-on-a-frame model for PAs that mimicked the antibody paratype [3]. The *E. coli* thioredoxin (TrxA) scaffold has proven to be an extremely versatile protein from which to display PA inserts from combinatorial libraries or rationally designed binding motifs. Thioredoxin-based PAs represent a new class of biologically active molecules that have the potential to bind a given target in solution, as well as extracellular and intracellular conditions to satisfy the demanding applications required in the biomedical, bioimaging, and bioanalytical fields.

References

1. Nimjee, S. M., Rusconi, C. P., and Sullenger, B. A. (2005). Aptamers: an emerging class of therapeutics, *Annu. Rev. Med.*, **56**, pp. 555–583.

2. Ng, E. W., Shima, D. T., Calias, P., Cunningham, E. T., Jr, Guyer, D. R., et al. (2006). Pegaptanib, a targeted anti-VEGF aptamer for ocular vascular disease, *Nat. Rev. Drug Discovery,* **5**, pp. 123–132.
3. Colas, P., Cohen, B., Jessen, T., Grishina, I., McCoy, J., and Brent, R. (1996). Genetic selection of peptide aptamers that recognize and inhibit cyclin-dependent kinase 2, *Nature,* **380**, pp. 548–550.
4. Huber, D., Cha, M. I., Debarbieux, L., Planson, A. G., Cruz, N., et al. (2005). A selection for mutants that interfere with folding of *Escherichia coli* thioredoxin-1 *in vivo, Proc. Natl. Acad. Sci. U S A,* **102**, pp. 18872–18877.
5. Tuerk, C., and Gold, L. (1990). Systematic evolution of ligands by exponential enrichment: RNA ligands to bacteriophage T4 DNA polymerase, *Science,* **249**, pp. 505–510.
6. Ellington, A. D., and Szostak, J. W. (1990). *In vitro* selection of RNA molecules that bind specific ligands, *Nature,* **346**, pp. 818–822.
7. Roberts, R. W., and Szostak, J. W. (1997). RNA-peptide fusions for the *in vitro* selection of peptides and proteins, *Proc. Natl. Acad. Sci. U S A,* **94**, pp. 12297–12302.
8. Georgiou, G., Stathopoulos, C., Daugherty, P. S., Nayak, A. R., Iverson, B. L., et al. (1997). Display of heterologous proteins on the surface of microorganisms: from the screening of combinatorial libraries to live recombinant vaccines, *Nat. Biotechnol.,* **15**, pp. 29–34.
9. Kanwar, J. R., Shankaranarayanan, J. S., Gurudevan, S., and Kanwar, R. K. (2014). Aptamer-based therapeutics of the past, present and future: from the perspective of eye-related diseases, *Drug Discovery Today,* **19**, pp. 1309–1321.
10. Santosh, B., and Yadava, P. K. (2014). Nuclei acid aptamers: research tools in disease diagnostics and therapeutics, *Biomed. Res. Int.,* **2014**, p. 540451.
11. Xing, H., Hwang, K., Li, J., Torabi, S. F., and Lu, Y. (2014). DNA aptamer technology for personalized medicine, *Curr. Opin. Chem. Eng.,* **4**, pp. 79–87.
12. Wang, C., Liu, B., Lu, J., Zhang, G., and Lu, A. (2014). Strategies for combination of aptamer and targeted drug delivery, *J Nanosci. Nanotechnol.,* **14**, pp. 501–512.
13. Zhou, J., and Rossi, J. J. (2014). Cell-type-specific, aptamer-functionalized agents for targeted disease therapy, *Mol. Ther. Nucleic Acids,* **3**, p. e169.

14. Xiang, D., Shigdar, S., Qiao, G., Wang, T., Kouzani, A. Z., Zhou, S., Kong, L., Li, Y., Pu, C., and Duan, W. (2015). Nucleic acid aptamer-guided cancer therapeutics and diagnostics: the next generation of cancer medicine, *Theranostics,* **5**, pp. 23–42.
15. Chen, Z. Y., Wang, Y. X., Lin, Y., Zhang, J. S., Yang, F., Zhou, Q. L., and Liao, Y. Y. (2014). Advance of molecular imaging technology and targeted imaging agent in imaging and therapy, *Biomed. Res. Int.,* **2014**, p. 819324.
16. Wang, A. Z., and Farokhzad, O. C. (2014). Current progress of aptamer-based molecular imaging. *J. Nucl. Med.,* **55**, pp. 353–356.
17. Kim, Y. S., and Gu, M. B. (2014). Advances in aptamer screening and small molecule aptasensors, *Adv. Biochem. Eng. Biotechnol.,* **140**, pp. 29–67.
18. MacKay, S., Wishart, D., Xing, J. Z., and Chen, J. (2014). Developing trends in aptamer-based biosensor devices and their applications, *IEEE Trans. Biomed. Circuits Syst.,* **8**, pp. 4–14.
19. Yoshida, W., Abe, K., and Ikebukuro, K. (2014). Emerging techniques employed in aptamer-based diagnostic tests, *Expert Rev. Mol. Diagn.,* **14**, pp. 143–151.
20. Zhou, W., Huang, P. J., Ding, J., and Liu, J. (2014). Aptamer-based biosensors for biomedical diagnostics, *Analyst,* **139**, pp. 2627–2640.
21. Mascini, M., Palchetti, I., and Tombelli, S. (2012). Nucleic acid and peptide aptamers: fundamentals and bioanalytical aspects, *Angew. Chem., Int. Ed. Engl.,* **51**, pp. 1316–1332.
22. Cohen, B. A., Colas, P., and Brent, R. (1998). An artificial cell-cycle inhibitor isolated from a combinatorial library, *Proc. Natl. Acad. Sci. U S A,* **95**, pp. 14272–14277.
23. Ladner, R. C. (1995). Constrained peptides as binding entities, *Trends Biotechnol.,* **13**, pp. 426–430.
24. Szeitner, Z., Andras, J., Gyurcsanyi, R. E., and Meszaros, T. (2014). Is more less? Lessons from aptamer selection strategies, *J. Pharm. Biomed. Anal.,* **101C**, pp. 58–65.
25. Uzawa, T., Tada, S., Wang, W., and Ito, Y. (2013). Expansion of the aptamer library from a "natural soup" to an "unnatural soup", *Chem. Commun. (Camb.),* **49**, pp. 1786–1795.
26. Santosh, B., and Yadava, P. K. (2014). Nucleic acid aptamers: research tools in disease diagnostics and therapeutics. *Biomed. Res. Int.,* **2014**, p. 540451.

27. Baines, I. C., and Colas, P. (2006). Peptide aptamers as guides for small-molecule drug discovery, *Drug Discovery Today,* **11**, pp. 334–341.
28. Reverdatto, S., Rai, V., Xue, J., Burz, D. S., Schmidt, A. M., and Shekhtman, A. (2013). Combinatorial library of improved peptide aptamers, CLIPs to inhibit RAGE signal transduction in mammalian cells, *PLoS One,* **8**, p. e65180.
29. Hamdi, A., and Colas, P. (2012). Yeast two-hybrid methods and their applications in drug discovery, *Trends Pharmacol. Sci.,* **33**, pp. 109–118.
30. Borghouts, C., Kunz, C., and Groner, B. (2008). Peptide aptamer libraries, *Comb. Chem. High Throughput Screen,* **11**, pp. 135–145.
31. Martin, J. L. (1995). Thioredoxin: a fold for all reasons, *Structure,* **3**, pp. 245–250.
32. LaVallie, E. R., DiBlasio, E. A., Kovacic, S., Grant, K. L., Schendel, P. F., and McCoy, J. M. (1993). A thioredoxin gene fusion expression system that circumvents inclusion body formation in the *E. coli* cytoplasm, *Biotechnology (N Y),* **11**, pp. 187–193.
33. Klevenz, B., Butz, K., and Hoppe-Seyler, F. (2002). Peptide aptamers: exchange of the thioredoxin-a scaffold by alternative platform proteins and its influence on target protein binding, *Cell. Mol. Life Sci.,* **59**, pp. 1993–1998.
34. Mondon, P., Dubreuil, O., Bouayadi, K., and Kharrat, H. (2008). Human antibody libraries: a race to engineer and explore a larger diversity, *Front. Biosci.,* **13**, pp. 1117–1129.
35. Bickle, M. B., Dusserre, E., Moncorge, O., Bottin, H., and Colas, P. (2006). Selection and characterization of large collections of peptide aptamers through optimized yeast two-hybrid procedures, *Nat. Protoc.,* **1**, pp. 1066–1091.
36. Crawford, M., Woodman, R., and Ko Ferrigno, P. (2003). Peptide aptamers: tools for biology and drug discovery, *Brief Funct. Genomic Proteomic,* **2**, pp. 72–79.
37. Sudha, G., Nussinov, R., and Srinivasan, N. (2014). An overview of recent advances in structural bioinformatics of protein-protein interactions and a guide to their principles, *Prog. Biophys. Mol. Biol.,* **116**, pp. 141–150.
38. Lane, M. D., and Seelig, B. (2014). Advances in the directed evolution of proteins, *Curr. Opin. Chem. Biol.,* **22C**, pp. 129–136.
39. Eijsink, V. G., Gaseidnes, S., Borchert, T. V., and van den Burg, B. (2005). Directed evolution of enzyme stability, *Biomol. Eng.,* **22**, pp. 21–30.

40. Romero, P. A., and Arnold, F. H. (2009). Exploring protein fitness landscapes by directed evolution, *Nat. Rev. Mol. Cell. Biol.,* **10**, pp. 866–876.

41. Schmid, F. X. (2011). Lessons about protein stability from *in vitro* selections, *ChemBioChem,* **12**, pp. 1501–1507.

42. Golynskiy, M. V., Haugner, J. C., 3rd, and Seelig, B. (2013). Highly diverse protein library based on the ubiquitous $(\beta/\alpha)_8$ enzyme fold yields well-structured proteins through in vitro folding selection, *ChemBioChem,* **14**, pp. 1553–1563.

43. Socha, R. D., and Tokuriki, N. (2013). Modulating protein stability-directed evolution strategies for improved protein function, *FEBS J.,* **280**, pp. 5582–5595.

44. Vanhee, P., van der Sloot, A. M., Verschueren, E., Serrano, L., Rousseau, F., and Schymkowitz, J. (2011). Computational design of peptide ligands, *Trends Biotechnol.,* **29**, pp. 231–239.

45. Hoffmann, A., Kovermann, M., Lilie, H., Fiedler, M., Balbach, J., Rudolph, R., and Pfeifer, S. (2012). New binding mode to TNF-alpha revealed by ubiquitin-based artificial binding protein. *PLoS One,* **7**, p. e31298.

46. Fabbrizio, E., Le Cam, L., Polanowska, J., Kaczorek, M., Lamb, N., Brent, R., and Sardet, C. (1999). Inhibition of mammalian cell proliferation by genetically selected peptide aptamers that functionally antagonize E2F activity, *Oncogene,* **18**, pp. 4357–4363.

47. Sardet, C., Le Cam, L., Fabbrizio, E., and Vidal, M. (1997). E2Fs and the retinoblastoma protein family, in *Progress in Gene Expression. Oncogenes as Transcriptional Regulators* (Yaniv, M., and Ghysdael, J., eds.) **2**, pp. 1–63, Birkhauser.

48. Blum, J. H., Dove, S. L., Hochschild, A., and Mekalanos, J. J. (1999). Isolation of peptide aptamers that inhibit intracellular processes, *Proc. Natl. Acad. Sci. U S A,* **97**(5), pp. 2241–2246.

49. Guzman, L. M., Belin, D., Carson, M. J., and Beckwith, J. (1995). Tight regulation, modulation, and high-level expression by vectors containing the arabinose PBAD promoter, *J. Bacteriol.,* **177**(14), pp. 1421–1430.

50. Butz, K., Denk, C., Ullmann, A., Scheffner, M., and Hoppe-Seyler, F. (2000). Induction of apoptosis in human papillomavirus-positive cancer cells by peptide aptamers targeting the viral E6 oncoprotein, *Proc. Natl. Acad. Sci. U S A,* **97**(12), pp. 6693–6697.

51. Pan, H., and Griep, A. E. (1995). Temporally distinct patterns of p53-dependent and p53-independent apoptosis during mouse lens development, *Genes Dev.,* 9, pp. 2157–2169.

52. Bates, S., Philips, A. C., Clark, P. A., Stott, F., Peters, G., Ludwig, R. L., and Vousden, K. H. (1998). p14ARF links the tumour suppressors RB and p53, *Nature (Lond.)*, **395**, pp. 124–125.

53. Buerger, C., Nagel-Wolfrum, K., Kunz, C., Wittig, I. Butz, K., Hoppe-Seyler, F., and Groner, B. (2003). Sequence-specific peptide aptamers, interacting with the intracellular domain of the epidermal growth factor eceptor, interfere with STat3 activation and inhibit the growth of tumor cells, *J. Biol. Chem.*, **278**(39), pp. 37610–37621.

54. Wender, P. A., Mitchell, D. J., Pattabiraman, K., Pelkey, E. T., Steinman, L., and Rothbard, J. B. (2000). The design, synthesis, and evaluation of molecules that enable or enhance cellular uptake: peptoid molecular transporters, *Proc. Natl. Acad. Sci. U S A*, **97**(24), pp. 13003–13008.

55. Nagel-Wolfrum, K., Buerger, C., Wittig, I., Butz, K., Hoppe-Seyler, F., and Groner, B. (2004). The interaction of specific peptide aptamers with the DNA binding domain and the dimerization domain of the transcription factor Stat3 inhibits transactivation and induces apoptosis in tumor cells, *Mol. Cancer Res.*, **2**, pp. 170–182.

56. Schmidt, S., Diriong, S., Méry, J., Fabbrizio, E., and Debant, A. (2002). Identification of the first Rho-GEF inhibitor, TRIPa, which targets the RhoA-specific GEF domain of TRIO, *FEBS Lett.*, **523**, pp. 35–42.

57. Lopez-Ochoa, L., Ramirez-Prado, J., and Hanley-Bowdoin, L. (2006). Peptide aptamers that bind to a geminivirus replication protein interfere with viral replication in plant cells, *J. Virol.*, **80**(12), pp. 5841–5853.

58. Kunz, C., Borghouts, C., Buerger, C., and Groner, B. (2006). Peptide aptamers with binding specificity for the intracellular domain of the ErbB2 receptor interfere with AKT signaling and sensitize breast cancer cells to taxol, *Mol. Cancer Res.*, **4**(12), pp. 983–998.

59. Gilch, S., Kehler, C., and Schätzl, H. M. (2007). Peptide aptamers expressed in the secretory pathway interfere with cellular PrP^{Sc} formation, *J. Mol. Biol.*, **371**, pp. 362–373.

60. Gronewold, T. M. A., Baumgartner, A., Weckmann, A., Knekties, J., and Egler, C. (2009). Selection process generating peptide aptamers and analysis of their binding to the TiO_2 surface of a surface acoustic wave sensor, *Acta Biomater.*, **5**, pp. 794–800.

61. Cobbert, J. D., DeMott, C., Majumder, S., Smith, E. A., Reverdattao, S., Burz, D. S., McDonough, K. A., and Shekhtman, A. (2014). Caught in action: selecting peptide aptamers against intrinsically disordered proteins in live cells, *Sci. Rep.*, **5**, p. #9402.

62. Haupt, Y., Maya, R., Kazaz, A., and Oren, M. (1997). Mdm2 promotes the rapid degradation of p53, *Nature,* **387**, pp. 296–299.

63. Kubbutat, M. H., Jones, S. N., and Vousden, K. H. (1997). Regulation of p53 stability by Mdm2, *Nature,* **387**, pp. 299–303.

64. Böttger, A., Böttger, V., Howard, S. F., Picksley, S. M., Chène, P., Garcia-Escheverria, C., Hochkeppel, H.-K., and Lane, D. P. (1996). Identification of novel mdm2 binding peptides by phage display, *Oncogene,* **13**, pp. 2141–2147.

65. Böttger, A., Böttger, V., Garcia-Escheverria, C., Chène, P., Hochkeppel, H.-K., Sampson, W., Ang, K., Howard, S. F., Picksley, S. M., and Lane, D. P. (1997). Molecular characterization of the hdm2-p53 interaction, *J. Mol. Biol.,* **269**, pp. 744–756.

66. Böttger, A., Böttger, V., Sparks, A., Liu, W. L., Howard, S. F., and Lane, D. P (1997a). Design of a synthetic Mdm2-binding mini protein that activates the p53 response *in vivo, Curr. Biol.,* **7**(11), pp. 860–869.

67. Brown, C. J., Dastidar, S. G., See, H. Y., Coomber, D. W., Ortiz-Lombardia, M., Verma, C., and Lane, D. P. (2009). Rational design and biophysical characterization of thioredoxin-based aptamers: insights into peptide grafting, *J. Mol. Biol.,* **395**, pp. 871–883.

68. Cui, Q., Lim, S. K., Zhao, B., and Hoffmann, F. M. (2005). Selective inhibition of TGF-β responsive genes by Smad-interacting peptide aptamers from FoxH1, Lef1 and CBP, *Oncogene,* **24**, pp. 3864–3874.

69. Zhao, B. M., and Hoffmann, F. M. (2006). Inhibition of transforming growth factor-β1-induced signaling and epithelial-to-mesenchymal transition by the Smad-binding peptide aptamer Trx-SARA, *Mol. Biol. Cell,* **17**, pp. 3819–3831.

70. Moretto, N., Bolchi, A., Rivetti, C., Imbimbo, B. P., Villetti, G., Pietrini, V., Polonelli, L., Del Signore, S., Smith, K. M., Ferrante, R. J., and Otonello, S. (2007). Conformation sensitive antibodies against Alzheimer amyloid-β by immunization with a thioredoxin-constrained B-cell epitope peptide, *J. Biol. Chem.,* **282**, pp. 11436–11445.

71. Rubio, I., Bolchi, A., Moretto, N., Canali, E., Gissmann, L., Tommasino, M., Müller, M., and Ottonello, S. (2009). Potent anti-HPV immune responses induced by tandem repeats of the HPV16 L2 (20-38) peptide displayed on bacterial thioredoxin, *Vaccine,* **27**, pp. 1949–1956.

72. Klevenz, B., Butz, K., and Hoppe-Seyler, F. (2002). Peptide aptamers: exchange of the thioredoxin-A scaffold by alternative platform proteins and its influence on target protein binding, *Cell. Mol. Life Sci.,* **59**(11), pp. 1993–1998.

73. Butz, K., Denk, C., Fitscher, B., Crnkovic-Mertens, I., Ullmann, A., Schröder, C. H., and Hoppe-Seyler, F. (2001). Peptide aptamers targeting the hepatitis B virus core protein: a new class of molecules with antiviral activity, *Oncogene,* **20**, pp. 6579–6586.
74. Huibregtse, J. M., Scheffner, M., and Howley, P. M. (1993). Cloning and expression of the cDNA for E6-AP, a protein that mediates the interaction of the human papillomavirus E6 oncoprotein with p53, *Mol. Cell. Biol.*, **13**(2), pp. 775–784.
75. Borghouts, C., Kunz, C., Delis, N., and Groner, B. (2008). Monomeric Recombinant peptide aptamers are required for efficient intracellular uptake and target inhibition, *Mol. Cancer Res.*, **6**(2), pp. 267–281.
76. Barrell, P. J., Liew, O. W., and Conner, A. J. (2004). Expressing an antibacterial protein in bacteria for raising antibodies, *Protein Expr. Purif.*, **33**, pp. 153–159
77. Canali, E., Bolchi, A., Spagnoli, G., Seitz, H., Rubio, I., Pertinhez, T. A., Muller, M., and Ottonello, S. (2014). A high-performance thioredoxin-based scaffold for peptide immunogen construction: proof-of-concept testing with a human papillomavirus epitope, *Sci. Rep.*, **4**, p. 4729.

Chapter 3

Aptamer Selection Methodologies

Penmetcha K. R. Kumar
Biomedical Research Institute, National Institute of Advanced Industrial Science and Technology (AIST), 1-1-1 Higashi, Central 6, Tsukuba City, Ibaraki 305-8566, Japan
pkr-kumar@aist.go.jp

Aptamers are rare functional nucleic acids that are isolated from a library of combinatorial nucleic acids by iterative rounds of selection and amplification. In this process, nucleic acid libraries directed against a predefined target (either biological or chemical) are created and screened to isolate nucleic acids that bind to the target with high specificity and affinity. Recently, aptamers have been isolated against a wide range of targets. The selection methodologies that are used during the selection cycles play a vital role in the successful isolation of aptamers against any target. In this chapter, several selection strategies employed for the selection of aptamers are described.

3.1 Introduction

Aptamers are isolated from completely random pools of nucleic acids in libraries that typically consist of 10^{13} molecules using a specific ligand. To enrich selective binders, iterative rounds of

Aptamers: Tools for Nanotherapy and Molecular Imaging
Edited by Rakesh N. Veedu

ISBN 978-981-4669-83-2 (Hardcover), 978-981-4669-84-9 (eBook)
www.panstanford.com

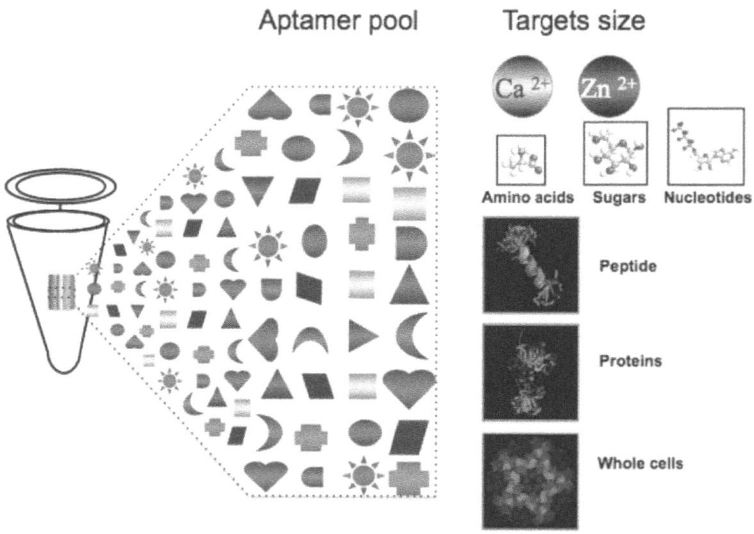

Figure 3.1 Aptamers selected against a wide range of target ligands.

selection and amplification are commonly employed. The selected aptamers display high specificity and affinity toward their cognate ligands. For this reason, aptamers are sometimes referred to as "rivals" of antibodies. However, aptamers have many advantages over antibodies, such as better molecular discrimination, smaller size, less temperature sensitivity, and no batch-to-batch variation. In addition, they are nonimmunogenic, and it is possible to isolate aptamers against a ligand of any size. Since the inception of aptamer methodologies in the 1990s by Ellington and Szostak [1] and by Tuerk and Gold [2], a number of aptamers have been isolated against a wide range of ligands, including simple ions, small molecules, peptides, proteins, organelles, viruses, and even entire cells (Fig. 3.1). Some of these selected aptamers were commercialized for both diagnostic and therapeutic applications. It was projected that many of these aptamers would attain a global market value $2.1 billion by 2018 (marketsandmarkets.com).

The general procedure for the selection of aptamers involves incubating a pool of nucleic acids with a target (ligand) of choice under an appropriate buffer condition, in which the ligand acquires its native conformation, followed by separating the free from

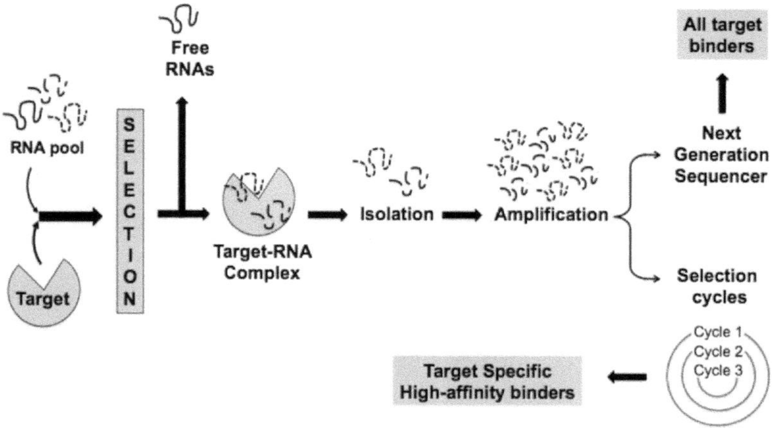

Figure 3.2 Aptamer selection process.

ligand-bound nucleic acids. The isolated ligand-bound aptamers are amplified to regenerate the ligand-bound nucleic acids and are then subjected to the next round of selection and amplification (Fig. 3.2). Alternatively, because of the availability of next-generation sequencers, it is possible to sequence the entire population of bound nucleic acids. The advantage of this approach is the ability to identify all binders that may recognize different epitopes or functional groups within a ligand. However, the bound pools may also contain both low and nonspecific binders such as matrix-bound species. Thus, a decision must be made whether or not to carry out additional selection cycles to eliminate nonspecific binders and to mitigate the postselection process of identifying the most efficient binders. The advantages of proceeding with additional rounds of selection and amplification are twofold: (1) the elimination of nonspecific binders and (2) the reduction of postselection analyses required for identification of efficient binders, as few sequences will dominate in the pool. Although methodologies for selection of aptamers have been around for more than two decades, aptamer applications in diagnostics and bioassays are limited compared to those of antibodies. Reasons for these discrepancies are partly due to the availability of suitable diagnostic platforms for aptamers but also because some of the reported aptamers bind either weakly or nonspecifically. Such weak or nonspecific aptamers might have originated from poor selection strategies applied during the aptamer selection process. The primary goal of the aptamer selection process should be to obtain the highest affinity (subpicomolar), specificity, and ability to discriminate between closely related targets so that they can be useful in medical diagnosis and therapeutic applications. Aptamers considered for these applications must meet all of the above conditions. Thus, the aptamer selection process, including the methods employed, plays a significant role in the performance of the selected aptamers in various biomedical applications. The present review is focused on aptamer selection strategies and methodologies that are employed to select aptamers with the highest affinities and specificities.

The selection of aptamers involves primarily three important steps: (1) random pool design and preparation, (2) selection of target-bound aptamers, and (3) amplification of the bound aptamers (Fig. 3.3). All three steps significantly contribute to the successful selection of aptamers with high affinity and specificity. The random pools that are used for the selection of aptamers all

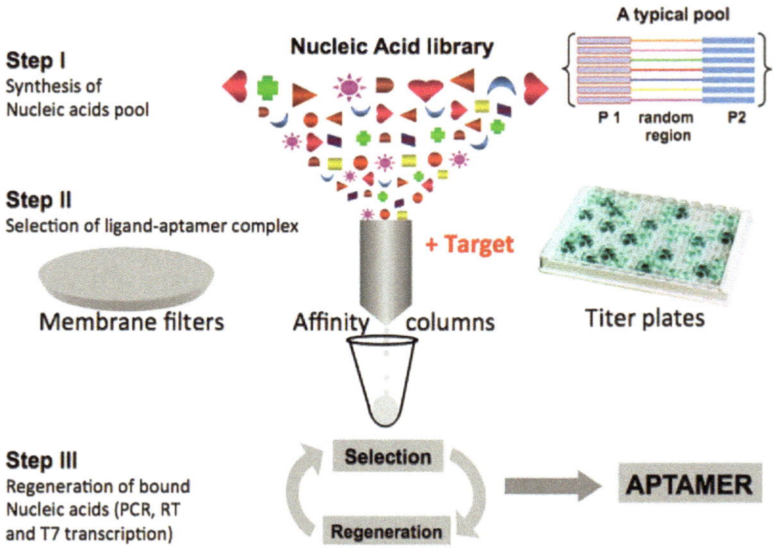

Figure 3.3 Important steps in aptamer selections.

have random sequences flanked by specific primer regions (Fig. 3.3, P1 and P2). In the pool design step, close attention is paid to four parameters: randomization (partial, segmental, or complete), length of the randomized region, the chemistry of the pool, and the primary primer sequences [3]. Successful and rapid aptamer selection (number of selection cycles required) primarily depends upon the approach by which separation and isolation of ligand-bound aptamers from the unbound nucleic acids are achieved. A number of approaches have been described for the separation of ligand-bound from free aptamers.

3.2 Selection Methods

Currently, several methodologies are available to separate the free, or target-unbound, aptamers from the complexed, or target-bound, aptamers. The selection of the method to be used generally depends upon the size of the ligand to which aptamers are directed. These methods have been adopted primarily from the biochemical techniques used for the analysis of nucleic acid–ligand interactions and from affinity chromatographic techniques. Listed below are methods that are widely used for the selection of aptamers against a wide range of ligands (Fig. 3.4).

3.2.1 *Affinity Column Selection*

An affinity column selection method is commonly employed when the target ligand molecular size is <1000 Daltons. In this strategy, the target ligand is linked to a matrix with the important, or target, functional groups of the ligand exposed and available for the selection of aptamers. The aptamer selection technology was demonstrated for the first time by Ellington and Szostak [1], who used an affinity column selection strategy. In their studies, the blue dyes (Cibacrone blue and reactive blue) were cross-linked to agarose and selection was then performed. To separate the matrix-bound aptamers from the blue dye–bound aptamers, counterselection was performed using cellulose columns. The following critical issues have been considered when affinity columns are used during the

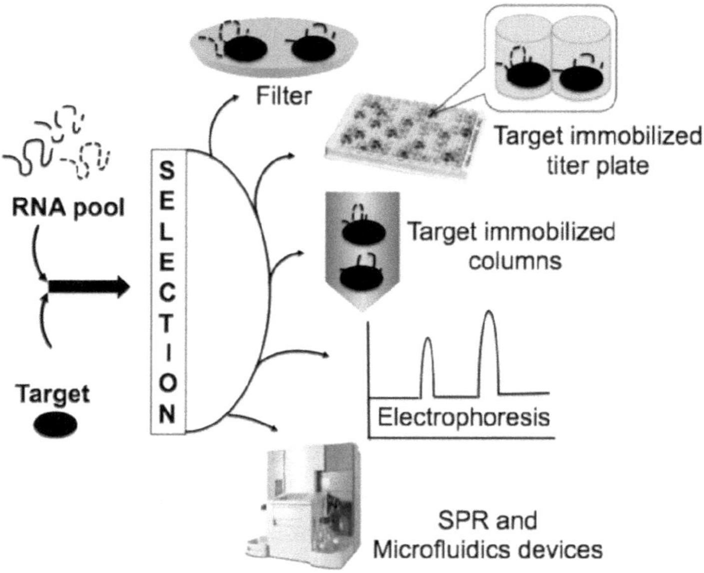

Figure 3.4 Commonly employed strategies for aptamer selection in vitro.

selection of aptamers against a wide variety of small molecules, such as metal ions, theophylline, adenosine triphosphate (ATP), porphyrins, antibiotics, mycotoxins, peptides, and other small molecules (Table 3.1):

- Stability of the matrix under various selection conditions, including buffers, pH, temperature, and detergent.
- Efficiency and density of the target ligand linked to the matrix. Currently, various types of activated matrices are commercially available for linking ligands.
- A spacer is generally considered between the matrix and the target ligand to facilitate accessibility of the functional group target for aptamer selection.
- Blocking of unreactive functional groups on the matrix and the use of monovalent and divalent cations in the selection buffer to prevent nonspecific binding.
- Other important factors, including the column size and the elution strategy for ligand-bound aptamers, also need to be considered.

Table 3.1 Aptamers selected using affinity columns

Target ligand	Matrix	References
Aminoglycosides	Agarose	[4]
ATP	Agarose	[5]
Cellulose	Cellulose	[6]
Cholic acid	Agarose	[7]
Cibaron blue	Agarose	[1]
Cocaine	Unknown	[8]
Gjanosine	Agarose	[9]
L-Arginine	Sepharose 6B	[10]
L-Citrulline	Sepharose 6B	[11]
Neomycin	Sepharose	[12]
Nicotinamide	Agarose	[13]
N-Methylmeso- porphyrin (NMM)	Acrylic	[14]
Ochratoxin	Agarose	[15]
Reactive blue	Agarose	[1]
Riboflavin	Agarose	[13]
Sephadex	Sephadex	[16]
Sphingolipid	SA-resin	[17]
Sphingosylphosphorylcholine	SA-resin	[18]
Tetracycline	Sepharose 68	[19]
Theophylline	EAH Sepharose	[20]
17β-estradiol	Sepharose	[21]

3.2.2 *Nitrocellulose Filter Selection*

The capture and analysis of RNA–protein complexes on a nitrocellulose filter were initially reported by Jones and Berg [22]. This method is based on the fact that nucleic acids do not bind to nitrocellulose filters, while many proteins can display a strong affinity toward the nitrocellulose filters without losing their binding affinity for their cognate ligands. This method involves the filtration of RNA–protein complexes in solution through nitrocellulose, followed by a wash step to remove nonspecifically retained free RNA. This method has been routinely used in the analysis of RNA–protein complexes, including thermodynamic characterization [23]. Tuerk and Gold [2] explored the use of cellulose filters for the selection of aptamers that bind to the T4 DNA polymerase. Nitrocellulose selection is commonly considered when the target ligand is a protein. Several aptamers that have been selected using nitrocellulose filters against a wide range of proteins are listed in Table 3.2. A prominent type of filter used for aptamer selection is the hydrophilic HAWP cellulose-mixed ester disk with a pore size of 0.45 μM and a thickness of 180 μm. Cellulose filters offer a convenient method for selecting aptamers without any modification of the target protein (under native conditions), and can be available for a wide range of selection conditions and used under a variety of laboratory conditions. Moreover, the method can be used to isolate multiple targets simultaneously by employing a 96-well filter manifold dot-blot apparatus.

Although a number of aptamers were selected using nitrocellulose filters, the method does pose problems during the selection process, including (1) the efficiency of the binding of the target protein to the cellulose filter and (2) the efficiency of removing certain free nucleic acids that bind specifically to the nitrocellulose filters, such as G-rich sequences. These nucleic acids must be removed effectively from the initial pool and also from successive pools in order to carry out successful selection cycles. A common approach for removing these specific filter-bound aptamers is a preselection or postselection procedure, where the nucleic acid pools are filtered through the prewetted nitrocellulose filters at least twice in the absence of target protein. This procedure facilitates the removal or reduction of the background nucleic acids—from the pool—that are retained on the nitrocellulose filters.

3.2.3 *Affinity Surfaces and Tag Selection*

A variety of surfaces and tags are currently available for immobilizing both small and macromolecular ligands. Commonly used tag proteins are GST, MBP, and His for immobilization on glutathione, amylose resins, and Ni-agarose, respectively. These tags can be added to the target proteins at either the N- or C-terminal ends. Using these tags, target proteins were immobilized specifically to the appropriate matrix, as mentioned earlier. For selection of aptamers, the nucleic acid pool was incubated with the target-immobilized matrix, followed by separation of the matrix and washing and elution of the fusion protein. Nucleic acids bound to the eluted protein are subjected to further rounds of amplification

Table 3.2 Aptamers selected using cellulose filters

Target ligand	References	Target ligand	References
Angiopoietin-2	[24]	Human coagulation factor IXa	[47]
Avian myeloblstosis and Moloney murine leukemia	[25]	Human cytomegalovirus	[48]
(Reverse transcriptase)		Human epidermal growth factor receptor-3	[49]
Bacteriophage MS2 coat protein	[26]	Human factor Vila	[50]
Bacteriophage R17 coat protein	[27]	Human IgE	[51]
Coat protein of alfalfa mosaic virus	[28]	Human rIFN-γ protein	[52]
Drosophila transformer 2 (Tra2)	[29]	Human nonpancreatic secretory phospholipase A2	[53]
Ebola Viral protein35 (C-terminal IFN inhibitory		Influenza A/H1N1/California/7/2009	
domain, IID)	[30]	Hemagglutinin	[54]
Elongation factor SelB	[31]	Influenza A/H3N2/Panama/2007/1999	
		Hemagglutinin	[55]
HCV NS3	[32–34]	Influenza A/H5N1/Indonesia/5/2005	
		Hemagglutinin	[56]
HCV NS3 helicase domain	[35]	Influenza A/H5N1/Vietnam/1194/2004	
		Hemagglutinin	[56]
Histidine utilizing protein (HutP)	[36]	Influenza A/H7N7/Netherlands/219/2003 Hemagglutinin	[56]
HIV-1 gag polyprotein	[37]	Influenza B/Johannesburg Hemagglutinin	[57]
HIV-1 Integrase	[38]	Keratinolytic growth factor	[58]
HIV-1 Nucleo capsid	[39]	Mouse prion protein	[59]
HIV-1 Rev	[40, 41]	Protein kinase C βII	[60]
HIV-1 Reverse transcriptase	[42]	Rex fusion protein	[61]
HIV-1 Tat	[43]	Ribosome-inactivating protein, Pepocin	[62]
HSV-1 gD	[44]	Rous sarcoma virus	[63]
Human activated protein (ARC)	[45]	T4 DNA polymerase	[2]
Human a-thrombin	[46]	Vascular endothelial growth factor (VEGF)	[64]

and selection. Alternatively, the fusion protein can be incubated with the nucleic acid pool, and the target protein can then be captured by the affinity matrix. Using these approaches, aptamers were selected against the ribosomal protein L22 [65], the hamster prion protein [66], and the nucleic acid protein of HIV [67]. As an alternative to using an affinity matrix in the absence of fusion tags, microtiter plates activated for covalent binding have been explored for immobilization of the target protein followed by in vitro selection [44, 55, 57]. For better recovery from the resins, magnetic beads have been utilized for immobilizing target proteins or other small ligands and used for in vitro selection [68–71]. This type of selection strategy is attractive because no special equipment is required, and it is suitable for multiple targets as well. As with other matrices, this approach also suffers from binding of free nucleic acid aptamers; thus a preselection strategy similar to that used for cellulose-bound aptamers described earlier is required.

3.2.4 *Electrophoresis Selection*

The use of a gel retardation assay, or gel shift assay, was initially reported for characterization of bacterial polyribosomes [72] and was later extended to study a variety of RNA–protein [34, 43, 73–75] and DNA–protein [76–78] complexes. It relies on the property that nucleic acids migrate through a gel matrix (either agarose or polyacrylamide) toward the anode when an electric field is applied. The migration of nucleic acids through gel is influenced by three factors: (1) the molecular weight of the nucleic acid, (2) the three-dimensional shape of the nucleic acid, and (3) the physical properties of the gel. In a nucleic acid–protein complex, the binding of protein modulates the conformation of the nucleic acid and the molecular weight of the nucleic acid–protein complex leads to a slower mobility compared to free nucleic acids in a gel matrix. The gel retardation assay has been used for various structure–function analyses of nucleic acid–protein interactions, including the evaluation of rate constants [78–80]. Since the gel shift clearly separates the free nucleic acids from the nucleic acid–protein complexes, this gel shift strategy has been employed during the selection of aptamers against three target proteins [81–83].

Although this method is used in the analysis of nucleic acid–protein complexes, the method has the following limitations as a common strategy for aptamer selection: (a) It is only suitable for nucleic acid–protein complexes that are well resolved from free nucleic acids in the gel matrix, (b) some complexes fail to enter the gel matrix or to be clearly resolved, (c) it is only suitable for complexes that have a long dissociation time, as the gel runtime is generally measured in hours, and (d) it is not suitable for high-throughput selection.

To circumvent the limitations of slab-gel electrophoresis, a capillary electrophoresis (CE) technique for the analysis of nucleic acid–protein complexes has been reported [84]. As CE has the ability to rapidly separate and quantitate the nucleic acid–protein complexes compared to slab-gel electrophoresis, the application of CE for aptamer selection has been explored, and it enables the selection of high-affinity aptamers against their cognate targets [84–88]. The advantage of the CE-based selection strategy is the in-solution aptamer–target binding and the possibility of expanding to high-throughput selection using multiple targets. However, the CE-based selection strategy poses some restrictions such as the size of the nucleic acid library, and despite the development of micro-free-flow electrophoresis [89], it is not suitable for targets that do not yield a shift during electrophoresis or larger targets such as whole cells.

3.2.5 *Centrifugation Selection*

A centrifugation strategy that settles particles suspended in fluids is commonly used for larger complexes. In this strategy, the aptamer library is incubated with macromolecular particles, followed by centrifugation of the reaction mixture to separate the particles; the particle-associated nucleic acids are isolated and multiple rounds of amplification and selection are performed. Using this strategy, aptamers to define the pro-head domain [90], African trypanosomes [91], rat brain tumor microvessels [92], disease-associated prion protein [93], and receptor tyrosine kinase [94] were selected. This method is convenient and suitable only for macromolecules, which can be recovered by simple centrifugation procedures. Thus, this methodology is suitable for, but limited to, targets associated with larger particles or whole cells.

3.2.6 *Surface Plasmon Resonance Selection*

Surface plasmon resonance (SPR) has emerged as a sensitive detection platform for a wide range of biomolecular interactions such as protein–protein, protein–ligand, protein–DNA, protein–RNA, and nucleic acid–small ligand interactions [95]. SPR is an optical detection platform that uses prism coupling and allows characterization of binding kinetics of biomolecular interactions in a real-time fashion in a label-free environment. Typically, in an SPR system, one interacting biomolecule is immobilized on the sensor surface using different matrixes, such as dextran. Beneath the dextran surface, a thin gold layer is present that is attached to a glass surface (Fig. 3.5A). This entire surface is placed within a flow cell. A cognate biomolecule (analyte) that interacts with an immobilized biomolecule is injected into the flow cell in the presence of a suitable buffer. Surface plasmons are generated when an incident beam of light is directed toward the gold surface at a critical angle. The critical angle depends on the refractive index near the surface, which changes when an analyte binds to the immobilized biomolecule (Fig. 3.5A). The resulting change in refractive index at the surface is measured, and the extent of that change correlates with the mass concentration near the surface and is displayed as a plot (sensogram) of resonance units (RUs) (Fig. 3.5B). The following three phases are observed in the sensogram when an analyte is injected on the immobilized protein surface: (1) an association phase, where the analyte binds to the immobilized protein during the injection; (2) an equilibrium phase, which is displayed at the end of the analyte injection, where the analyte binding rate reaches

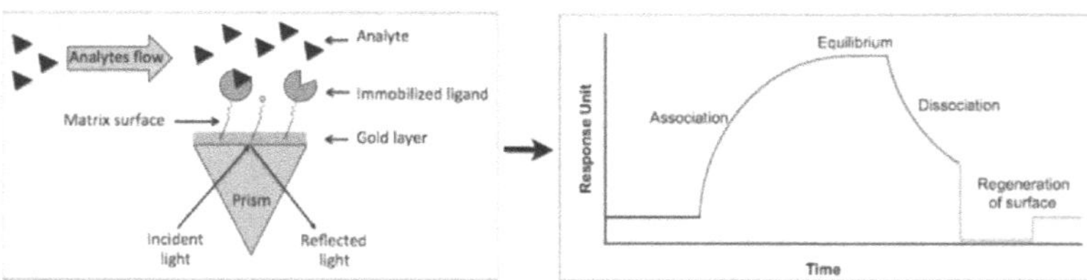

Figure 3.5 Surface plasmon resonance (SPR) platform for biomolecular interaction analyses. (A) SPR biomolecular analyses system and (B) sensogram generated upon ligand interaction with immobilized biomolecule.

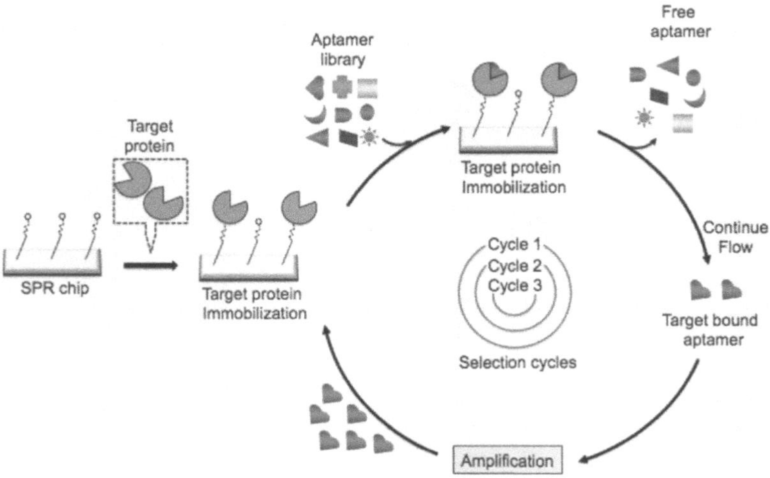

Figure 3.6 Aptamer selection scheme adopted for the SPR system.

a steady state; and (3) a dissociation phase, which is observed when the analyte dissociates from the immobilized protein. This occurs during the continuous flow of buffer over the immobilized surface (Fig. 3.5B). These events can all be monitored in real time. The residual analyte remaining on the sensor surface can be removed, and the surface can be regenerated without damaging the activity of the immobilized protein by using various strategies.

The SPR system (Biacore) was first explored for isolating high-affinity binding peptides from a phage antibody library that efficiently recognizes the tumor antigen c-erbB-2 [96]. Similarly, aptamer selection using Biacore was performed for RNase H1 [97], HIV-1 gp120 [98], and influenza hemagglutinin (HA) [99]. A general scheme employed for the selection of aptamers from a nucleic acid library using the SPR system is summarized in Fig. 3.6 [99]. First, HA is immobilized on the surface of a CM4 chip (step a, in Fig. 3.7), followed by the injection of an RNA pool that contains approximately 10^{13} sequences. The injected RNA pool flows over the immobilized HA surface and facilitates binding to the HA (step b, in Fig. 3.7). Upon completion of aptamer binding, the binding buffer (10 mM Hepes, pH 7.4, 150 mM NaCl, 3 mM EDTA and 0.005% Tween 20) is injected into the same flow cell. During this process, aptamers dissociate

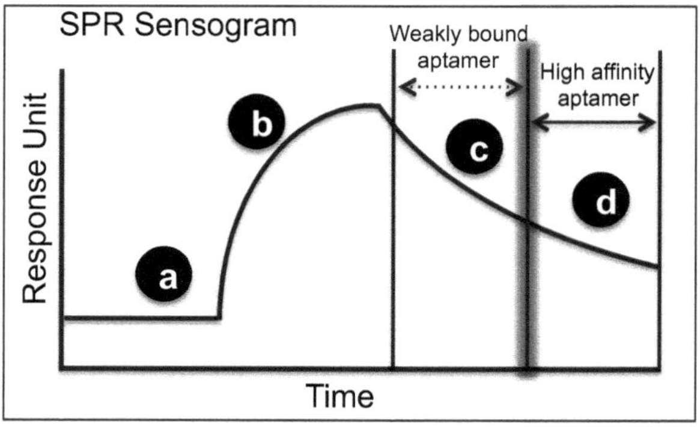

Figure 3.7 A simplified representation of sensogram observed during aptamer selection using the SPR system.

from the HA and are collected in fractions instead of being commonly directed to the waste bottle. It is believed that the initial fractions from the flow cell contain a pool of aptamers that represents either unbound aptamers or weakly HA-interacting aptamers (step c, in Fig. 3.7). High-affinity aptamers are then collected in later fractions (step d, in Fig. 1.7). Therefore, in each selection cycle, the first few fractions are not used for amplification. A total of five selection cycles are performed using a single flow cell (requiring only 1.2 μg of HA). Interestingly, high-affinity aptamers that bind to the HA were recovered from the RNA library near the completion of two selection cycles using the SPR system. These results suggest that the SPR system provides an improved methodology for rapidly enriching high-affinity binding species of multiple targets. Moreover, the great advantage of SPR is that it provides information about the target-binding abilities of molecules before they are selected. This procedure can be extended to new SPR systems, such as Biacore 4000, for use in high-throughput selection for multiple targets.

3.2.7 *Microfluidics Selection*

Affinity chromatography and cellulose filter selection methodologies have been used successfully for selecting a number of aptamers against a wide range of biomolecules. However, these conventional methods are tedious and time-consuming processes that typically

Table 3.3 Aptamers selected using microfluidics

Microfluidics election method	Target	Reference
Capillar Electrophresis	Human IgE	[85, 86, 89]
	Nueropeptide Y	[100]
	HIV Reverse transcriptase	[101]
	rh VEGF165	[102]
	N-methyl mesoporphyrin	[103]
	Protein fernesyltransferase	[104]
	MutS protein	[88]
	h-Ras	[105]
Sol-Gel	Yeast tarnscription factor MB	[106]
	Xanthine	[107]
	TATA-binding protein	[108]
Magnetic beads	Streptavidin	[109]
	BoNt/A-rLc	[110]
	PDGF-BB protein	[111]
	Angiopoietin-2	[112]
Integrated system	C-recative protein	[113]
	α-fetoprotein	[114]
	A549	[102]
	Myoglobin	[115]
Agarose droplet	Shp2	[116]
Monoclonal surface display	EpCam	[118]
	AFB	[118]

require more than 10 cycles of selection and amplification. Considering these limitations, microfluidic-based selection methodologies, including the SPR methodology, were explored to improve both the efficiency as well as the separation of free and complexed (target-bound) aptamers and to provide automation capability. Aptamers that have been selected using microfluidics are listed in Table 3.3.

3.2.8 *Alternating Two Selection Methods*

A critical step in the above-listed strategies for the selection of aptamers is the complete or near-complete separation of the matrix-bound nucleic acid species from the target-bound species before proceeding to downstream applications of aptamers. All of the above strategies are successful in selecting aptamers mainly because a negative selection step is employed to remove species that bound

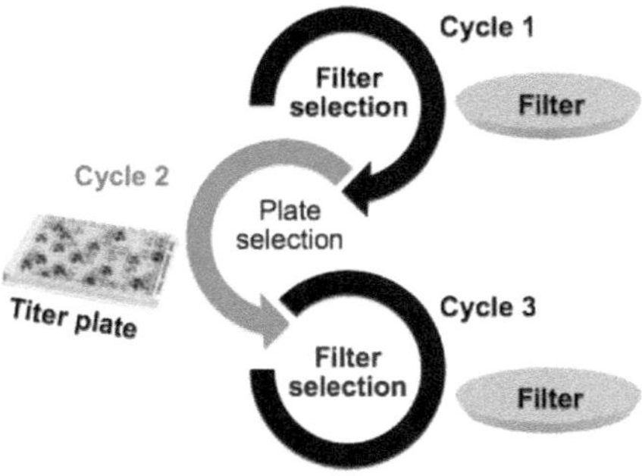

Figure 3.8 Aptamer selections using alternating two selection methods.

to the matrix. Alternatively, two types of selection strategies can be exploited during selection cycles. In this scenario, the matrix-bound species that accumulate in the first selection are eliminated in the next round of selection when a different matrix is used. If more than two cycles are required for aptamer selection, then rotating two matrices is recommended instead of using a single matrix (Fig. 3.8). This approach appears to be more convenient and has a greater success rate for aptamer selection.

Acknowledgments

Aptamer research in the author's lab was supported by funds from the National Institute of Advanced Industrial Science and Technology (AIST).

References

1. Ellington, A. D., and Szostak, J. W. (1990). In vitro selection of RNA molecules that bind specific ligands, *Nature*, **346**, pp. 818–822.
2. Tuerk, C., and Gold, L. (1990). Systematic evolution of ligands by exponential enrichment: RNA ligands to bacteriophage T4 DNA polymerase, *Science*, **249**, pp. 505–510.
3. Marshall, K. A., and Ellington, A. D. (2000). *In vitro* selection of RNA aptamers, *Methods Enzymol.*, **318**, pp. 93–214.

4. Lato, S. M., Boles, A. R., and Ellington, A. D. (1995). *In vitro* selection of RNA lectins: using combinatorial chemistry to interpret ribozyme evolution, *Chem. Biol.*, **2**, pp. 291–303.
5. Huizenga, D. E., and Szostak, J. W. (1995). A DNA aptamer that binds adenosine and ATP, *Biochemistry*, **34**, pp. 656–665.
6. Boese, B. J., and Breaker, R. R. (2007). *In vitro* selection and characterization of cellulose-binding DNA aptamer, *Nucleic Acids Res.*, **35**, pp. 6378–6388.
7. Kato, T., Takemura, T., Yano, K., Ikebukuro, K., and Karube, I. (2000). *In vitro* selection of DNA aptamers, which binds to cholic acid, *Biochim. Biophs. Acta*, **1493**, pp. 12–18.
8. Stojanovic, M. N., de Prada, P., and Landry, D. W. (2000). Fluorescent sensor based on aptamer self-assembly, *J. Am. Chem. Soc.*, **122**, pp. 11547–11548.
9. Connell, G. J., and Yarus, M. (1994). RNAs with dual specificity and dual RNAs with similar specificity, *Science*, **264**, pp. 1137–1141.
10. Connell, G. J., Illangesekare, M., and Yarus, M. (1993). Three small ribooligonucleotides with specific arginine sites, *Biochemistry*, **32**, pp. 5497–5502.
11. Famulok, M. (1994). Molecular recognition of amino acids by RNA aptamers: an L-citrulline binding RNA motif and its evolution into an L-arginine, binder, *J. Am. Chem. Soc.*, **116**, pp. 1698–1706.
12. Wallis, M. G., Vonahsen, U., Schroeder, R., and Famulok, M. (1995). A novel RNA motif for neomycin recognition, *Chem. Biol.*, **2**, pp. 543–552.
13. Lauhon, C. T., and Szostak, J. W. (1995). RNA aptamers that bind flavin and nicotinamide redox cofactors, *J. Am. Chem. Soc.*, **117**, pp. 1246–1257.
14. Li, Y. F., Geyer, C. R., and Sen, D. (1996). Recognition of anionic porphyrins by DNA aptamers, *Biochemistry*, **35**, pp. 6911–6922.
15. Cruz-Aguado, J. A., and Penner, G. (2008). Determination of ochratoxin a with a DNA aptamer, *J. Agric. Food Chem.*, **56**, pp. 10456–10461.
16. Srisawat, C., Goldstein, I. J., and Engelke, D. R. (2001). Sephadex-binding RNA ligands: rapid affinity purification of RNA from complex RNA mixtures, *Nucleic Acids Res.*, **29**, p. e4.
17. Purschke, W. G., Hoehlig, K., Buchner, K., Zboralski, D., Schwoebel, F., Vater, A., and Klussmann, S. (2014). Identification and characterization of a mirror-image oligonucleotide that binds and neutralizes sphingosine 1-phosphate, a central mediator of angiogenesis, *Biochem. J.*, **462**, pp. 153–162.

18. Hori, K., Omi, K., Yoshida, Y., Imai, Y., Sakai, N., Oka, A., Masuda, H., Furuichi, M., Tanimoto, T., and Waga, I. (2010). Development of a sphingosylphosphorylcholine detection system using aptamers, *Molecules*, **15**, pp. 5742–5755.
19. Berens, C., Thain, A., and Schroeder, R. (2001). A tetracycline-binding RNA aptamer, *Bioorg. Med. Chem.*, **9**, pp. 2549–2556.
20. Jenison, R. D., Gill, S. C., Pardi, A., and Polisky, B. (1994). High resolution molecular discrimination by RNA, *Science*, **263**, pp. 1425–1429.
21. Vanschoenbeek, K., Vanbrabant, J., Hosseinkhani, B., Vermeeren, V., and Michiels, L. (2015). Aptamers targeting different functional groups of 17b-estradiol, *J. Steroid Biochem. Mol. Biol.*, **147**, pp. 10–16.
22. Krug, M., de Haseth, P. L., and Uhlenbeck, O. C. (1982). Enzymatic synthesis of a21-nucletide coat protein binding fragment of R17 ribonucleic acid, *Biochemistry*, **21**, pp. 4713–4729.
23. White, R. R., Shan, S., Rusconi, C. P., Shetty, G., Dewhirst, M. W., Kontos, K. D., and Sullenger, B. A. (2003). Inhibition of rat corneal angiogenesis by a nuclease-resistant RNA aptamer specific for angiopoietin-2, *Proc. Natl. Acad Sci. U S A*, **100**, pp. 5028–5033.
24. Chen, H., and Gold, L. (1994). Selection of high-affinity RNA ligands to reverse transcriptase: inhibition of cDNA synthesis and RNase H activity, *Biochemistry*, **33**, pp. 8746–8756.
25. Shtatland, T., Gill, S. C., Javornik, B. E., Johansson, H. E., Singer, B. S., Uhlenbeck, O. C., Zichi, D. A., and Gold, L. (2000). Interactions of Escherichia coli RNA with bacteriophage MS2 coat protein: genomic SELEX, *Nucleic Acids Res.*, **28**, p. e93.
26. Schneider, D., Tuerk, C., and Gold, L. (1992). Selection of high affinity RNA ligands to the bacteriophage R17 coat protein, *J. Mol. Biol.*, **228**, pp. 862–869.
27. Houser-Scott, F., Ansel-Mckinney, P., Cai, J., and Gehrke, L. (1997). *In vitro* genetic selection analysis of alfalfa mosaic virus coat protein binding to 3’-terminal AUGC repeats in the viral RNAs, *J. Virol.*, **71**, pp. 2310–2319.
28. Brunel, C., Ehresmann, B., Ehresmann, C., and McKeown, M. (2001). Selection of genomic target RNAs by iterative screening, Bioorganic, *Med. Chem.*, **9**, pp. 2533–2541.
29. Binning, J. M., Wang, T., Luthra, P., Shabman, R. S., Borek, D. M., Liu, G., Xu, W., Leung, D. W., Basler, C. F., and Amarasinghe, G. K. (2013). Development of RNA aptamers targeting Ebola virus VP35, *Biochemistry*, **52**, pp. 8406–8419.

30. Klug, S. J., Hüttenhofer, A., Kromayer, M., and Famulok, M. (1997). *In vitro* and *in vivo* characterization of novel mRNA motifs that bind special elongation factor SelB, *Proc. Natl. Acad Sci. U S A*, **94**, pp. 6676–6681.
31. Kumar, P. K. R., Machida, K., Urvil, P. T., Kakiuchi, N., Vishnuvardhan, D., Shimotohno, K., Taira, K., and Nishikawa, S. (1997). Isolation of RNA aptamers specific to the NS3 protein of hepatitis C virus from a pool of completely random RNA, *Virology*, **237**, pp. 270–282.
32. Fukuda, K., Vishnuvardhan, D., Sekiya, S., Hwang, J., Kakiuchi, N., Taira, K., Shimotohno, K., Kumar, P. K. R., and Nishikawa, S. (2000). Isolation and characterization of RNA aptamers specific for the hepatitis C virus nonstructural protein 3 protease, *Eur. J. Biochem.*, **267**, pp. 3685–3694.
33. Urvil, P. T., Kakiuchi, N., Zhou, D., Shimotohno, K., Kumar, P. K. R., and Nishikawa, S. (1997). Selection of RNA aptamers that bind specifically to the NS3 protease of hepatitis C virus, *Eur. J. Biochem.*, **248**, pp.130–138.
34. Nishikawa, F., Funaji, K., Fukuda, K., and Nishikawa, S. (2004). *In vitro* selection of RNA aptamers against the HCV NS3 helicase domain, *Oligonucleotides*, **14**, pp. 114–129.
35. Kumarevel, T. S., Gopinath, S. C. B., Nishikawa, S., Mizuno, H., and Kumar, P. K. R. (2004). Identification of important chemical groups of the hut mRNA for HutP interactions that regulate the *hut* operon in *Bacillus subtilis*, *Nucleic Acids Res.*, **32**, pp. 3904–3912.
36. Lochrie, M. A., Waugh, S., Pratt, D. G. Jr., Clever, J., Parslow, G., and Polisky, B. (1997). *In vitro* selection of RNAs that bind to the human immunodeficiency virus type-1 gag polyprotein, *Nucleic Acids Res.*, **24**, pp. 2902–2910.
37. Allen, P., Worland, S., and Gold, L. (1995). Isolation of high-affinity RNA ligands to HIV-1 integrase from a random pool, *Virology*, **209**, pp. 327–336.
38. Berglund, J. A., Charpentier, B., and Rosbash, M. (1997). A high affinity binding site for the HIV-1 nucleocapsid protein, *Nucleic Acids Res.*, **25**, pp. 1042–1049.
39. Bartel, D. P., Zapp, M. L., Green, M. R., and Szostak, J. W. (1991). HIV-1 Rev regulation involves recognition of non-Watson-Crick base pairs in viral RNA, *Cell*, **67**, pp. 529–536.
40. Jensen, K. B., Atkinson, B. L., Willis, M. C., Koch, T. H., and Gold, L. (1995). Using in vitro selection to direct the covalent attachment of

human immunodeficiency virus type 1 Rev protein to high-affinity RNA ligands, *Proc. Natl. Acad. Sci. U S A*, **92**, pp. 12220–12223.

41. Tuerk, C., Macdougal. S., and Gold, L. (1992). RNA pseudoknots that inhibit human immunodeficiency virus type 1 reverse transcriptase, *Proc. Natl. Acad. Sci. U S A*, **89**, pp. 6988–6992.
42. Yamamoto, R., Murakami, K., Taira, K., and Kumar, P. K. R. (1998). Isolation and characterization of an RNA that binds with high affinity to Tat protein of HIV-1 from a completely random pool of RNA, *Gene Ther. Mol. Biol.*, **1**, pp. 451–466.
43. Gopinath, S. C. B, Hayashi, K., and Kumar, P. K. (2012). Aptamer that binds to the gD protein of herpes simplex virus 1 and efficiently inhibits viral entry, *J. Virol.*, **86**, pp. 6732–6744.
44. Gal, S. W., Amontov, S., Urvil, P. T., Vishnuvardhan, D., Nishikawa, F., Kumar P. K. R., and Nishikawa, S. (1998). Selection of a RNA aptamer that binds to human activated protein C and inhibits its protease function, *Eur. J. Biochem.*, **252**, pp. 553–562.
45. Kubik, M. F., Stephens, A. W., Schneider, D., Marlar, R., and Tasset, D. (1994). High-affinity RNA ligands to human alpha-thrombin, *Nucleic Acids Res.*, **22**, pp. 2619–2626.
46. Rusconi, C. P., Scardino, E., Layzer, J., Pitoc, G. A., Ortel, T. L., Monroe, D., and Sullenger, B. A. (2002). RNA aptamers as reversible antagonists of coagulation factor IXa, *Nature*, **419**, pp. 90–94.
47. Wang, J., Jiang, H., and Liu, F. (2000). *In vitro* Selection of novel RNA ligands that bind human cytomegalovirus and block viral infection, *RNA*, **6**, pp. 571–583.
48. Chen, C. B., Chernis, G. A., Hoang, V. Q., and Landgraf, R. (2003). Inhibition of heregulin signaling by an aptamer that preferentially binds to the oligomeric form of human epidermal growth factor receptor-3, *Proc. Natl. Acad. Sci. U S A*, **100**, pp. 9226–9231.
49. Rusconi, C. P., Yeh, A., Lyerly, H. K., Lawson, J. H., and Sullenger, B. A. (2000). Blocking the initiation of coagulation by RNA aptamers to factor VIIa, *Thromb. Haemost.*, **84**, pp. 841–848.
50. Wiegand, T. W., Williams, B. P., Dreskin, S. C., Jouvin, M., Kinet, J., and Tasset, D. (1996). High-affinity oligonucleotide ligands to human IgE inhibit binding to Fc epsilon receptor I, *J. Immunol.*, **157**, pp. 221–230.
51. Kubik, M. F., Bell, C., Fitzwater, T., Watson, S. R., and Tasset, D. M. (1997) Isolation and characterization of 2'-fluoro-, 2'-amino-, and 2'-fluoro-/amino-modified RNA ligands to human IFN-gamma that inhibit receptor binding, *J. Immunol.*, **159**, pp. 259–267.

52. 53.Bridonneau, P., Chang, Y., O'Connell, D., Gill, S. C., Snyder, D. W., Johnson, L., Goodson, T., Herron, D. K., and Parma, D. H. (1998). High-affinity aptamers selectively inhibit human nonpancreatic secretory phospholipase A2 (hnps-PLA2), *J. Med. Chem.,* **41**, pp. 778–786.
53. Gopinath, S. C. B., and Kumar, P. K. (2013). Aptamer that binds to the hemagglutinin of the recent pandemic influenza virus H1N1 and efficiently inhibit agglutination, *Acta Biomater.,* **9**, pp. 8932–8941.
54. Gopinath, S. C. B., Misono, T., Kawasaki, K., Mizuno, T., Imai, M., Odagiri, T., and Kumar, P. K. R. (2006). An RNA aptamer that distinguishes between closely related human influenza viruses and inhibits haemagglutinin-mediated membrane fusion, *J. Gen. Virol.,* **87**, pp. 479–487.
55. Suenaga, E., and Kumar, P. K. (2014). An aptamer that binds efficiently to the hemagglutinins of highly pathogenic avian influenza viruses (H5N1 and H7N7) and inhibits hemagglutinin-glycan interactions, *Acta Biomater.,* **10**, pp. 1314–1323.
56. Gopinath, S. C. B., Sakamaki, Y., Kawasaki, K., and Kumar, P. K. R. (2006). An efficient RNA aptamer against human influenza B virus hemagglutinin, *J. Biochem.,* **139**, pp. 837–846.
57. Pagratis, N. C., Bell, C., Chang, Y., Jennings, S., Fitzwater, T., Jellinek, D., and Dang, C. (1997). Potent 2'-amino-, and 2'-fluoro-2'-deoxyribonucleotide RNA inhibitors of keratinocyte growth factor, *Nat. Biotechnol.,* **15**, pp. 68–73.
58. Sekiya, S., Noda, K., Nishikawa, F., Yokoyama, T., Kumar, P. K. R., and Nishikawa, S. (2006). Characterization and application of a novel RNA aptamer against the mouse prion protein, *J. Biochem.,* **139**, pp. 383–390.
59. Conrad, R., Keranen, L. M., Ellington, A. D., and Newton, A. C. (1994). Isozyme-specific inhibition of protein kinase C by RNA aptamers, *J. Biol. Chem.,* **269**, pp. 32051–32054.
60. Baskerville, S., Zapp, M., Ellington, A. D. (1999). High-resolution mapping of the human T-cell leukemia virus type 1 Rex-binding element by in vitro selection, *J. Virol.,* **73**, pp. 4962–4971.
61. Hirao, I., Madin, K., Endo, Y., Yokoyama, S., and Ellington, A. D. (2000). RNA aptamers that bind to and inhibit the ribosome-inactivating protein, pepocin, *J. Biol. Chem.,* **275**, pp. 4943–4948.
62. Pan, W., Craven, R. C., Qiu, D., Wilson, C. B., Wills, J. W., Golovine, S., and Wang, J. (1995). Isolation of virus-neutralizing RNAs from a large pool of random sequences, *Proc. Natl. Acad. Sci. U S A,* **92**, pp. 11509–11513.

63. Ruckman, J., Green, L. S., Beeson, J., Waugh, S., Gillette, W. L., Henninger, D. D., Claesson-Welsh, L., and Janjić, N. (1998). 2'-Fluoropyrimidine RNA-based aptamers to the 165-amino acid form of vascular endothelial growth factor (VEGF165). Inhibition of receptor binding and VEGF-induced vascular permeability through interactions requiring the exon 7-encoded domain, *J. Biol. Chem.*, **273**, pp. 2056–20567.
64. Dobbelstein, M., and Shenk, T. (1995). *In vitro* selection of RNA ligands for the ribosomal L22 protein associated with Epstein-Barr virus-expressed RNA by using randomized and cDNA-derived RNA libraries, *J. Virol.*, **69**, pp. 8027–8034.
65. Rhodes, A., Deakin, A., Spaull, J., Coomber, B., Aitken, A., Ligfe, P., and Rees, S. (2000). The generation and characterization of antagonist RNA aptamers to human oncostatin M, *J. Biol. Chem.*, **275**, pp. 28555–28561.
66. Kim, S. J., Kim, M. Y., Lee, J. H., You, J. C., and Jeong, S. (2002). Selection and stabilization of the RNA aptamers against the human immunodeficiency virus type-1 nucelocapsid protein, *Biochem. Biophys. Res. Commun.*, **291**, pp. 925–931.
67. Cox, J. C., and Ellington, A. D. (2001). Automated selection of anti-protein aptamers, *Bioorg. Med. Chem.*, **9**, pp. 2525–2531.
68. Bruno, J. G., and Kiel, J. L. (2002). Use of magnetic beads in selection and detection of biotoxin aptamers by electrochemiluminescence and enzymatic methods, *Biotechniques*, **32**, pp. 178–180.
69. Bryant, K. F., Cox, J. C., Wang, H., Hogle, J. M., Ellington, A. D., and Coen, D. M. (2005). Binding of herpes simplex virus-1 US11 to specific RNA sequences, *Nucleic Acids Res.*, **33**, pp. 6090–6100.
70. Stoltenburg, R., Reinemann, C., and Strehlitz, B. (2005). FluMag-Selex as an advantageous method for DNA aptamer selection, *Anal. Bioanal. Chem.*, **383**, pp. 83–91.
71. Dahlberg, A. E., Dingman, C. W., and Peacock, A. C. (1969). Electrophoretic characterization of bacterial polyribosomes in agarose-acrylamidecomposite gels, *J. Mol. Biol.*, **41**, pp. 139–147.
72. Schaup, H. W., Green, M., and Kurland, C. G. (1970). Molecular interactions of ribosomal components. I. Identification of RNA binding sites for individual 30S ribosomal proteins, *Mol. Gen. Genet.*, **109**, pp. 193–205.
73. Yakhnin, A. V., Trimble, J. J., Chiaro, C. R., and Babitzke, P. (2000). Effects of mutations in the L-tryptophan binding pocket of the *Trp* RNA-binding attenuation protein of *Bacillus subtilis*, *J. Biol. Chem.*, **275**, pp. 4519–4524.

74. Kumarevel, T., Mizuno, H., and Kumar, P. K. R. (2005). Structural basis of HutP-mediated anti-termination and role of the Mg^{2+} ion and L-histidine ligand, *Nature*, **434**, pp. 183–191.

75. Garner, M. M., and Revzin, A. (1981). A gel electrophoresis method for quantifying the binding of proteins to specific DNA regions: application to components of the Escherichia coli lactose operon regulatory system, *Nucleic Acids Res.*, **9**, pp. 3047–3060.

76. Fried, M., and Crothers, D. M. (1981). Equilibria and kinetics of lac repressor-operator interactions by polyacrylamide gel electrophoresis, *Nucleic Acids Res.*, **9**, pp. 6505–6525.

77. Fried, M., and Crothers, D. M. (1984). Equilibrium studies of the cyclic AMP receptor protein-DNA interaction, *J. Mol. Biol.*, **172**, pp. 241–262.

78. Esteban, R., Fujimura, T., and Wickner, R. (1988). Site-specific binding of viral plus single-stranded RNA to replicase-containing open virus-like particles of yeast, *Proc. Natl. Acad. Sci. U S A,* **85**, pp. 4411–4415.

79. Shen, Y., and Bruenn, J. A. (1993). RNA structural requirements for RNA binding, replication, and packaging in the yeast double-stranded RNA virus, *Virology,* **195**, pp. 481–491.

80. Smith, D., Kirschenheuter, G.P., Charlton, J., Guidot, D. M., and Repine, J. E. (1993). *In vitro* selection of RNA-based irreversible inhibitors of human neutrophil elastase, *Chem. Biol.,* **2**, pp. 741–750.

81. Yao, W., Adelman, K., and Bruenn, J. A. (1997). In vitro selection of packaging sites in a double-stranded RNA virus, *J. Virol.,* **71**, pp. 2157–2162.

82. Goodman, S. D. Velten, N. J., Gao, Q., Robinson, S., and Segall, A. M. (1999). *In vitro* selection of integration host factor binding sites, *J. Bacteriol.,* **181**, pp. 3246–3255.

83. Stebbins, M. A. Hoyt, A. M., Sepaniak, M. J., and Hurlburt, B. K. (1996). Design and optimization of a capillary electrophoretic mobility shift assay involving *trp* repressor-DNA complexes, *J. Chromatogr. B Biomed. Appl.,* **683**, pp. 77–84.

84. Mendonsa, S. D., and Bowser, M. T. (2004). *In vitro* evolution of functional DNA using capillary electrophoresis, *J. Am. Chem. Soc.,* **126**, pp. 20–21.

85. Mendonsa, S. D., and Bowser, M. T. (2004). *In vitro* selection of high-affinity DNA ligands for human IgE using capillary electrophoresis, *Anal. Chem.,* **76**, pp. 5387–5392.

86. Berezovski, M., Drabovich, A., Krylova, S. M., Musheev, M., Okhonin, V., Petrov, A., and Krylov, S. N. (2005). Nonequilibrium capillary elec-

trophoresis of equilibrium mixtures: a universal tool for development of aptamers, *J. Am. Chem. Soc.,* **127**, pp. 3165–3171.

87. Drabovich, A. Berezovski, M., Krylov, S. N. (2005). Selection of smart aptamers by equilibrium capillary electrophoresis of equilibrium mixtures (ECEEM), *J. Am. Chem. Soc.,* **127**, pp. 11224–11225.
88. Jing, M., and Bowser, M. T. (2011). Isolation of DNA aptamers using micro free flow electrophoresis, *Lab Chip,* **11**, pp. 3703–3709.
89. Zhang, F., and Anderson, D. (1998). *In vitro* selection of bacteriophage Φ29 prohead RNA aptamers for prohead binding, *J. Biol. Chem.,* **273**, pp. 2947–2953.
90. Homann, M., and Goringer, H. U. (1999). Combinatorial selection of high affinity RNA ligands to live African trypanosomes, *Nucleic Acids Res.,* **27**, pp. 2006–2014.
91. Blank, M., Weinschenk, T., Priemer, M., and Schluesener, H. (2001). Systematic evolution of a DNA aptamer binding to rat brain tumor micro vessels, *J. Biol. Chem.,* **276**, pp. 16464–16468.
92. Rhie, A., Kirby, L., Sayer, N., Wellesley, R., Disterer, P., Sylvester, I., Gill, A., Hope, J., James, W., and Tahiri-Alaoui, A. (2003). Characterization of 2'-Fluoro-RNA aptamers that bind preferentially to disease-associated conformations of prion protein and inhibit conversion, *J. Biol. Chem.,* **278**, pp. 39697–39706.
93. Cerchia, L., Duconnge, F., Pestourie, C., Boulay, J., Aissouni, Y., Gombert, K., Tavitian, B., de Franciscis, V., and Libri, D. (2005). Neutralizing aptamers from whole-cell selex inhibit the RET receptor tyrosine kinase, *PLoS Biol.,* **3**, p. e123.
94. Long, S. D., and Myszka, D. G. (2006). Affinity-based optical biosensors, in *Handbook of Affinity Chromatography,* 2nd Ed. (Hage, D. S., and Cazes, J., eds.), pp. 685–696, CRC Press, Boca Raton, FL.
95. Schier, R., and Marks, J. D. (1996). Efficient in vitro affinity maturation of phage antibodies using BIAcore guided selections, *Hum. Antibodies Hybridomas,* **7**, pp. 97–105.
96. Pileur, F., Andreola, M. L., Dausse, E., Michel, J., Moreau, S., Yamada, H., Gaidamakov, S. A., Crouch, R. J., Toulme, J. J., and Cazenave, C. (2003). Selective inhibitory DNA aptamers of the human RNaseH1, *Nucleic Acids Res.,* **31**, pp. 5776–5788.
97. Khati, M., Schuman, M., Ibrahim, J., Sattentau, D., Gordon, S., and James, W. (2003). Neutralization of infectivity of diverse R5 clinical isolates of human immunodeficiency virus type 1 by gp120-binding 2'F-RNA aptamers, *J. Virol.,* **77**, pp. 12692–12698.

98. Misono T. S., and Kumar, P. K. R. (2005). Selection of RNA aptamers against human influenza virus hemagglutinin using surface plasmon resonance, *Anal. Biochem.,* **342**, pp. 312–317.

99. Mendonsa, S. D., and Bowser, M. T. (2005). *In vitro* evolution of functional aptamers with affinity for neuropeptide Y using capillary electrophoresis, *J. Am. Chem. Soc.,* **127**, pp. 9382–9383.

100. Mosing, R. K., Mendonsa, S. D., and Bowser, M. T. (2005). Capillary electrophoresis-SELEX selection of aptamers with high affinity for HIV-1 reverse transcriptase, *Anal. Chem.,* **77**, pp. 6107–6112.

101. Weng, C. H., Hsieh, I. S., Hung, L. Y., Lin, H. I., Shiesh, S. C., Chen, Y. L., and Lee, G. B. (2012). An automatic microfluidic system for rapid screening of cancer stem-like cell-specific aptamers, *Microfluid. Nanofluidics,* **14**, pp. 753–765.

102. Yang, J., and Bowser, M. T. (2013). Capillary electrophoresis-SELEX selection of catalytic DNA aptamers for a small-molecule porphyrin target, *Anal. Chem.,* **85**, pp. 1525–1530.

103. Berezovski, M., and Krylov, S. N. (2002). Nonequilibrium capillary electrophoresis equilibrium mixtures- a single experiment reveals equilibrium and kinetic parameters of protein-DNA interactions, *J. Am. Chem. Soc.,* **124**, pp. 13674–13675.

104. Berezovski, M., Musheev, M., Drabovich, A., and Krylov, S. N. (2006). Non-SELEX selection of aptamers, *J. Am. Chem. Soc.,* **128**, pp. 1410–1411.

105. Park, S. M., Ahn, J. Y., Jo, M., Lee, D. K., Lis, J. T., Craighead, H. G., and Kim, S. (2009). Selection and elution of aptamers using nanoporous sol-gel arrays with integrated microheads, *Lab Chip,* **9**, pp. 1206–1212.

106. Hyunjung, B., Shuo, R., Jeehye, K., Minjung, K., Yuanyuan, J., Moonsoo, J. M., Irene, M. M., and Kim, S. (2013). Sol-gel SELEX circumventing chemical conjugation of low molecular weight metabolites discovers aptamers selective to xanthine, *Nucleic Acid Ther.,* **23**, pp. 443–449.

107. Ahn, J. Y., Jo, M., Dua, P., Lee, D. K., and Kim, S. (2011). Sol-gel-based microfluidics system enhances the efficiency of RNA aptamer selection, *Oligonucleotides,* **21**, pp. 93–100.

108. Quin, J., Lou, X., Zhang, Y., Xiao, Y., and Soh, H. T. (2009). Generation of highly specific aptamers via micromagnetic selection, *Anal. Chem.,* **81**, pp. 5490–5495.

109. Lou, X., Quin, J., Xiao, Y., Viel, L., Gerdon, A., Lagally, E. T., Atzberger, P., Tarasow, T. M., Heeger, A. J., and Soh, H. T. (2009). Micromagnetic

selection of aptamers in microfluidics, *Proc. Natl. Acad. Sci. U S A,* **106**, pp. 2989–2994.

110. Cho, M., Xiao, Y., Nie, J., Stewart, R., Csordas, A. T., Oh, S. S., Thomson, J. A., and Soh, H. T. (2009). Quantitative selection of DNA aptamers through microfluidic selection and high-throughput sequencing, *Proc. Natl. Acad. Sci. U S A,* **107**, pp. 15373–15378.

111. Cho, M., Oh, S. S., Nie, J., Stewart, R., Eisenstein, M., Chambers, J., Marth, J. D., Walker, F., Thomson, J. A., and Soh, H. T. (2009). Quantitative selection and parallel characterization of aptamers, *Proc. Natl. Acad. Sci. U S A,* **110**, pp. 18460–18465.

112. Huang, C. H., Lin, H. I., Shiesh, S. C., and Lee, G. B. (2010). Integrated microfluidic system for rapid screening of CRP aptamers utilizing systematic evolution of ligands by exponential enrichment (SELEX), *Biosens. Bioelectron.,* **25**, pp. 1716–1766.

113. Huang, C. H., Lin, H. I., Shiesh, S. C., and Lee, G. B. (2012). An Integrated microfluidic system for rapid screening of alpha-fetoprotein-specific aptamers, *Biosens. Bioelectron.,* **35**, pp. 50–55.

114. Weng, Q., Liu, W., XIng, Y., Yang, X., Wang, K., Jiang, R., Wang, P., and Zhao, Q. (2014). Screening of DNA aptamers against myoglobin using a positive and negative selection units integrated microfluidic chip and its biosensing application, *Anal. Chem.,* **86**, pp. 6572–6579.

115. Hybarger, G., Bynum, J., Williams, R. F., Valdes, J. J., and Chambers, J. P. (2006). A microfluidic SELEX prototype, *Anal. Bioanal. Chem.,* **384**, pp. 191–198.

116. Zhu, Z., Song, Y., Li, C., Zou, Y., Zhu, L., An, Y., and Yang, C. J. (2014). Monoclonal surface display SELEX for simple, rapid, efficient, an cost-effective aptamer enrichment and identification, *Anal. Chem.,* **86**, pp. 5881–5888.

Chapter 4

Selection and Application of Catalytically Active Oligonucleotides

Rachel Gysbers, Kha Tram, Sepehr Manochehry, Dingran Chang, and Yingfu Li

Department of Biochemistry and Biomedical Sciences, McMaster University, 1280 Main Street West, Hamilton, Ontario, Canada L8S 4K1

liying@mcmaster.ca

4.1 Nucleic Acids: Beyond Genetic Storage

Nucleic acids are the genetic blueprint for all life on earth. They are responsible for storing and passing on critical information from one generation to the next. This hereditary material was first recognized for its passive role in faithfully transferring the template of life, as it is the starting point of molecular biology's central dogma. However, with the discovery of RNA molecules that are capable of catalyzing reactions at a rate comparable to that of protein enzymes [1, 2], many believed that this would change how we fundamentally perceive nucleic acids [3]. DNA and RNA molecules functioning beyond their role as genetic carriers have shaped our understanding of modern biology.

Aptamers: Tools for Nanotherapy and Molecular Imaging
Edited by Rakesh N. Veedu

ISBN 978-981-4669-83-2 (Hardcover), 978-981-4669-84-9 (eBook)
www.panstanford.com

In the 1990s, pioneering research for probing DNA or RNA molecules that can serve purposes beyond information storage has led to the development of a technique known as in vitro selection or systematic evolution of ligands by exponential enrichment (SELEX) [4–6]. It is a combinatorial approach that begins with a large randomized library of 10^{14}–10^{16} DNA or RNA molecules. These pools of sequences are subjected to successive rounds of selection whereby sequences that express the desired function are enriched, while inactive sequences are discarded. The result of this test-tube evolution approach will favor synthetic molecules that can either catalyze chemical reactions (termed "DNAzymes" or "ribozymes") or function as molecular receptors (aptamers). Collectively, these unique molecules are referred to as functional nucleic acids (FNAs).

The invention of in vitro selection has led to the generation of many FNAs with fascinating properties. It has sparked the interest of the scientific community and has recruited researchers from different disciplines to answer various fundamental questions ranging from the origins of life to chemistry and biology. The contributions made by these communities have been productive over the past 25 years. We have witnessed the discoveries of several naturally occurring ribozymes and RNA aptamers [7–10] (e.g., riboswitches) in biology and the discovery of a substantial number of man-made FNAs.

These discoveries sparked a significant number of initiatives to employ FNAs for a broad spectrum of applications such as therapeutics, biosensing, molecular imaging, and drug discovery. FNAs have increasingly been explored within these areas for several reasons. First, FNAs are chemically synthesized. This process is readily scalable, produces excellent batch-to-batch consistency, and eliminates biological contaminations from viruses and bacteria. Second, FNAs can be chemically modified to increase their stability and bioavailability. Introducing functional groups also allows access to interesting conjugation chemistries, such as linking nucleic acids to small molecules (dyes, carbohydrates, amino acids), proteins, and various nanomaterials. Third, they are nonimmunogenic. Lastly, FNAs are resistant to thermal denaturation and are able to refold into their native structure. These attractive features offered by FNAs are the principal motivation behind many innovative FNA-based technologies.

4.2 Isolating Catalytically Active Oligonucleotides

Unlike in vitro selection of aptamers, nucleic acid enzymes (NAEs) such as DNAzymes and ribozymes are often more challenging to isolate. Aptamer selection experiments are now routinely performed, and over the past few years, we have even seen the emergence of companies offering aptamer selection services. Conversely, commercially available services for the selection of NAEs are not readily seen. Fundamentally, the requirement for aptamers is to simply bind to its cognate ligand, but an NAE must not only bind its substrates but also facilitate a chemical reaction. The experimental design for selecting catalytically active oligonucleotides is often more dynamic and varied. The selection protocol will change depending on the chemical reaction of interest, and the challenges will be dictated by the complexity of the chemical transformation.

4.2.1 *In vitro Selection Using DNA or RNA*

Synthetic NAEs can be composed of RNA (ribozymes) or DNA (DNAzymes). The choice of using either biopolymer is mostly decided by personal preference. However, it is worth considering that the selection design will be different. For the selection of DNAzymes, the randomized single-stranded DNA (ssDNA) library is obtained through solid-phase synthesis. This library is then subjected to a function-based selection step in which the active and nonactive species are separated from one another. The functionally desired sequences are then amplified using polymerase chain reaction (PCR). Lastly, the double-stranded DNA (dsDNA) needs to be denatured and the strand that contains the catalytic sequence retrieved for the next cycle of selection (Fig. 4.1A).

The selection of ribozymes begins with the generation of a dsDNA library that was synthesized from an ssDNA pool using PCR. This preparation step is required for in vitro transcription

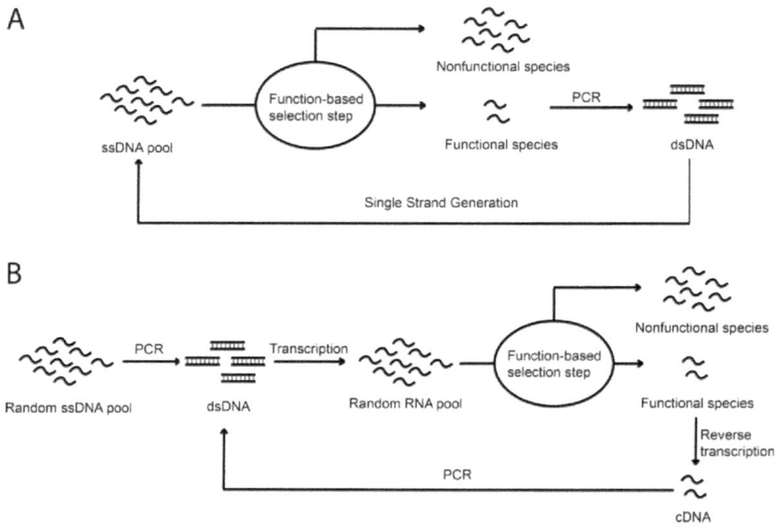

Figure 4.1 A general in vitro selection approach for isolating nucleic acid enzymes that are composed of (A) DNA or (B) RNA.

that will yield the ssRNA library used for selection. Similar to the selection of DNAzymes, a function-based step is applied to isolate sequences that are capable of catalyzing the reaction of interest. These sequences are then reverse-transcribed into an ssDNA pool that will be amplified by PCR. The cycle then begins again except with an enriched library (Fig. 4.1B). These additional preparation steps may not necessarily affect the outcome of isolating a ribozyme, but the process is certainly more laborious and tedious than DNAzyme selection. Furthermore, extra care needs to be taken when handling RNA. The ubiquitous presence of RNases may contaminate the RNA library and substantially reduce the chance of success.

4.2.2 *Effects of Library Design during in vitro Selection*

The process of enriching catalytically active oligonucleotides begins with a large library of diverse sequences. A key decision is to be made at the start of an in vitro selection experiment to determine the length of a random region, which would dictate the diversity of the starting library. A longer random region would result in a greater number of unique sequences that would have access to intricate secondary and tertiary structures. However, it has been previously demonstrated that longer random regions may not necessarily be the most optimal solution [11, 12]. Long random regions (with $N>80$) will also have the ability to form inhibitory motifs that negatively affect the enrichment of desired catalysts. In addition, longer and more complex motifs are often outcompeted by shorter motifs simply through having higher copy numbers in a randomized library [13–15]. As a result, there will be a bias for the shorter motif sequence, even though the longer motif may have better activity. The propensity for this bias is often dubbed "the tyranny of short motifs" [16, 17]. Conversely, a shorter randomized library may reduce this bias but will also limit the possible structural complexity required to perform certain chemical reactions.

Another consideration for library design is the practical limit of sequence sampling. With only four nucleotide building blocks available, the diversity of sequence space is expressed as 4^N. Although a typical starting library will have random regions that range from $N40$ to $N80$, there are libraries with regions as low as $N20$ to as high as $N228$ being utilized for the isolation of catalytically active oligonucleotides [18–21]. The first challenge is the physical limit in which the starting library is prepared. A library can begin with 10^{14} (or up to 10^{16}) molecules, but the size of the 4^N sequence space can greatly outnumber the initial pool. From Fig. 4.2, as the size of N increases past $\sim N25$, the starting library becomes undersampled. For example, a library with $N80$ will contain approximately 10^{48} unique sequences, of which only a very small fraction is sampled initially ($10^{14}/10^{48}$).

4.2.3 *Incorporating Chemically Modified Nucleotides into in vitro Selection*

There are two key reasons for using chemically modified oligonucleotides. The first is to expand the potential chemical repertoire that can be carried out by NAEs. Using substrates composed of only nucleic acids provides a great starting point as it is predisposed to have an affinity for the enzyme; however, it also limits the

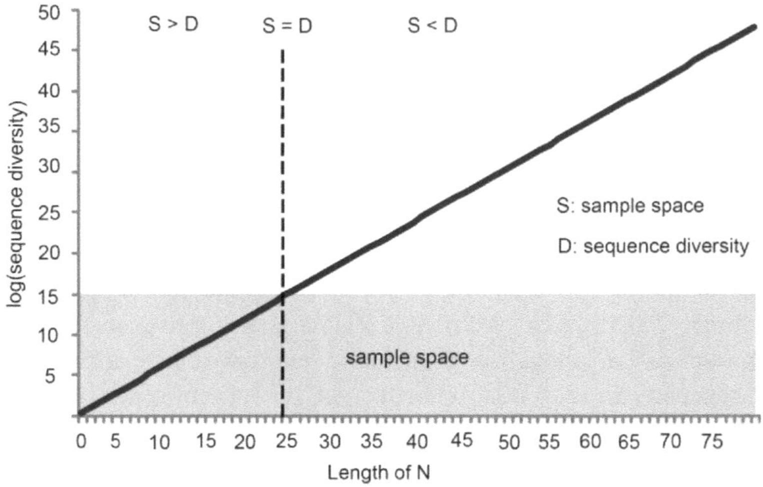

Figure 4.2 Relationship between sequence diversity and length of the randomized region. Extending the length of the random region will exponentially increase the diversity of the starting library. After ~*N*25, the initial library becomes undersampled.

scope of available chemical transformation. To extend beyond the known nucleic acid manipulations, nucleic acid substrates can be modified with auxiliary functional groups to explore alternative chemical reactions. These modifications play a key role in the design of the selection experiment since it will be the focal point of catalysis. Secondly, chemical modifications also provide a solution for separating catalytically active molecules from the inactive ones. Modifications such as 5′-biotinylated sequences have been routinely used as a separation technique for isolating many NAEs [22].

4.2.4 *In vitro Selection Design for Catalytic Specificity*

In vitro selection experiments need to be well planned and designed for successful isolation of NAEs that are highly efficient and specific. Even with every precaution taken to ensure enrichment for the catalyst of interest, the inherent process of isolation and enrichment may introduce complications. As an example, in some in vitro selection experiments, identification of an RNA-cleaving catalyst relies on a gel shift to separate cleaved and uncleaved sequences. However, RNA truncation can proceed through either transesterification or hydrolysis [23], both of which will lead to a shortened sequence that will have approximately the same gel shift. If attention is not given to the selection design, control for enrichment of the desired sequence will be lost. The library will simply favor the more active catalyst and not the catalyst of interest.

4.2.5 *Negative Selection*

To remove undesired catalytic species, a negative selection step may be incorporated into the selection design. Undesired sequences can often survive the in vitro selection process through various mechanisms, and if left unchecked, these sequences can be selectively amplified over the sequence of interest. For example, selection of an RNA-cleaving enzyme that utilizes a specific metal ion for catalysis may result in an RNA-cleaving catalyst that nonspecifically responds to other metal cofactors. These catalysts will survive the selection process and can be competitively more active than the catalyst of interest. As a result, they become selectively enriched due to better activity, but the end result yields a nonspecific RNA-cleaving catalyst. To enhance selectivity for the desired metal ion, the library can first be exposed to a broad spectrum of similar cofactors that can be used to eliminate these sequences. Although this technique certainly aids in directing the selection process, a single negative selection step does not remove all unwanted sequences. Conversely, a catalyst with low activity may also be enriched and biased over the desired sequence. These "opportunistic" sequences can survive by inhibiting the activity of more efficient catalysts. Over time, these inhibitions substantially suppress enrichment of faster-acting catalysts as the population of opportunistic sequences grows.

4.2.6 *Other Considerations for Selection of Catalytically Active Oligonucleotides*

Depending on the application and intended use of the NAE, the choice of using DNA or RNA is an important consideration and there are certain situations in which one may be better suited than the other. For example, there is a growing interest in using DNA over RNA for bioanalytical applications. From the perspective of stability, DNA is chemically more stable than RNA due to the lack of a 2′-OH. The presence of a 2′-OH enables RNA to undergo spontaneous hydrolysis. Consequently, oligonucleotides that are made of RNA will have limited shelf life. In addition, RNA is susceptible to ribonucleases that will readily degrade RNA sequences. Although loss of this 2′-OH has been speculated to restrict the functionality of DNA-based enzymes [24, 25], a substantial number of DNAzymes have been isolated that are as effective as their RNA counterparts. After all, DNA and RNA are very similar chemically and both are capable of creating intricate tertiary structures through similar molecular interactions such as hydrogen bonding, charge–charge interactions, and π-stacking.

4.3 Demonstrating the Capabilities of Nucleic Acid Enzymes

Compared to protein enzymes that are well versed in chemical diversity, nucleic acids lack the many useful functional groups, which would seem to limit their capabilities to function as catalysts. Despite this disadvantage, nucleic acids have their own set of advantages. First, nucleic acids can be chemically modified in various ways to append additional functional groups that would reduce the limitation for lack of chemical diversity. Second, NAEs are typically smaller and can be easily manipulated and altered. Lastly, NAEs are characteristically more stable than their proteinaceous counterparts, making them easier to handle. Despite the challenges faced by nucleic acids, a diverse number of artificial NAEs have been isolated from in vitro selection experiments for catalyzing chemical reactions that are important in biology, as listed in Table 4.1.

4.3.1 *DNAzymes for RNA Cleavage*

The very first catalytic DNA, a Pb^{2+}-dependent RNA-cleaving DNAzyme, was isolated by the Joyce group in 1994. A DNA sequence

Table 4.1 Artificial FNA catalysts

Artificial ribozymes		Artificial deoxyribozymes	
Reactions catalyzed	Reference	Reactions catalyzed	Reference
RNA cleavage	[26–28]	RNA cleavage	[29–35]
RNA ligation	[14, 36–38]	RNA ligation	[39–42]
RNA branching	[43]	RNA branching	[44, 45]
RNA phosphorylation	[46, 47]	RNA hydrolysis	[23]
RNA triphosphorylation	[48]	DNA phosphorylation	[49]
RNA capping	[50–52]	DNA capping	[53]
RNA polymerization	[54–56]	DNA ligation	[57, 58]
Diels–Alder	[59–61]	DNA deglycosylation	[62]
Amino acylation	[63–65]	DNA hydrolysis	[66–68]
Peptide formation	[69]	Thymine dimer repair	[70]
Porphyrin metalation	[71]	Phosphoramidate cleavage	[72]
Amide synthesis	[73]	Porphyrin metalation	[74]
Aldol reaction	[75]	Depurination	[76]
Cross-chiral polymerization	[77]	Ester and amide hydrolysis	[78]

that could cleave the RNA via transesterification with a rate enhancement of 10^5 over the uncatalyzed reaction was isolated after five rounds of in vitro selection. A wide assortment of DNAzymes has since been isolated to cleave RNA substrates or chimeric DNA-RNA substrates under a wide range of conditions. Many of these RNA-cleaving DNAzymes typically relied on divalent metal ion cofactors [29, 33, 79–82].

The first DNAzyme of the Joyce group cleaved a single ribonucleotide in an all-DNA substrate, so subsequently they developed a DNAzyme that could cleave an all-RNA substrate under physiological conditions [34]. Two classes of DNAzymes, each distinctive in structure, were isolated by slight modifications to their previous protocols. Named 8-17 and 10-23 after their, respective, in vitro selection cycles and clones, these DNAzymes have since been well studied (Fig. 4.3). 10-23 can facilitate cleavage of any RNA substrate that contains a purine–pyrimidine junction as a cleavage site under physiological conditions, and several variants of 10-23 have been developed to target messenger and viral RNA targets [83]. The Joyce group additionally isolated 8-17, which has a catalytic core consisting of a small three-base pair stem

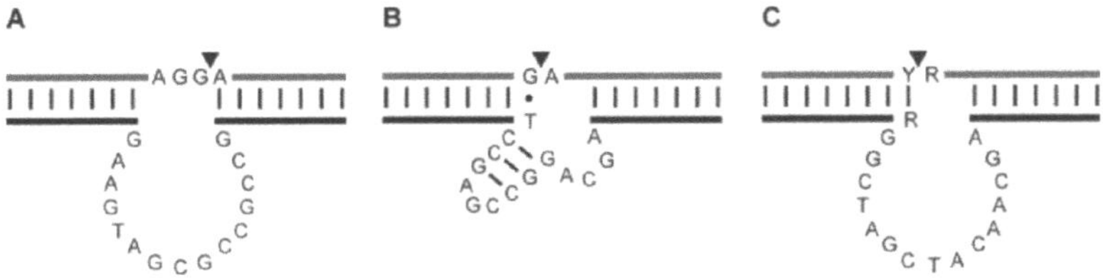

Figure 4.3 RNA-cleaving DNAzymes. (A) The first Pb^{2+}-dependent DNAzyme, (B) the 8-17 DNAzyme, and (C) the 10-23 DNAzyme. The filled triangle represents the cleavage site, R is a purine-based nucleotide, and Y is a pyrimidine-based nucleotide.

and eight conserved nucleotides and has been used extensively in biosensor development. This DNAzyme has been isolated in multiple independent in vitro selection experiments, suggesting that it could be the simplest structure for DNA to cleave RNA [84, 85].

More recent developments have led to the isolation of RNA-cleaving DNAzymes that did not conform to conventional expectations. The Perrin group appended imidazoles and amides to DNA to isolate a DNAzyme that does not require divalent metal ions; however, the transesterification reaction still requires imidazole assistance [86]. The Hollinger group isolated nucleic acid–derived synthetic polymers (termed "XNA") capable of cleaving RNA; these were called FANAzyme, ANAzyme, HNAzyme, and CeNAzyme [87].

In addition, the reaction mechanism of RNA cleavage can differ depending on the DNAzyme used to catalyze the reaction. Transesterification is the common reaction mechanism followed by most DNAzymes that cleave RNA, although not by all [84, 88, 89]. The Silverman group selected a DNAzyme that cleaved RNA via hydrolysis as opposed to transesterification [23]. Hydrolysis of RNA involves an attack of a water molecule to the phosphodiester bond, creating a 5′-phosphate end and a 3′-OH terminus. When compared to the mechanisms of transesterification, RNA hydrolysis is considered a highly unfavorable reaction [90, 91] (Fig. 4.4). However, the result of a 5′-phosphate is unique and can be utilized by enzymes such as T4 DNA ligase, which makes this reaction useful.

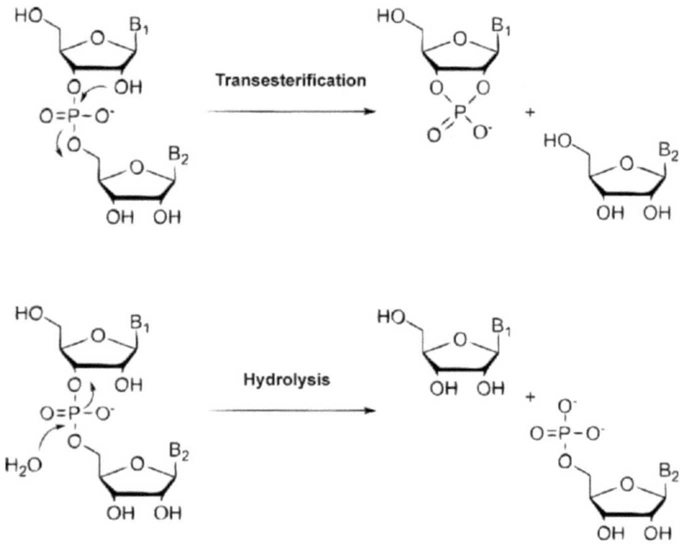

Figure 4.4 Sequence truncation can be catalyzed through a transesterification reaction (top panel) or hydrolysis (bottom panel).

4.3.2 *DNAzymes for Phosphorylation*

To participate in a ligation reaction, the 5′-hydroxy terminal of DNA needs to be phosphorylated. Additionally, radioisotopic phosphorylation of nucleic acids is often carried out for purposes of labeling, isolation, and sequencing. Traditionally, the 5′-phosphorylation of DNA molecules has been performed by protein enzymes, such as polynucleotide kinases, which transfer the γ-phosphate of ATP to the 5′-terminus of the DNA strand. An in vitro selection method was designed by Li and Breaker, which resulted in the isolation of self-phosphorylating DNAzymes that could utilize different nucleoside triphosphates (NTPs) and deoxyribonucleoside triphosphates (dNTPs) for each (d)NTP (with the exception of dTTP) [92]. In addition, a reselection effort resulted in DNAzymes that could differentiate between guanosine-5′-triphosphate (GTP) and deoxyguanosine triphosphate (dGTP) as a phosphate source [92]. These DNAzymes transfer a phosphate to the 5′-end of the sequence, using the γ-phosphate of either NTPs or dNTPs as a phosphate source. The reaction proceeds via nucleophilic attack

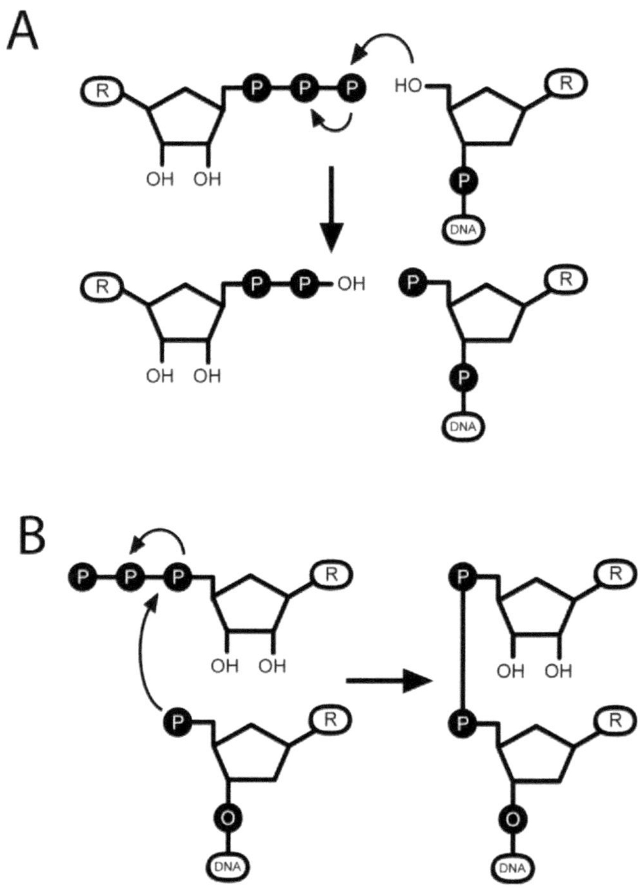

Figure 4.5 DNAzymes for (A) phosphorylation reaction and (B) capping reaction.

of the 5′-hydroxyl on the γ-phosphorous of a molecule of NTP (Fig. 4.5A). The result is phosphate transfer and the formation of a 5′-phosphorylated DNA and a nucleoside diphosphate (NDP) molecule.

Another separate series of parallel in vitro selection experiments performed by Wang et al. isolated self-phosphorylating DNAzymes that required different divalent metal ion cofactors [49]. The divalent metal cations used included Mn^{2+}, Cu^{2+}, Ca^{2+}, and Mg^{2+};

some broad classes emerged that could use any cofactors, whereas others were specific in requiring one or two divalent metal ions [49].

4.3.3 *DNAzymes for Self-Adenylation*

Li et al. isolated 12 classes of DNAzymes capable of self-adenylating themselves, also known as self-capping. These DNAzymes mimic the first reaction catalyzed by the protein enzyme T4 DNA ligase, which is the formation of AppDNA via the addition of adenosine monophosphate (AMP) to a 5′-phosphorylated DNA molecule [53]. An oxygen from 5′-phosphorylated DNA undergoes a nucleophilic attack on the ATP molecule's α-phosphate, resulting in a 5′ pyrophosphate cap (Fig 4.5B). This reaction results in a DNA molecule with an adenyl 5′-5′- pyrophosphate cap, allowing ligation to proceed via T4 DNA ligase. The dominant deoxyribozyme class isolated was called the class I capase, which has a predicted four-tier guanine quartet structure [53].

The second reaction catalyzed by T4 DNA ligase was also mimicked by an additional DNAzyme, isolated by the Breaker group, in which class I capase was used to generate AppDNA, and DNAzymes were selected that could form 3′–5′ linkages between two DNA molecules [58]. This system was configured into a trimolecular arrangement for ligation.

4.3.4 *DNAzymes for Ligation*

Due to the difficulties involved in manufacturing lengthy nucleic acid polymers, the ligation or joining together of separate sequence components is an important method in generating extensive sequences. Traditionally, ligase enzymes are used to join RNA and DNA pieces with the aid of a complementary template or "splint" oligonucleotide. An alternative to the established methods is the use of FNAs to ligate nucleic acids. Ligase deoxyribozymes were first generated soon after the isolation of the very first DNAzyme. The first ligase, developed by Cuenoud and Szostack, is capable of joining two DNA substrates by the reaction of the 5′-hydroxyl of one substrate with an activated 3′-phosphorimidazolide of the other [57]. A proposed secondary structure for the E47 DNA ligase

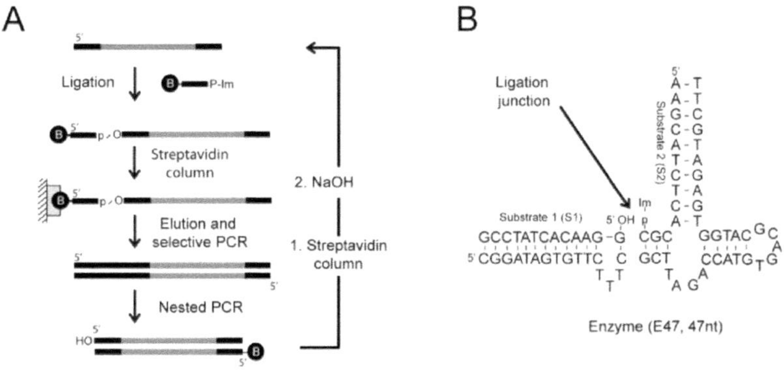

Figure 4.6 Isolating DNA-ligating DNAzymes.

DNAzyme isolated by Cuenoud and Szostak (1995) is shown in Fig 4.6B. The formation of a phosphodiester bond between S2 (phosphorimidazolide substrate) and S1 (the 5′-hydroxyl substrate) is catalyzed by the 47-nt E47. This reaction requires the presence of either Zn^{2+} or Cu^{2+}. However, a limitation of the use of this deoxyribozyme is that the activated 3′-phosphorimidazolide termini of the substrates is not obtained easily.

To isolate a DNAzyme capable of ligation, a combinatorial pool of ss sequences is incubated with substrate molecules, activated with a 5′-biotin moiety and a 3′-phosphorimidazolide (P-Im) (Fig 4.6A). The ligation reaction is carried out for a designated time period in a buffer containing divalent metal ions. This mixture is then passed over a streptavidin column; ligated molecules now contain the biotin tag and therefore are retained on the column. Denaturing conditions (e.g., NaOH, urea) are used to wash away the nonligated molecules. PCR amplification is performed with one primer being the same sequence as the substrate, selecting only for ligated molecules. A second nested PCR is performed to regenerate DNA of the correct size using a biotinylated primer; these double strands are passed over a strepdavidin column to yield single strands, used during the next round of selection.

Silverman's group isolated the first deoxyribozyme capable of ligating RNA by joining a 2′,3′-cyclic phosphate with the 5′-hydroxyl of the second substrate, but this enzyme created nonnative

2′-5′ linkages using Mg^{2+} as a cofactor [39]. Subsequently, the group developed a DNAzyme that could link a 2′,3′-diol with a 5′-triphosphate terminal, which resulted in both native 3′–5′ linkages as well as nonnative 2′–5′ linkages [44]. In addition, another deoxyribozyme from Silverman's group, 9F7, provided the first synthetic means of producing branched RNA [45]. This DNAzyme was able to form 2′,5′-branched RNA by catalyzing the reaction of an internal 2′-hydroxyl group of an unpaired adenosine with a 5′-triphosphate [45]. This is an exciting development, as this branching reaction is comparable to the first step of RNA splicing, which has a branched intermediate.

4.4 Nucleic Acid Enzymes for Biosensing Applications

A particularly useful application of NAEs is their incorporation as the biological components of biosensors, which are analytical devices capable of detecting specific analytes. Biosensors consist of two components: a detector, known as a molecular recognition element (MRE), and a transducer, which translates the detection target into a quantitative signal. The world's first biosensor, an enzyme electrode, developed in 1962 by Leland Clark, used an immobilized layer of glucose oxidase. The oxidase detector generated signals recognized by an oxygen electrode [93]. Initially developed for use in surgery to monitor blood oxygen, the decrease in glucose concentration proportional to oxygen levels allows for inadvertent glucose monitoring [94].

The conventional biosensor design has employed proteins, especially antibodies, receptors, and enzymes, as MREs. Recent use of FNAs, including ribozymes, DNAzymes, and aptazymes (NAE–aptamer conjugates in which the aptamer domain binds to the target, regulating the catalytic activity of the enzyme domain), has shown that FNAs have certain advantages over traditional protein MREs.

Commercially available biosensors possess specific limitations as they utilize proteins as their biological component for catalysis or recognition [95]. Enzymes are evolved biological machineries with a very specific target—their associated ligand, and not much else, leaving no opportunity for adaptability or modifications. Antibodies are more flexible as they can be isolated to target new analytes, but isolating antibodies is an expensive process, difficult to scale for mass production [96].

The demand for biosensor technology is steadily rising [97], with a startling swell in FNAs in analytical applications [98]. Rigorous study toward understanding the fundamental structure of FNAs is elucidating how they could be used to replace proteins for developing biosensors. Currently, the FNA equivalent of protein enzymes and antibodies are NAEs and aptamers, respectively. Although FNA-based biosensors are still in their infancy, researchers are beginning to understand how to manipulate and modify FNAs to engineer the most ideal biosensor.

Both RNA- and DNA-based FNEs have been used in biosensors, and many existing DNA-based NAEs have proven to be as effective as their RNA counterparts. These DNAzymes are flexible and have been isolated to carry out a wide array of different chemical transformations. Several comprehensive reviews have been published, which discuss the practical niche of FNAs in biosensing applications [99–101]. NAEs meet a vast majority of the characteristics of a good target recognition element: high affinities and specificities, lengthy shelf life, rapid responses, and broad ranges of detection [101]. NAEs as components of biosensors have been extensively reviewed; both ribozymes [102] and DNAzymes [103] have been explored. RNA-cleaving DNAzymes in particular have been studied extensively [104].

RNA cleaving is a favored reaction for developing DNA-based biosensors on several grounds. First, the RNA transesterification reaction carried out by DNAzymes has been extensively studied, with well-characterized structural and catalytic properties. As a result, the available information on this class of DNAzymes significantly facilitates biosensor development. Second, the rates of RNA cleavage performed by these DNAzymes are among the best of all chemical reactions catalyzed by NAEs. One of the key traits of an ideal biosensor is to generate results as quickly as possible; thus, a faster catalyst is able to process more substrates, which results in a quicker response time. Lastly, RNA cleavage results in two short strands that can be easily monitored using various separation techniques. The mechanism of two departing strands can be readily coupled to various transduction platforms to generate an electrochemical, colorimetric, or fluorescent signal.

RNA-cleaving DNAzymes can be engineered to detect a vast variety of targets, ranging from metal ions to complex biological mixtures. In addition to in vitro selection, another approach called rational design can be used to create FNA sensors employing a known aptamer for a specific target. An aptamer, which acts as the MRE, is integrated with an NAE so that its catalytic activity is regulated by target binding [105]. In addition, FNAs can ultimately be manipulated to produce unique ways of signal amplification. For example, Yi Lu's group used a Pb^{2+}-dependent RNase DNAzyme to cleave an RNA-containing fluorescence quenching molecular beacon, which was able to generate an amplified signal from a low concentration of DNAzyme, increasing sensitivity [106]. Due to these advantages, RNase DNAzyme-based biosensors are thus expected to be used in biosensing devices for many upcoming applications [107].

4.5 Nucleic Acid Enzymes as Potential Therapeutics

Catalytically active nucleic acids have been exploited in various research fields ranging from the development of biosensors to drug discovery. Over the past decade, DNAzymes and ribozymes have also been explored as potential therapeutic agents. Although there has been tremendous progress in these fields, NAEs are still faced with many challenges with regard to utility and effectiveness within a biological system. Factors such as delivery, cellular uptake, subcellular localization, and stability are key issues that need to be addressed in order for NAEs to be viable therapeutic agents. In addition, the cellular environment must also be favorable for NAE activity. For example, temperature, concentrations of metal ions, and the presence of potential inhibitors will influence the effectiveness of an NAE.

4.5.1 *Delivery of Nucleic Acid Enzymes*

Despite promising progress in the therapeutic applications of NAEs for a variety of cancers and numerous noncancerous diseases (most commonly cardiovascular disease models), there remains a significant challenge in finding reliable and efficient delivery methods that maintain low cytotoxicity. Certain drug delivery systems (DDSs) have shown promising results, but a model capable of being used clinically has yet to be developed [108]. Prior to the discussion of therapeutic applications, it is vital to consider some of the DDSs that are currently under development.

One of the earliest DDS studies explored encapsulation of ribozymes and DNAzymes by copolymers of poly(lactic acid) (PLA) and poly(glycolic acid) (PLGA) microspheres [109]. The PLGA microspheres were shown to have a biphasic effect, with an initial burst in the first five days followed by a second phase of sustained release the following 40 days [109]. Additional attempts include complexation of DNAzymes with cyclodextrin-containing polycation (CDP) via electrostatic interactions that caused the formation of uniformly sized 100 nm particles known as polyplexes [110]. Polyplexes were rapidly internalized by cells within 15 minutes of exposure. While free DNAzymes lack cellular uptake, DNAzymes that were protected in a polyplex experienced intracellular uptake into three different cell lines: HeLa cells, colorectal carcinoma HT29 cells, and ovarian carcinoma A2780 cells [110].

4.5.1.1 Dendrimers

Dendrimers have also been introduced as a potential DDS. They are a synthetic macromolecule with a well-defined structure and are commercially available in various forms such as the poly(propylene imine) (PPI) dendrimers [108]. Modified PPI dendrimers, termed G4, were assessed for their DNAzyme-binding capacity, toxicity, in vivo transfection efficiency, and in vivo delivery characteristics. G4 demonstrated reversible binding capabilities and high transfection efficiency when compared to the commercially available agent *N*-[1-(2,3-dioleoyloxy)propyl]-*N*,*N*,*N*-trimethylammonium methylsulfate (DOTAP), while having lower cytotoxicity than DOTAP at high concentrations [111]. The study revealed that the contrastingly low transfection efficiency of free DNAzymes, capable of only entering 5%–10% of cells, was dramatically enhanced (>80%) when using G4 [111].

Another example of a dendritic DDS that can be used as a carrier in living cells is the polyethylene-cationic poly(p-phenyleneethynlene) (PPE). Zhang et al. used this nanocarrier to deliver a Pb^{2+}-dependent DNAzyme for imaging lead ions in living cells [112].

4.5.1.2 Vesicles and nanoparticles

Multilamellar vesicle liposomes have been successfully employed for the encapsulation of Dz13, a DNAzyme targeting *c-Jun* mRNA, a proto-oncogene that is overexpressed in osteosarcoma cells [113]. These liposomes had better transfection efficiency than the commercially available transfection reagent Fugene-6. Following transfection using liposomes, Liang et al. demonstrated that an 80% decrease in target proteins was observed in human pancreatic carcinoma cells in which mRNAs were being cleaved by DNAzymes in a time- and dose- dependent manner [114].

Another alternative DDS for transporting NAEs relies on carriers such as nanoparticles. Tack et al. demonstrated that immobilizing DNAzymes onto gold nanoparticles was capable of effectively transfecting >77% of HT29 carcinoma cells [115]. DNAzymes have also been encapsulated into nanoparticles made of chitostan, a polysaccharide that serves as a useful biodegradable sustained release depot [116]. These DNAzyme-loaded nanocarriers were shown to down-regulate a proto-oncogene *c-Jun* [113]. Chitostan is considerably cheaper than other vehicles such as liposomes or synthetic particles [108].

4.5.2 *NAEs in Therapeutic Applications*

As an intermediate in the production of proteins and the genetic material of many pathogenic viruses, RNA presents an attractive target for both biological and therapeutic manipulation. The most established nucleic acid–based approaches to gene suppression at the RNA level is through antisense oligonucleotides and small interfering RNAs. Alternatively, RNA can be targeted by catalytic RNA such as the hammerhead ribozyme. Ribozymes have the advantage of being equipped with their own RNA cleavage apparatus and are therefore independent of the host nuclear protein activity. However, the utility of ribozyme oligonucleotides is restricted by the relative difficulty of synthesizing active molecules with sufficient resistance to nuclease degradation. DNAzymes have more desirable properties for therapeutic application because of their small size, ease of synthesis, and increased resistance to both chemical and enzymatic degradation. Since all NAEs are susceptible to nucleases, their stability in any in vivo system is jeopardized. Modifications have been made to enhance the stability of mRNA-targeting DNAzymes, in particular by introducing at both 5′- and 3′-ends either two phosphorothioate linkages or two 2′-O-methyl-substituted residues [117]. However, these modifications are prone to nonspecific interactions [118], as well as immunological side effects and toxicity [119, 120].

4.5.2.1 NAEs targeting cancer

In recent years, DNAzymes have been repeatedly demonstrated to be capable of acting as efficient tools for the modulation of cancer. DNAzymes modified with nuclease-resistant properties have been successfully used to target the expression of mRNA that codes for cellular components critical to cell growth. We will highlight some examples of cancer therapy using DNAzymes; however, there are a substantial number of studies regarding this topic and many comprehensive reviews are available [121–123].

The targeting of DNAzymes against cancer cells was first tested with a line of proto-oncogene variant, *bcr-abl,* known to code for the Bcr-Abl fusion protein. The Bcr-Abl fusion protein is encoded by the chimeric mRNA that results from chromosome 22 translocations [124]. This chimeric mRNA is commonly detected in patients with leukemia [125], making it a potential target for therapeutic applications involving RNA-cleaving DNAzymes. Cleavage of this chimeric mRNA at its junction has been successfully conducted using DNAzymes to inactivate the oncogene, while leaving intact the normal *bcr* or *abl* mRNA [126]. The DNAzyme-induced cleavage of the mRNA inhibited Bcr-Abl protein expression in chronic myeloid leukemia cells by almost 40% [125].

4.5.2.2 DNAzymes combating angiogenesis

Angiogenesis is the process by which new blood vessels are formed and has long been recognized to be required for growth and eventual metastasis of solid tumors [127]. Angiogenesis involves an intricate combination of cellular processes that are tightly regulated in a healthy cell and are mandatory for tumor progression as the processes ultimately supply oxygen and nutrients to the growing tumor [128]. Consequently, using DNAzymes to target mRNA of genes responsible for angiogenesis is a therapeutic approach that has been adopted by a number of researchers.

DNAzymes partially modified with phosphorothioate and 2′-O-methyl groups have been used to target integrins B_1 and B_2, which are important in cell–cell and cell–matrix interactions [129]. The DNAzymes' capability to eliminate microvascular endothelial cell capillary tube formation led authors to propose the potential use of DNAzymes for targeting tumor-induced angiogenesis [129].

An example of a protein implicated in biological processes leading toward angiogenesis is the urokinase-type plasminogen activator (uPA) receptor (uPAR) [130]. Targeting of uPAR is another case where the DNAzyme designed to perform cleavage of the mRNA coding for this protein successfully decreased mRNA levels and protein concentrations, ultimately resulting in suppression of osteosarcoma cell invasion [130].

4.5.2.3 Growth factors as DNAzyme targets

A growth factor known to influence tumor angiogenesis is the vascular endothelial growth factor (VEGF), released by malignant tumors. These growth factors bind their receptors on endothelial cells and induce angiogenesis [131, 132]. One such receptor, VEGFR-1, has been selected as the target for a modified ribozyme known as ANGIOZYME, which is currently in phase 2 trials for several tumor types [116].

NAEs have also been used to cleave the mRNA of growth factor receptors [133]. A DNAzyme and a ribozyme designed to silence the mRNA of the epidermal growth factor receptor (EFGR), a receptor that is abnormally activated in epithelial tumors, inhibited both EFGR expression and growth of EFGR overexpressing cells [133].

4.5.3 *NAEs Targeting Pathogenic Bacteria*

Controlling protein expression through mRNA modulation has been a topic of interest that extends beyond cancer therapy. DNAzymes have been also been employed to target pathogenic bacteria, particularly antibiotic-resistant strains. Antibiotic resistance typically occurs when the infectious pathogen obtains a genetic mutation that permits it to be resistant to the effects of one or more antibiotics. Since treatment with various antibiotics can result in new antibiotic-resistant strains, new therapeutic approaches are needed. By using RNA-cleaving DNAzymes, mRNAs that encode the proteins responsible for antibiotic resistance can be regulated.

Luo's group demonstrates the viability of DNAzymes that targets the mRNAs for β-lactamase [134] and penicillin-binding protein [135] in drug-resistant *Staphylococcus aureus*. DNAzyme treatment reduced the concentration of protein expression in a dose-dependent manner. This result suggests that incorporating DNAzymes in the treatment process may be a strategy to inhibit antibiotic resistance and restore the pathogen's sensitivity to antibiotic therapy. Although this field is still in its infancy, the prospect of using DNAzymes as a tool to potentially answer the growing resistance of antibiotics is certainly an effort worth investigating. Table 4.2 lists some of the current research made toward using DNAzymes as anti-bacteria agents.

Table 4.2 DNAzymes used as antibacterial reagents

Bacteria	Gene target	Reference
Escherichia coli	*ftsZ*	[136]
Mycobacterium tuberculosis	*icl*	[137]
Staphylococcus aureus	*blaR1*	[134]
Staphylococcus aureus	*mecR1*	[135]

Table 4.3 DNAzymes used as antiviral reagents

Virus	Gene target	Reference
Epstein–Barr virus	*LMP1*	[139]
Hepatitis B virus	*HBsAg, HBeAg*	[140, 141]
Hepatitis B virus	*HBx*	[142]
Hepatitis C virus	Core protein	[143]
HIV-1	P24 Gag, Nef regions	[144]
HIV-1	TAR	[145]
HIV-1	TAT	[146]
HIV-1	Integrase	[138]
SARSassociated corona virus	5′-UTR	[147]
Respiratory syncytial virus	Nucleocapsid	[148]

4.5.4 *Nucleic Acid Enzymes Targeting Viruses*

DNAzymes are also being developed as therapeutics for viral diseases such as HIV and hepatitis (Table 4.3). Targeting different regions of the HIV genome with DNAzymes allows for inhibition at different stages of the HIV life cycle. For example, HIV integrase is an enzyme that is responsible for integration of the proviral genome into the human genome and this integration step is the first step of the virus hijacking the human cell machinery for its propagation and replication. RNA-cleaving DNAzymes were shown to suppress integrase expressions through sequence-specific mRNA cleavage [138].

4.6 Nucleic Acid Enzymes in Imaging Applications

There has been significant progress made in the development of numerous NAEs with selective detection capabilities for a variety of biological targets, showing especially great promise in the detection of heavy metal ions [149]. Despite the promising detection capabilities, nucleic acids are limited to functioning in extracellular environments due to their poor cellular uptake [150, 151]. Without this barrier, detection of a variety of important biological targets could be performed in live-cell assays using fluorescent-linked FNA probes. Thus a number of groups have focused on the development of high-efficiency delivery systems that maintain low cytotoxicity.

One of the successful attempts at delivery of DNAzymes to cells for imaging involved fluorescein isothiocyanate (FITC)–labeled DNAzymes complexed with dendrimers, a synthetic macromolecule with a well-defined structure [111]. Whole-body imaging of ovarian carcinoma-bearing mice following intravascular injection of the complex containing DNAzymes specific for mRNA implicated in cancer progression led to externally visible fluorescence near the tumor, followed by accumulation in the tumor and within the cancer cell nucleus [111]. Zhang et al. used PPE dendrimers as a nanocarrier to deliver DNAzymes specific to lead ions for imaging in live cells [112].

Tumors in mice have been visualized using fluorescence through employment of transferrin-modified PEGylated polyplexes (TFPs) made of DNAzymes, specific for a gene that is highly expressed in cancerous cells, and CDP [110]. While unformulated DNAzymes are eliminated from the body within 24 hours after intravenous injection, DNAzymes delivered through TFPs were retained and concentrated in tumors for more than 24 hours [110].

4.6.1 *NAEs for Imaging Metal Ion Targets*

Determining the levels of harmful toxins in living cells and biological samples is of tremendous value to human health. Environmental toxins such as lead, a small amount of which is capable of causing serious health problems [152], are targets that have raised the demand for imaging in living cells. A DNAzyme highly sensitive to lead has been used to cause induction of fluorescence in the presence of lead, in murine models [112]. By employing a dendritic nanocarrier delivery system to transport the DNAzyme across cell membranes, Zhang et al. demonstrated the ability to image Pb^{2+} within living cells. In addition to the detection of Pb^{2+} in live cells, uranyl was detected using a UO_2^{2+}-dependent DNAzyme that relied on gold nanoparticles for intracellular delivery [112, 153].

DNAzymes sensitive to heavy metals have been applied as contrast agents for magnetic resonance imaging (MRI) [154]. MRI is a noninvasive method of 3D imaging in vivo, which involves paramagnetic compounds as contrast agents that respond differently to magnetization in different tissues, thereby producing a black-and-white contrast image. DNAzymes have been conjugated with such paramagnetic compounds as gadolinium to produce a larger DNAzyme–Gd complex that is capable of enhancing the contrast in MRI images in the presence of uranium [154].

4.6.2 *NAEs Used for Imaging Nucleic Acid Targets*

There is a growing interest for molecular imaging technologies using NAEs to monitor endogenous gene expressions as an indicator for disease progression. Hong et al. demonstrated the use of ribozymes for the imaging of RNA expression in mouse models [155]. A *trans*-splicing ribozyme was used for imaging the expression of mouse cytoskeleton-associated protein 2 (mCKAP2) RNA in intrahepatic tumors, which has been proposed as a potential method for monitoring of gene therapy.

Fluorescent imaging of ribozyme mRNA splicing activity has been monitored through the expression of a reporter protein β-lactamase [156]. The level of reporter production is reflected in the splicing ribozymes activity and efficiency. An increase in fluorescence occurred when the reporter protein catalyzed a membrane-permeable fluorescent substrate. These reporters combined with flow cytometry enabled screening of large combinatorial libraries of ribozymes for highly efficient splicing activity [156].

In general, the use of RNAs for imaging in vivo is complex [157, 158]. To visualize RNA in vivo, one unique approach is to use RNA aptamers that can significantly enhance the fluorescent output on binding to their cognate fluorophores. Fusion of these RNA–fluorophore complexes to targeted RNA can be conducted to probe live-cell imaging of RNAs [159, 160].

4.7 Conclusion

Nucleic acids are indisputably adept in many functions beyond basic genetic storage and transfer. A vast assortment of NAEs that are capable of carrying out complex chemistries have been developed using a method that mimics natural selection at a molecular scale. These polymers are tractable and can undergo modifications, which make them even more proficient.

Thus, it comes as no surprise that there is a veritable market for commercial and medical applications of NAEs. These molecules have distinct properties that make them advantageous for in vivo applications, and have been developed for a large array of purposes in therapeutics and imaging. As new technologies develop, the future looks promising for further expansion of the rapidly growing employment of NAEs.

References

1. Kruger, K., Grabowski, P. J., Zaug, A. J., Sands, J., Gottschling, D. E., et al. (1982). Self-splicing RNA: autoexcision and autocyclization of the ribosomal RNA intervening sequence of Tetrahymena, *Cell*, **31**, pp. 147–157.
2. Guerrier-Takada, C., Gardiner, K., Marsh, T., Pace, N., and Altman, S. (1983). The RNA moiety of ribonuclease P is the catalytic subunit of the enzyme, *Cell*, **35**, pp. 849–857.
3. Breaker, R. R., and Joyce, G. F. (2014). The expanding view of RNA and DNA function, *Chem. Biol.*, 21: 1059–1065.
4. Ellington, A. D., and Szostak, J. W. (1990). In vitro selection of RNA molecules that bind specific ligands, *Nature*, **346**, pp. 818–822.
5. Robertson, D. L., and Joyce, G. F. (1990). Selection in vitro of an RNA enzyme that specifically cleaves single-stranded DNA, *Nature*, **344**, pp. 467–468.
6. Tuerk, C., and Gold, L. (1990). Systematic evolution of ligands by exponential enrichment: RNA ligands to bacteriophage T4 DNA polymerase, *Science*, **249**, pp. 505–510.
7. Mandal, M., and Breaker, R. R. (2004). Gene regulation by riboswitches, *Nat. Rev. Mol. Cell Biol.*, **5**, pp. 451–463.
8. Tucker, B. J., and Breaker, R. R. (2005). Riboswitches as versatile gene control elements, *Curr. Opin. Struct. Biol.*, **15**, pp. 342–348.
9. Winkler, W. C., and Breaker, R. R. (2005). Regulation of bacterial gene expression by riboswitches, *Annu. Rev. Microbiol.*, **59**, pp. 487–517.

10. Wickiser, J. K., Cheah, M. T., Breaker, R. R., and Crothers, D. M. (2005). The kinetics of ligand binding by an adenine-sensing riboswitch, *Biochemistry*, **44**, pp. 13404–13414.
11. Sabeti, P. C., Unrau, P. J., and Bartel, D. P. (1997). Accessing rare activities from random RNA sequences: the importance of the length of molecules in the starting pool, *Chem. Biol.*, **4**, pp. 767–774.
12. Wedel, A. B. (1996). Fishing the best pool for novel ribozymes, *Trends Biotechnol.*, **14**, pp. 459–465.
13. Salehi-Ashtiani, K., and Szostak, J. W. (2001). In vitro evolution suggests multiple origins for the hammerhead ribozyme, *Nature*, **414**, pp. 82–84.
14. Bartel, D. P., and Szostak, J. W. (1993). Isolation of new ribozymes from a large pool of random sequences, *Science*, **261**, pp. 1411–1418.
15. Legiewicz, M., Lozupone, C., Knight, R., and Yarus, M. (2005). Size, constant sequences, and optimal selection, *RNA*, **11**, pp. 1701–1709.
16. Cowperthwaite, M. C., and Ellington, A. D. (2008). Bioinformatic analysis of the contribution of primer sequences to aptamer structures, *J. Mol. Evol.*, **67**, pp. 95–102.
17. Ellington, A. D. (1994). The RNA world. Empirical explorations of sequence space: host-guest chemistry in the RNA world, *Ber. Bunsenges. Phys. Chem.*, **98**, pp. 1115–1121.
18. Velez, T. E., Singh, J., Xiao, Y., Allen, E. C., Wong, O. Y., et al. (2012). Systematic evaluation of the dependence of deoxyribozyme catalysis on random region length, *ACS Comb. Sci.*, **14**, pp. 680–687.
19. Coleman, T. M., and Huang, F. (2002). RNA-catalyzed thioester synthesis, *Chem. Biol.*, **9**, pp. 1227–1236.
20. Coleman, T. M., and Huang, F. (2005). Optimal random libraries for the isolation of catalytic RNA, *RNA Biol.*, **2**, pp. 129–136.
21. Unrau, P. J., and Bartel, D. P. (1998). RNA-catalysed nucleotide synthesis, *Nature*, **395**, pp. 260–263.
22. Singer, B. S., Shtatland, T., Brown, D., and Gold, L. (1997). Libraries for genomic SELEX, *Nucleic Acids Res.*, **25**, pp. 781–786.
23. Parker, D. J., Xiao, Y., Aguilar, J. M., and Silverman, S. K. (2013). DNA catalysis of a normally disfavored RNA hydrolysis reaction, *J. Am. Chem. Soc.*, **135**, pp. 8472–8475.
24. Cech, T. R. (1987). The chemistry of self-splicing RNA and RNA enzymes, *Science*, **236**, pp. 1532–1539.

25. Cate, J. H., Gooding, A. R., Podell, E., Zhou, K., Golden, B. L., et al. (1996). RNA tertiary structure mediation by adenosine platforms, *Science,* **273**, pp. 1696–1699.
26. Pan, T., and Uhlenbeck, O. C. (1992). A small metalloribozyme with a two-step mechanism, *Nature,* **358**, pp. 560–563.
27. Williams, K. P., Ciafre, S., and Tocchini-Valentini, G. P. (1995). Selection of novel Mg(2+)-dependent self-cleaving ribozymes, *EMBO J.*, **14**, pp. 4551–4557.
28. Jayasena, V. K., and Gold, L. (1997). In vitro selection of self-cleaving RNAs with a low pH optimum, *Proc. Natl. Acad. Sci. U S A,* **94**, pp. 10612–10617.
29. Breaker, R. R., and Joyce, G. F. (1994). A DNA enzyme that cleaves RNA, *Chem. Biol.,* **1**, pp. 223–229.
30. Santoro, S. W., Joyce, G. F., Sakthivel, K., Gramatikova, S., and Barbas, C. F., 3rd (2000). RNA cleavage by a DNA enzyme with extended chemical functionality, *J. Am. Chem. Soc.,* **122**, pp. 2433–2439.
31. Perrin, D. M., Garestier, T., and Helene, C. (2001). Bridging the gap between proteins and nucleic acids: a metal-independent RNAseA mimic with two protein-like functionalities. *J. Am. Chem. Soc.,* **123**, pp. 1556–1563.
32. Sidorov, A. V., Grasby, J. A., and Williams, D. M. (2004). Sequence-specific cleavage of RNA in the absence of divalent metal ions by a DNAzyme incorporating imidazolyl and amino functionalities, *Nucleic Acids Res.*, **32**, pp. 1591–1601.
33. Breaker, R. R., and Joyce, G. F. (1995). A DNA enzyme with Mg(2+)-dependent RNA phosphoesterase activity, *Chem. Biol.,* **2**, pp. 655–660.
34. Santoro, S. W., and Joyce, G. F. (1997). A general purpose RNA-cleaving DNA enzyme, *Proc. Natl. Acad. Sci. U S A,* **94**, pp. 4262–4266.
35. Geyer, C. R., and Sen, D. (1997). Evidence for the metal-cofactor independence of an RNA phosphodiester-cleaving DNA enzyme, *Chem. Biol.,* **4**, pp. 579–593.
36. Teramoto, N., Imanishi, Y., and Ito, Y. (2000). In vitro selection of a ligase ribozyme carrying alkylamino groups in the side chains, *Bioconjug. Chem.*, **11**, pp. 744–748.
37. Ekland, E. H., Szostak, J. W., and Bartel, D. P. (1995). Structurally complex and highly active RNA ligases derived from random RNA sequences, *Science,* **269**, pp. 364–370.
38. Hager, A. J., and Szostak, J. W. (1997). Isolation of novel ribozymes that ligate AMP-activated RNA substrates, *Chem. Biol.,* **4**, pp. 607–617.

39. Flynn-Charlebois, A., Wang, Y., Prior, T. K., Rashid, I., Hoadley, K. A., et al. (2003). Deoxyribozymes with 2′-5′ RNA ligase activity, *J. Am. Chem. Soc.*, **125**, pp. 2444–2454.
40. Hoadley, K. A., Purtha, W. E., Wolf, A. C., Flynn-Charlebois, A., and Silverman, S. K. (2005). Zn^{2+}-dependent deoxyribozymes that form natural and unnatural RNA linkages, *Biochemistry*, **44**, pp. 9217–9231.
41. Purtha, W. E., Coppins, R. L., Smalley, M. K., and Silverman, S. K. (2005). General deoxyribozyme-catalyzed synthesis of native 3′-5′ RNA linkages, *J. Am. Chem. Soc.*, **127**, pp. 13124–13125.
42. Kost, D. M., Gerdt, J. P., Pradeepkumar, P. I., and Silverman, S. K. (2008). Controlling the direction of site-selectivity and regioselectivity in RNA ligation by Zn^{2+}-dependent deoxyribozymes that use 2′,3′-cyclic phosphate RNA substrates, *Org. Biomol. Chem.*, **6**, pp. 4391–4398.
43. Tuschl, T., Sharp, P. A., and Bartel, D. P. (1998). Selection in vitro of novel ribozymes from a partially randomized U2 and U6 snRNA library, *EMBO J.*, **17**, pp. 2637–2650.
44. Wang, Y., and Silverman, S. K. (2003). Deoxyribozymes that synthesize branched and lariat RNA, *J. Am. Chem. Soc.*, **125**, pp. 6880–6881.
45. Coppins, R. L., and Silverman, S. K. (2005). A deoxyribozyme that forms a three-helix-junction complex with its RNA substrates and has general RNA branch-forming activity, *J. Am. Chem. Soc.*, **127**, pp. 2900–2907.
46. Lorsch, J. R., and Szostak, J. W. (1994). In vitro evolution of new ribozymes with polynucleotide kinase activity, *Nature*, **371**, pp. 31–36.
47. Curtis, E. A., and Bartel, D. P. (2005). New catalytic structures from an existing ribozyme, *Nat. Struct. Mol. Biol.*, **12**, pp. 994–1000.
48. Moretti, J. E., and Müller, U. F. (2014). A ribozyme that triphosphorylates RNA 5′-hydroxyl groups, *Nucleic Acids Res.*, **42**, pp. 4767–4778.
49. Wang, W., Billen, L. P., and Li, Y. (2002). Sequence diversity, metal specificity, and catalytic proficiency of metal-dependent phosphorylating DNA enzymes, *Chem. Biol.*, **9**, pp. 507–517.
50. Chapman, K. B., and Szostak, J. W. (1995). Isolation of a ribozyme with 5′-5′ ligase activity, *Chem. Biol.*, **2**, pp. 325–333.
51. Huang, F., and Yarus, M. (1997). 5′-RNA self-capping from guanosine diphosphate, *Biochemistry*, **36**, pp. 6557–6563.
52. Huang, F., Yang, Z., and Yarus, M. (1998). RNA enzymes with two small-molecule substrates, *Chem. Biol.*, **5**, pp. 669–678.
53. Li, Y., Liu, Y., and Breaker, R. R. (2000). Capping DNA with DNA, *Biochemistry*, **39**, pp. 3106–3114.

54. Ekland, E. H., and Bartel, D. P. (1996). RNA-catalysed RNA polymerization using nucleoside triphosphates, *Nature,* **383**, p. 192.
55. Attwater, J., Wochner, A., and Holliger, P. (2013). In-ice evolution of RNA polymerase ribozyme activity, *Nat. Chem.*, **5**, pp. 1011–1018.
56. Wochner, A., Attwater, J., Coulson, A., and Holliger, P. (2011). Ribozyme-catalyzed transcription of an active ribozyme, *Science,* **332**, pp. 209–212.
57. Cuenoud, B., and Szostak, J. W. (1995). A DNA metalloenzyme with DNA ligase activity, *Nature,* **375**, pp. 611–614.
58. Sreedhara, A., Li, Y., and Breaker, R. R. (2004). Ligating DNA with DNA, *J. Am. Chem. Soc.,* **126**, pp. 3454–3460.
59. Seelig, B., Keiper, S., Stuhlmann, F., and Jaschke, A. (2000). Enantioselective ribozyme catalysis of a bimolecular cycloaddition reaction, *Angew. Chem., Int. Ed. Engl.*, **39**, pp. 4576–4579.
60. Tarasow, T. M., Tarasow, S. L., and Eaton, B. E. (1997). RNA-catalysed carbon-carbon bond formation, *Nature,* **389**, pp. 54–57.
61. Tarasow, T. M., Kellogg, E., Holley, B. L., Nieuwlandt, D., Tarasow, S. L., et al. (2004). The effect of mutation on RNA Diels-Alderases, *J. Am. Chem. Soc.,* **126**, pp. 11843–11851.
62. Sheppard, T. L., Ordoukhanian, P., and Joyce, G. F. (2000). A DNA enzyme with N-glycosylase activity, *Proc. Natl. Acad. Sci. U S A,* **97**, pp. 7802–7807.
63. Lee, N., Bessho, Y., Wei, K., Szostak, J. W., and Suga, H. (2000). Ribozyme-catalyzed tRNA aminoacylation, *Nat. Struct. Biol.*, **7**, pp. 28–33.
64. Saito, H., Kourouklis, D., and Suga, H. (2001). An in vitro evolved precursor tRNA with aminoacylation activity, *EMBO J.*, **20**, pp. 1797–1806.
65. Illangasekare, M., Sanchez, G., Nickles, T., and Yarus, M. (1995). Aminoacyl-RNA synthesis catalyzed by an RNA, *Science,* **267**, pp. 643–647.
66. Chandra, M., Sachdeva, A., and Silverman, S. K. (2009). DNA-catalyzed sequence-specific hydrolysis of DNA, *Nat. Chem. Biol.*, **5**, pp. 718–720.
67. Fekry, M. I., Gates, K. S. (2009). DNA-catalyzed hydrolysis of DNA phosphodiesters, *Nat. Chem. Biol.*, **5**, pp. 710–711.
68. Gu, H., Furukawa, K., Weinberg, Z., Berenson, D. F., and Breaker, R. R. (2013). Small, highly active DNAs that hydrolyze DNA, *J. Am. Chem. Soc.,* **135**, pp. 9121–9129.
69. Zhang, B., and Cech, T. R. (1997). Peptide bond formation by in vitro selected ribozymes, *Nature,* **390**, pp. 96–100.

70. Chinnapen, D. J., and Sen, D. (2004). A deoxyribozyme that harnesses light to repair thymine dimers in DNA, *Proc. Natl. Acad. Sci. U S A,* **101**, pp. 65–69.

71. Conn, M. M., Prudent, J. R., and Schultz, P. G. (1996). Porphyrin metalation catalyzed by a small RNA molecule, *J. Am. Chem. Soc.,* **118**, pp. 7012–7013.

72. Burmeister, J., von Kiedrowski, G., and Ellington, A. D. (1997). Cofactor-Assisted Self-Cleavage in DNA Libraries with a 3′–5′-Phosphoramidate Bond, *Angew. Chem., Int. Ed.*, **36**, pp. 1321–1324.

73. Wiegand, T. W., Janssen, R. C., and Eaton, B. E. (1997). Selection of RNA amide synthases, *Chem. Biol.,* **4**, pp. 675–683.

74. Li, Y., and Sen, D. (1996). A catalytic DNA for porphyrin metallation, *Nat. Struct. Biol.*, **3**, pp. 743–747.

75. Fusz, S., Eisenfuhr, A., Srivatsan, S. G., Heckel, A., and Famulok, M. (2005). A ribozyme for the aldol reaction, *Chem. Biol.,* **12**, pp. 941–950.

76. Hobartner, C., Pradeepkumar, P. I., and Silverman, S. K. (2007). Site-selective depurination by a periodate-dependent deoxyribozyme, *Chem. Commun. (Camb.),* pp. 2255–2257.

77. Sczepanski, J. T., and Joyce, G. F. (2014). A cross-chiral RNA polymerase ribozyme, *Nature,* **515**, pp. 440–442.

78. Brandsen, B. M., Hesser, A. R., Castner, M. A., Chandra, M., and Silverman, S. K. (2013). DNA-catalyzed hydrolysis of esters and aromatic amides, *J. Am. Chem. Soc.,* **135**, pp. 16014–16017.

79. Faulhammer, D., Famulok, M. (1996). The Ca^{2+} ion as a cofactor for a novel RNA-cleaving deoxyribozyme, *Angew. Chem., Int. Ed.*, **35**, pp. 2837–2841.

80. Roth, A., and Breaker, R. R. (1998). An amino acid as a cofactor for a catalytic polynucleotide, *Proc. Natl. Acad. Sci. U S A,* **95**, pp. 6027–6031.

81. Feldman, A. R., and Sen, D. (2001). A new and efficient DNA enzyme for the sequence-specific cleavage of RNA, *J. Mol. Biol.*, **313**, pp. 283–294.

82. Cruz, R. P., Withers, J. B., and Li, Y. (2004). Dinucleotide junction cleavage versatility of 8-17 deoxyribozyme, *Chem. Biol.,* **11**, pp. 57–67.

83. Isaka, Y. (2007). DNAzymes as potential therapeutic molecules, *Curr. Opin. Mol. Ther.*, **9**, pp. 132–136.

84. Schlosser, K., and Li, Y. (2009). Biologically inspired synthetic enzymes made from DNA, *Chem. Biol.,* **16**, pp. 311–322.

85. Liu, J., and Lu, Y. (2003). A colorimetric lead biosensor using DNAzyme-directed assembly of gold nanoparticles, *J. Am. Chem. Soc.,* **125**, pp. 6642–6643.

86. Lermer, L., Roupioz, Y., Ting, R., and Perrin, D. M. (2002). Toward an RNaseA mimic: a DNAzyme with imidazoles and cationic amines, *J. Am. Chem. Soc.,* **124**, pp. 9960–9961.

87. Taylor, A. I., Pinheiro, V. B., Smola, M. J., Morgunov, A. S., Peak-Chew, S., et al. (2015). Catalysts from synthetic genetic polymers, *Nature,* **518**, pp. 427–430.

88. Doudna, J. A., and Cech, T. R. (2002). The chemical repertoire of natural ribozymes, *Nature,* **418**, pp. 222–228.

89. Silverman, S. K. (2010). DNA as a versatile chemical component for catalysis, encoding, and stereocontrol, *Angew. Chem., Int. Ed. Engl.*, **49**, pp. 7180–7201.

90. Li, Y., and Breaker, R. R. (1999). Kinetics of RNA degradation by specific base catalysis of transesterification involving the 2′-hydroxyl group, *J. Am. Chem. Soc.,* **121**, pp. 5364–5372.

91. Schroeder, G. K., Lad, C., Wyman, P., Williams, N. H., and Wolfenden, R. (2006). The time required for water attack at the phosphorus atom of simple phosphodiesters and of DNA, *Proc. Natl. Acad. Sci. U S A,* **103**, pp. 4052–4055.

92. Li, Y., and Breaker, R. R. (1999). Phosphorylating DNA with DNA, *Proc. Natl. Acad. Sci. U S A,* **96**, pp. 2746–2751.

93. Yoo, E.-H., and Lee, S.-Y. (2010). Glucose biosensors: an overview of use in clinical practice, *Sensors*, **10**, pp. 4558–4576.

94. Clark, L. C., Jr., and Lyons, C. (1962). Electrode systems for continuous monitoring in cardiovascular surgery, *Ann. N. Y. Acad. Sci.*, **102**, pp. 29–45.

95. Luong, J. H., Male, K. B., and Glennon, J. D. (2008). Biosensor technology: technology push versus market pull, *Biotechnol. Adv.*, **26**, pp. 492–500.

96. Chames, P., Van Regenmortel, M., Weiss, E., and Baty, D. (2009). Therapeutic antibodies: successes, limitations and hopes for the future, *Br. J. Pharmacol.*, **157**, pp. 220–233.

97. Thusu, R. (2010). Strong Growth Predicted for Biosensor Market. *Sensors Online*. www.sensorsmag.com: Questex Media Group.

98. Li, Y., and Lu, Y. (2009). *Functional Nucleic Acids for Analytical Applications*, Springer.

99. Navani, N. K., and Li, Y. (2006). Nucleic acid aptamers and enzymes as sensors, *Curr. Opin. Chem. Biol.*, **10**, pp. 272–281.

100. Mok, W., and Li, Y. (2008). Recent progress in nucleic acid aptamer-based biosensors and bioassays, *Sensors*, **8**, pp. 7050–7084.

101. Liu, J., Cao, Z., and Lu, Y. (2009). Functional nucleic acid sensors, *Chem. Rev.*, **109**, pp. 1948–1998.

102. Kuwabara, T., Warashina, M., and Taira, K. (2000). Allosterically controllable ribozymes with biosensor functions, *Curr. Opin. Chem. Biol.*, **4**, pp. 669–677.

103. Achenbach, J., Chiuman, W., Cruz, R., and Li, Y. (2004). DNAzymes: from creation in vitro to application in vivo, *Curr. Pharm. Biotechnol.*, **5**, pp. 321–336.

104. Silverman, S. K. (2005). In vitro selection, characterization, and application of deoxyribozymes that cleave RNA, *Nucleic Acids Res.*, **33**, pp. 6151–6163.

105. Tang, J., and Breaker, R. R. (1997). Rational design of allosteric ribozymes, *Chem. Biol.*, **4**, pp. 453–459.

106. Zhang, X.-B., Wang, Z., Xing, H., Xiang, Y., and Lu, Y. (2010). Catalytic and molecular beacons for amplified detection of metal ions and organic molecules with high sensitivity, *Anal. Chem.*, **82**, pp. 5005–5011.

107. Tram, K., Kanda, P., and Li, Y. (2012). Lighting up RNA-cleaving DNAzymes for biosensing, *J. Nucleic Acids*, **2012**, Article ID 958683, 8 pages.

108. Tan, M. L., Choong, P. F., and Dass, C. R. (2009). Cancer, chitosan nanoparticles and catalytic nucleic acids, *J. Pharm. Pharmacol.*, **61**, pp. 3–12.

109. Khan, A., Benboubetra, M., Sayyed, P. Z., Ng, K. W., Fox, S., et al. (2004). Sustained polymeric delivery of gene silencing antisense ODNs, siRNA, DNAzymes and ribozymes: in vitro and in vivo studies, *J. Drug Target.*, **12**, pp. 393–404.

110. Pun, S. H., Bellocq, N. C., Liu, A., Jensen, G., Machemer, T., et al. (2004). Cyclodextrin-modified polyethylenimine polymers for gene delivery, *Bioconjug. Chem.*, **15**, pp. 831–840.

111. Tack, F., Bakker, A., Maes, S., Dekeyser, N., Bruining, M., et al. (2006). Modified poly (propylene imine) dendrimers as effective transfection agents for catalytic DNA enzymes (DNAzymes), *J. Drug Target.*, **14**, pp. 69–86.

112. Zhang, L., Huang, H., Xu, N., and Yin, Q. (2014). Functionalization of cationic poly(p-phenylene ethynylene) with dendritic polyethylene enables efficient DNAzyme delivery for imaging Pb^{2+} in living cells, *J. Mater. Chem. B*, **2**, p. 4935.

113. Dass, C. R., Friedhuber, A. M., Khachigian, L. M., Dunstan, D. E., and Choong, P. F. (2008). Downregulation of c-jun results in apoptosis-

mediated anti-osteosarcoma activity in an orthotopic model, *Cancer Biol. Ther.,* **7**, pp. 1033–1036.

114. Liang, Z., Wei, S., Guan, J., Luo, Y., Gao, J., et al. (2005). DNAzyme-mediated cleavage of survivin mRNA and inhibition of the growth of PANC-1 cells, *J. Gastroenterol. Hepatol.*, **20**, pp. 1595–1602.
115. Tack, F., Noppe, M., Van Dijck, A., Dekeyzer, N., Van Der Leede, B.-J., et al. (2008). Delivery of a DNAzyme targeting c-myc to HT29 colon carcinoma cells using a gold nanoparticulate approach, *Pharmazie*, **63**, pp. 221–225.
116. Fahmy, R. G., Waldman, A., Zhang, G., Mitchell, A., Tedla, N., et al. (2006). Suppression of vascular permeability and inflammation by targeting of the transcription factor c-Jun, *Nat. Biotechnol.*, **24**, pp. 856–863.
117. Lin Tan, M., Choong, P. F., and Dass, C. R. (2009). DNAzyme delivery systems: getting past first base, *Expert Opin. Drug Deliv.*, **6**, pp. 127–138.
118. Rockwell, P., O'Connor, W. J., King, K., Goldstein, N. I., Zhang, L., et al. (1997). Cell-surface perturbations of the epidermal growth factor and vascular endothelial growth factor receptors by phosphorothioate oligodeoxynucleotides, *Proc. Natl. Acad. Sci.*, **94**, pp. 6523–6528.
119. Wahlestedt, C., Salmi, P., Good, L., Kela, J., Johnsson, T., et al. (2000). Potent and nontoxic antisense oligonucleotides containing locked nucleic acids, *Proc. Natl. Acad. Sci.*, **97**, pp. 5633–5638.
120. Fluiter, K., ten Asbroek, A. L., de Wissel, M. B., Jakobs, M. E., Wissenbach, M., et al. (2003). In vivo tumor growth inhibition and biodistribution studies of locked nucleic acid (LNA) antisense oligonucleotides, *Nucleic Acids Res.*, **31**, pp. 953–962.
121. Dass, C. R., Choong, P. F., and Khachigian, L. M. (2008). DNAzyme technology and cancer therapy: cleave and let die, *Mol. Cancer Ther.*, **7**, pp. 243–251.
122. Burgess, D. J. (2012). Antitumour potential of catalytic DNA, *Nat. Rev. Cancer,* **12**, pp. 508–509.
123. Xu, Z., Yang, L., Sun, L., and Cao, Y. (2012). Use of DNAzymes for cancer research and therapy, *Chin. Sci. Bull.*, **57**, pp. 3404–3408.
124. Warashina, M., Kuwabara, T., Nakamatsu, Y., and Taira, K. (1999). Extremely high and specific activity of DNA enzymes in cells with a Philadelphia chromosome, *Chem. Biol.,* **6**, pp. 237–250.
125. Wu, Y., Yu, L., McMahon, R., Rossi, J. J., Forman, S. J., et al. (1999). Inhibition of bcr-abl oncogene expression by novel deoxyribozymes (DNAzymes), *Hum. Gene Ther.,***10**, pp. 2847–2857.

126. Kuwabara, T., Warashina, M., Tanabe, T., Tani, K., Asano, S., et al. (1997). Comparison of the specificities and catalytic activities of hammerhead ribozymes and DNA enzymes with respect to the cleavage of BCR-ABL chimeric L6 (b2a2) mRNA, *Nucleic Acids Res.,* **25**, pp. 3074–3081.

127. Weidner, N., Semple, J. P., Welch, W. R., and Folkman, J. (1991). Tumor angiogenesis and metastasis—correlation in invasive breast carcinoma, *New Engl. J. Med.*, **324**, pp. 1–8.

128. Hanahan, D., and Folkman, J. (1996). Patterns and emerging mechanisms of the angiogenic switch during tumorigenesis, *Cell*, **86**, pp. 353–364.

129. Cieslak, M., Niewiarowska, J., Nawrot, M., Koziolkiewicz, M., Stec, W. J., et al. (2002). DNAzymes to beta 1 and beta 3 mRNA down-regulate expression of the targeted integrins and inhibit endothelial cell capillary tube formation in fibrin and matrigel, *J. Biol. Chem.*, **277**, pp. 6779–6787.

130. de Bock, C. E., Lin, Z., Itoh, T., Morris, D., Murrell, G., et al. (2005). Inhibition of urokinase receptor gene expression and cell invasion by anti-uPAR DNAzymes in osteosarcoma cells, *FEBS J.*, **272**, pp. 3572–3582.

131. Masood, R., Cai, J., Zheng, T., Smith, D. L., Hinton, D. R., et al. (2001). Vascular endothelial growth factor (VEGF) is an autocrine growth factor for VEGF receptor–positive human tumors, *Blood*, **98**, pp. 1904–1913.

132. Herold-Mende, C., Steiner, H.-H., Andl, T., Riede, D., Buttler, A., et al. (1999). Expression and functional significance of vascular endothelial growth factor receptors in human tumor cells, *Lab. Invest.*, **79**, pp. 1573–1582.

133. Beale, G., Hollins, A. J., Benboubetra, M., Sohail, M., Fox, S. P., et al. (2003). Gene silencing nucleic acids designed by scanning arrays: anti-EGFR activity of siRNA, ribozyme and DNA enzymes targeting a single hybridization-accessible region using the same delivery system, *J. Drug Target.*, **11**, pp. 449–456.

134. Hou, Z., Meng, J. R., Zhao, J. R., Hu, B. Q., Liu, J., et al. (2007). Inhibition of beta-lactamase-mediated oxacillin resistance in Staphylococcus aureus by a deoxyribozyme, *Acta Pharmacol. Sin.*, **28**, pp. 1775–1782.

135. Hou, Z., Meng, J. R., Niu, C., Wang, H. F., Liu, J., et al. (2007). Restoration of antibiotic susceptibility in methicillin-resistant Staphylococcus

aureus by targeting mecr1 with a phosphorothioate deoxyribozyme, *Clin. Exp. Pharmacol. Physiol.*, **34**, pp. 1160–1164.

136. Tan, X. X., Rose, K., Margolin, W., and Chen, Y. (2004). DNA enzyme generated by a novel single-stranded DNA expression vector inhibits expression of the essential bacterial cell division gene ftsZ, *Biochemistry*, **43**, pp. 1111–1117.

137. Li, J., Zhu, D., Yi, Z., He, Y., Chun, Y., et al. (2005). DNAzymes targeting the icl gene inhibit ICL expression and decrease Mycobacterium tuberculosis survival in macrophages, *Oligonucleotides*, **15**, pp. 215–222.

138. Singh, N., Ranjan, A., Sur, S., Chandra, R., and Tandon, V. (2012). Inhibition of HIV-1 Integrase gene expression by 10-23 DNAzyme, *J. Biosci.*, **37**, pp. 493–502.

139. Lu, Z. X., Ye, M., Yan, G. R., Li, Q., Tang, M., et al. (2005). Effect of EBV LMP1 targeted DNAzymes on cell proliferation and apoptosis, *Cancer Gene Ther.*, **12**, pp. 647–654.

140. Wo, J. E., Wu, X. L., Zhou, L. F., Yao, H. P., Chen, L. W., et al. (2005). Effective inhibition of expression of hepatitis B virus genes by DNAzymes, *World J. Gastroenterol.*, **11**, pp. 3504–3507.

141. Wo, J. E., Wu, X. L., Zhu, H. H., Zhou, L. F., Yao, H. P., et al. (2003). [DNAzymes in vitro inhibit the expression of hepatitis B virus genes]. *Zhejiang Da Xue Xue Bao Yi Xue Ban*, **32**, pp. 112–115.

142. Hou, W., Ni, Q., Wo, J., Li, M., Liu, K., et al. (2006). Inhibition of hepatitis B virus X gene expression by 10-23 DNAzymes, *Antiviral Res.*, **72**, pp. 190–196.

143. Trepanier, J., Tanner, J. E., Momparler, R. L., Le, O. N., Alvarez, F., et al. (2006). Cleavage of intracellular hepatitis C RNA in the virus core protein coding region by deoxyribozymes, *J. Viral Hepat.*, **13**, pp. 131–138.

144. Dash, B. C., and Banerjea, A. C. (2004). Sequence-specific cleavage activities of DNA enzymes targeted against HIV-1 Gag and Nef regions, *Oligonucleotides*, **14**, pp. 41–47.

145. Chakraborti, S. (2003). Inhibition of HIV-1 gene expression by novel DNA enzymes targeted to cleave HIV-1 TAR RNA: potential effectiveness against all HIV-1 isolates, *Mol. Ther.*, **7**, pp. 817–826.

146. Unwalla, H., Chakraborti, S., Sood, V., Gupta, N., and Banerjea, A. C. (2006). Potent inhibition of HIV-1 gene expression and TAT-mediated apoptosis in human T cells by novel mono- and multitarget anti-

TAT/Rev/Env ribozymes and a general purpose RNA-cleaving DNA-enzyme, *Antiviral Res.*, **72**, pp. 134–144.

147. Wu, S., Xu, J., Liu, J., Yan, X., Zhu, X., et al. (2007). An efficient RNA-cleaving DNA enzyme can specifically target the 5′-untranslated region of severe acute respiratory syndrome associated coronavirus (SARS-CoV), *J. Gene Med.*, **9**, pp. 1080–1086.

148. Zhou, J., Yang, X. Q., Xie, Y. Y., Zhao, X. D., Jiang, L. P., et al. (2007). Inhibition of respiratory syncytial virus of subgroups A and B using deoxyribozyme DZ1133 in mice, *Virus Res.*, **130**, pp. 241–248.

149. Zhang, G., Dass, C. R., Sumithran, E., Di Girolamo, N., Sun, L. Q., et al. (2004). Effect of deoxyribozymes targeting c-jun on solid tumor growth and angiogenesis in rodents, *J. Natl. Cancer Inst.*, **96**, pp. 683–696.

150. Juliano, R., and Yoo, H. (2000). Aspects of the transport and delivery of antisense oligonucleotides, *Curr. Opin. Mol. Ther.*, **2**, pp. 297–303.

151. Yanze, M., Lee, W.-S., Poon, K., Piquette-Miller, M., and Macgregor, R. (2003). Cellular uptake and metabolism of DNA frayed wires, *Biochemistry*, **42**, pp. 11427–11433.

152. Godwin, H. A. (2001). The biological chemistry of lead, *Curr. Opin. Chem. Biol.*, **5**, pp. 223–227.

153. Wu, P., Hwang, K., Lan, T., and Lu, Y. (2013). A DNAzyme-gold nanoparticle probe for uranyl ion in living cells, *J. Am. Chem. Soc.*, **135**, pp. 5254–5257.

154. Xu, W., Xing, H., and Lu, Y. (2013). A smart T(1)-weighted MRI contrast agent for uranyl cations based on a DNAzyme-gadolinium conjugate, *Analyst*, **138**, pp. 6266–6269.

155. Hong, S. H., Jeong, J. S., Lee, Y. J., Jung, H. I., Kim, K. T., et al. (2007). Molecular imaging of endogenous mRNA expression in a mouse tumor model by adenovirus harboring trans-splicing ribozyme, *FEBS Lett.*, **581**, pp. 5396–5400.

156. Hasegawa, S., Jackson, W. C., Tsien, R. Y., and Rao, J. (2003). Imaging Tetrahymena ribozyme splicing activity in single live mammalian cells, *Proc. Natl. Acad. Sci. U S A*, **100**, pp. 14892–14896.

157. Armitage, B. A. (2011). Imaging of RNA in live cells, *Curr. Opin. Chem. Biol.*, **15**, pp. 806–812.

158. Gillingham, D., and Shahid, R. (2015). Catalysts for RNA and DNA modification, *Curr. Opin. Chem. Biol.*, **25**, pp. 110–114.

159. Paige, J. S., Wu, K. Y., and Jaffrey, S. R. (2011). RNA mimics of green fluorescent protein, *Science,* **333**, pp. 642–646.

160. Dolgosheina, E. V., Jeng, S. C., Panchapakesan, S. S. S., Cojocaru, R., Chen, P. S., et al. (2014). RNA mango aptamer-fluorophore: a bright, high-affinity complex for rna labeling and tracking, *ACS Chem. Biol.*, **9**, pp. 2412–2420.

Chapter 5

Stem-Cell-Specific Aptamers for Targeted Cancer Therapy

Hadi Al. Shamaileh, Dongxi Xiang, Tao Wang, Wang Yin, Wei Duan, and Sarah Shigdar

School of Medicine, Deakin University, Pigdons Road, Waurn Ponds Geelong, Victoria 3216, Australia

sarah.shigdar@deakin.edu.au

Cancer is a malignant disease that affects millions of people worldwide and is one of the leading causes of death. Early stages of the disease can be treated with surgical resection with overall positive prognosis. However, advanced and metastatic stages require additional therapeutic intervention such as chemo-/radiotherapy. Despite significant progress in conventional cancer medicine and its effectiveness in eradicating the rapidly proliferating cancer cells, patients remain victims to tumor regrowth regardless of positive treatment outcomes. A wave of evidence suggests the existence of a population of cells with stem cell–like qualities that are resistant to conventional therapies and are capable of initiating tumor growth. These cells have come to be known as cancer stem cells (CSCs). An emerging new class of drugs known as aptamers shows promising potential in targeting CSCs for therapeutic and diagnostic applications. Much like antibodies, aptamers show high binding

Aptamers: Tools for Nanotherapy and Molecular Imaging
Edited by Rakesh N. Veedu

ISBN 978-981-4669-83-2 (Hardcover), 978-981-4669-84-9 (eBook)
www.panstanford.com

specificity and affinity in the nanomolar range. Many aptamers have been developed to recognize CSCs and have been implemented as drug delivery agents by direct or indirect conjugation with anticancer drugs. Aptamer-mediated drug delivery to CSCs showed great potential in tumor penetration, reduced off-target effect, and reduction in stem cell properties, making aptamers a novel therapeutic strategy in targeting CSCs.

5.1 Introduction

Despite current progress in anticancer treatments, a high proportion of patients fall victim to tumor regrowth and metastasis regardless of a positive response to initial treatment [1, 2]. Accumulating evidence suggests that a specific population of cancer cells possesses stemness qualities and plays a key role in tumor growth and developing resistance to anticancer treatments [3–5]. These tumor-initiating cells, sometimes referred to as cancer stem cells (CSCs), possess stem cell–like properties of self-renewal and pluripotency that contribute to the maintenance and growth of tumors, as well as forming new tumors of histological similarity to the parent tumor when transplanted into mice, thus giving them the description as the "root of cancer" [6–8]. The significance of the CSC model of cancer is that it provides a valuable explanation for several poorly understood clinical phenomena: How can tumors regrow in patients despite having a positive and robust response to therapy? And how can cancer cells remain dormant and metastasize after nearly a decade of initial treatment, as observed with breast cancer [9]? As a result of these questions and findings, CSCs have proven themselves to be vital targets for effective cancer treatment.

Most clinical anticancer drugs rely on passive diffusion, which has a risk of exposure to healthy tissues, whereas the use of drug nanocarriers relies on passive targeting exploited by the enhanced permeability and retention (EPR) effect, which is limited by poor tumor penetration. Consequently, active targeting has attracted wide attention for improved tumor penetration and cellular uptake of drugs [10–12]. Active targeting operates by guiding therapeutic agents to their specific targets with the aid of a binding ligand and the target cell's ability for receptor-mediated endocytosis [13–16]. Development of molecular probes or binding ligands that bind to specific cell surface markers as active targeting devices represents one of the major advances in next-generation tumor therapeutics. Aptamers are an emerging new means of CSC targeting and drug delivery for more effective cancer treatment. They offer significant advantages over antibodies in terms of size, tissue penetration, and production. In addition, aptamers allow for increased delivery of a wide range of therapeutic agents as well as reduce their associated toxicity and side effects [17–20].

CSCs can be identified by their overexpression of certain surface markers. Although the majority of such markers have been identified from cell line studies, confirmation of the expression of these markers in human primary tumors is necessary. Breast cancer cells expressing $CD44^{+}CD24^{-/low}$ and $CD133^{+}$ were found to initiate tumor growth at much higher frequencies than the majority of the tumor cell population [6, 21]. Similarly, colon cancer cells with epithelial cell adhesion molecule ($EpCAM^{+}$), $CD44^{+}$, $CD166^{+}$, and $CD133^{+}$ were able to initiate tumor growth at a much higher frequency [7, 22]. Additionally, leukemic cells with $CD133^{+}$ could self-renew and produce engraftments in vivo in xenograft models with a similar immunophenotype and karyotype as the parent cells [23]. Table 5.1 shows a list of commonly used surface markers to identify CSCs in various cancer types.

5.1.1 *Cancer Stem Cells*

In the first half of the twentieth century, the long-standing cancer model of clonal evolution was challenged by a new wave of evidence that supported a new model of cancer that is now known as the CSC model. This model contradicted the clonal evolution model, which postulates that all the cells within a tumor had equal capacity for tumor initiation (Fig. 5.1). In 1937, Furth and Khan demonstrated that single cells extracted from a mouse tumor were capable of forming new tumors when transplanted into a different mouse, resembling the functionality of stem cells in creating new cell progeny [42]. The frequency of these tumor-initiating cells found within solid tumors and leukemias were

Table 5.1 CSC markers in solid tumors

Tumor type	Marker	Reference
Breast cancer	$CD44^{+}/CD24^{-}/^{low}$	[6]
	$CD133^{+}$	[24]
	$EpCAM^{+}$	[25]
	$CD133^{+}/CXCR4^{+}$	[26]
Colon cancer	$CD44^{+}/CD166^{+}/EpCAM^{+}$	[7]
	$CD44^{+}/CD24^{+}$	[27]
	$CD133^{+}/CD24^{+}$	[28]
Brain cancer	$CD133^{+}$	[29, 30]
Liver cancer	$CD133^{+}/CD90^{+}/CD44^{+}/OV6^{+}/EpCAM^{+}$	[31]
Lung cancer	$CD133^{+}/ALDH\text{-}1^{+}$	[32]
Ovarian cancer	$CD133^{+}/ALDH\text{-}1^{+}$	[33]
Pancreatic cancer	$CD44^{+}/CD24^{+}/EpCAM^{+}$	[34]
	$CD133^{+}$	[35, 36]
	$CXCR4^{+}/CD133^{+}$	[37]
Prostate cancer	$CD133^{+}$	[38]
	$CD44^{+}$/integrin $\alpha 2\beta 1^{+}/CD133^{+}$	[39]
	$CD44^{+}$	[40]
	$CD133^{+}/Trop\text{-}2^{+}$	[41]

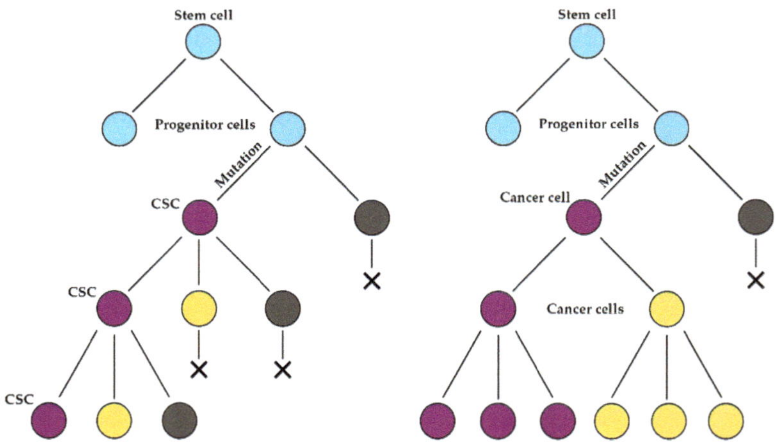

Figure 5.1 The CSC model (left) shows the hierarchical relationship in which only a specific population of cells has tumorigenic capacity, which self-renews and differentiates into a heterogeneous population of tumor cells, while recapitulating the original tumor heterogeneity. The clonal evolution model (right) shows that each cell of the tumor population has equal capacity for tumorigenesis and further proliferates in a nonhierarchical relationship.

variably low but were sufficiently capable of forming new tumors of similar morphological heterogeneity as the parent tumors [43–45]. With the emergence of cell radiolabeling and autoradiography techniques in the midtwentieth century, labels in undifferentiated cells later appeared in differentiated cells, which in turn were not capable of forming new tumors [46]. This finding aided in giving a formal definition of the CSC models by describing two separate cell populations, a malignant stem cell population with a proliferative capacity and a limited differentiation capacity and a population of differentiated cells that are the progeny of the malignant stem cell population [47].

The support for the CSC model began to decline in the 1970s after the discovery of mutations in oncogenes and tumor suppressor genes and their role in cancer, which later led to the description of acquired genetic variability in tumor progression, allowing for the sequential selection of more aggressive cells and resulting in advanced malignancies composed of cell populations that are karyotypically and biologically similar [48]. Despite the prevalence of the clonal evolution model, advances in hematopoietic stem cell research and the development of fluorescence-activated cell sorting (FACS) led to the revival of the CSC model in leukemias [49]. Only a specific subset of cells, such as $CD34^+CD38^-$ cells in acute myeloid leukemia (AML), were successful in initiating leukemia in mice models, and the frequency of these tumor-initiating cells was found to be in the order of one cell per million tumor cells [50, 51]. Subsequently, cells expressing $EpCAM^+CD44^+CD24^{-/low}$ and Lin^- in breast cancer showed a higher tumor initiation ability in xenograft models and retained the parent tumor heterogeneity [6]. Similar findings were reported in colon and brain tumors in which only a small subset of cells were capable of initiating tumorigenesis and retaining the parent tumor heterogeneity [7, 8, 52, 53].

5.1.2 *CSC Markers and Association with Stemness and Survival*

The CSC model attributes tumor initiation and growth to a specific population of cells within the tumor mass. These CSCs are best identified by their functional properties of self-renewal, which

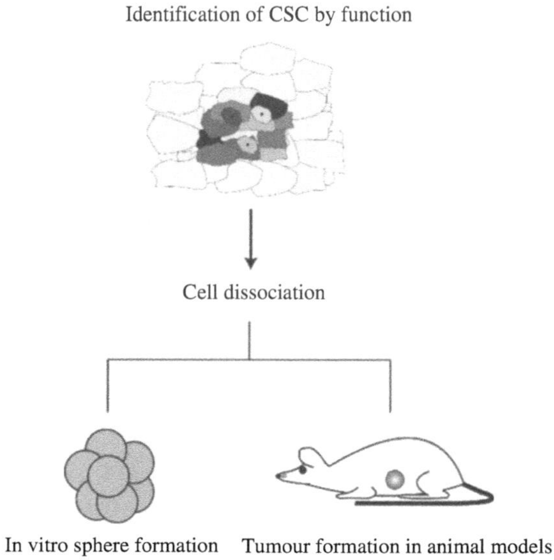

Figure 5.2 Functional identification of a CSC. The self-renewal property of CSCs allows for the formation of tumor spheres in nonadherent culture, while the differentiated cells undergo cell death. The tumorigenic property of CSCs can be assessed by tumor formation in xenografts.

can be evaluated by in vitro sphere formation; the ability to differentiate into different cell progeny; and tumorigenesis—the ability to form tumors in xenograft models (Fig. 5.2). Lineage tracing has also been used to identify and support the CSC model by tracking CSC activity throughout tumor development with inducible fluorescent proteins [54]. CSCs can also be identified by their frequent overexpression of specific cell surface markers, but it is constantly stressed that such markers are not exclusive to tumor cells, and expression patterns may differ between different or even similar cancer types. However, these surface markers allow for enrichment and prospective isolation of CSCs. One of the most common cell surface expression patterns observed in CSCs is $CD44^{+}CD24^{-/low}$, which can be found in breast, prostate, and ovarian cancers [6, 21, 55, 56]. These cells showed high sphere formation and tumor initiation capacity in xenografts at cell densities as low as 100 cells compared to $CD44^{-}CD24^{+}$ cells. Similarly, colorectal cancer cells with high $EpCAM^{+}CD44^{+}CD166^{+}$ displayed high stem cell characteristics and high tumor-initiating capacity in xenograft models that recapitulated the parent tumor [7, 22, 52]. $CD133^+$ is another example of a cell surface marker that allows for the enrichment of CSCs from a wide range of cancers, which includes breast, colon, glioma, liver, lung, melanoma, ovarian, pancreatic, and prostate cancers [57]. The frequent overexpression of such cell surface markers has been linked to signaling pathways that lead to stemness, cell proliferation, drug resistance, and survival. CD44 plays an important role in cell–cell and cell–extracellular matrix communications and is also linked to pathways essential for stem cell maintenance and self-renewal, such as the stem cell genes *BMI-1* and *Oct-3/4* [55]. EpCAM is a transmembrane protein that is crucial for cell–cell adhesion, and its intracellular domain contributes to the formation of transcription factor complexes that induces expression of stem cell–associated genes, such as *Klf4, Oct4, Nanog*, and *Myc* [58]. CD133 expression is found in many CSCs of various cancer types, as mentioned earlier. However, its role as a CSC marker has been under much debate, as both $CD133^+$ and $CD133^-$ populations were shown to have equal tumorigenic potential, and its functional role for the maintenance of stemness isn't as clearly elucidated as that of CD44 and EpCAM [59]. Its role as a CSC marker was finally clarified when it was shown that only a specific epitope, the AC133 epitope, rather than the whole CD133, serves as a more reliable CSC marker due to the loss of the epitope as the cell differentiates [60, 61]. The expression of CD133, specifically the AC133 epitope, was found to be 1000-fold higher in leukemic stem cells as opposed to normal cells, and its overexpression was highly associated with poor prognosis in acute leukemic patients [62, 63]. Additionally, cells that are $CD133^+$ showed increased expression of survival protein involved in the Akt/PKB and Bcl-2 pathways, which contributes to chemoresistance to conventional chemotherapeutic drugs, including doxorubicin and 5-fluorouracil [64].

Although conventional chemo- or radiotherapy are successful at eradicating the majority of tumor cells, the CSC population within often develops resistance to anticancer treatments and is spared from eradication, which later contributes to tumor regrowth, if left unchecked. Colon CSCs were shown to be resistant to 5-fluorouracil, oxaliplatin, and irinotecan-based treatments [65]. In fact, these chemotherapeutic agents have led to the enrichment of CSCs with high tumorigenicity in xenograft models [65–67]. In breast cancer, residual tumor cells that survive conventional anticancer treatment were found to be enriched for cells of high tumorigenic potential. Evidence suggests that the breast CSC population exhibits resistance to docetaxel or doxorubicin and cyclophosphamide treatments, leading to the enrichment of the $CD44^{+}CD24^{-/low}$ cell population [68, 69]. CSCs in glioblastoma multiforme (GBM) were also resistant to radiotherapy and led to an increase in the $CD133^{+}$ cell population [70]. Such resistance to conventional treatments has been attributed to many cellular pathways, including the elevated expression of ABC transporters that efflux the drugs out of the cells, active DNA repair capacity, and the up-regulation of antiapoptotic proteins [71, 72]. Indeed, both CD44 and EpCAM markers were reported to be associated with signaling pathways, such as PI3K/E2F1 and wnt/β-catenin, which led to the expression of the antiapoptotic protein survivin (BIRC5) [58, 73]. In addition, a new wave of evidence supports the role of autophagy as a survival mechanism of CSCs in response to DNA damage [74–76]. The overexpression of these markers in CSCs and their association with cell growth and stemness pathways, as well as therapy resistance, suggest a strategy of overcoming therapy resistance of cancer by targeting the CSC population (Fig. 5.3).

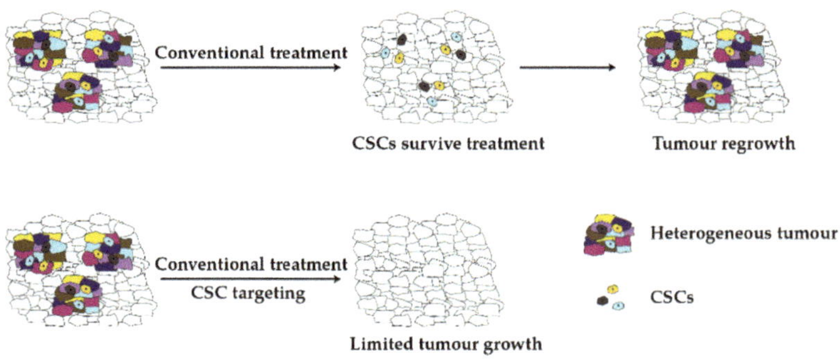

Figure 5.3 Conventional anticancer treatments are effective against bulk tumor cells but do not affect the resistant CSCs, which eventually leads to tumor regrowth. However, a combination of conventional treatment and CSC-targeted therapy inhibits tumor regrowth potential and leads to a more desired anticancer result.

5.1.3 *Aptamers in Anticancer Therapy*

Targeted anticancer therapy aims to improve the treatment response and overall survival of patients. Conventional cancer treatment mostly relies on passive targeting by the EPR effect by utilizing the differences in vasculature organization between normal tissues and tumors to achieve preferential accumulation of anticancer agents in tumors [10]. However, efficacy of anticancer drugs guided by EPR is limited due to poor tumor penetration and specific drug uptake [11, 77]. Active targeting of tumor cells has gained wide attention as a promising method of guiding anticancer drugs to their target cells and inducing cellular uptake of the drugs through receptor-mediated endocytosis [13–15]. Therefore, development of a novel molecular binding ligand that binds specifically to tumor cells for targeted drug delivery represents one of the major advances in the next generation of tumor therapeutics.

Using tumor-specific binding ligands, such as antibodies, peptides, and aptamers, as guides for anticancer agents can potentially enhance specific drug accumulation, internalization, and retention in tumors, thus increasing the therapeutic index [78, 79]. To achieve this effect, targeting specificity and drug-delivering capacity need to be considered [13]. Both antibodies and aptamers offer high binding affinity and specificity to their target molecules. However, aptamers offer favorable features over antibodies as an anticancer delivery system. Unlike antibodies, aptamers can be chemically synthesized in vitro with a relatively low-cost advantage and without labor-intensive antibody production techniques [17, 80]. Additionally, chemical synthesis of aptamers greatly reduces batch-to-batch variation and allows for large-scale manufacturing with minimal risk of contamination of microorganisms and endotoxins [18–20]. Antibodies are immunogenic as even humanized antibodies may elicit immune responses in patients [17, 18, 81]. In contrast, aptamers are generally nonimmunogenic or low immunogenic [82, 83]. But most importantly, aptamers can penetrate into tumor cores much more efficiently than antibodies due to their smaller sizes compared to monoclonal antibodies [84, 85]. A CD133 aptamer demonstrated effective penetration of tumor spheres and was retained for a minimum of 24 hours. In contrast, the antibody counterpart (AC133) was not capable of penetrating tumor spheres, even at a 300-fold higher concentration than the aptamer [85]. Despite the advantages of aptamers, they also face several limitations, especially when applied in in vivo applications. Consisting of nucleotides, aptamers are susceptible to enzymatic degradation by exo- and/or endonucleases, leading to a reduced in vivo circulatory half-life. Such limitations can be alleviated by introducing chemical modifications to the backbone or side chain of the aptamer, as well as incorporating unnatural nucleotide bases (locked and unlocked nucleic acids) and capping the aptamer ends, thus minimizing the susceptibility to endonuclease and exonuclease attacks [86–88]. The small size means the aptamer faces the challenge of blood residence time, as aptamers are susceptible to renal filtration, being smaller than the renal filtration threshold of 40 kDa [88]. To achieve the desired serum half-life, aptamers can be conjugated to polyethylene glycol (PEG) to increase the overall size, but this may compromise the tumor penetration efficacy [87, 89]. Post–systematic evolution of ligands by exponential enrichment (SELEX) modifications may risk the 3D structure of the aptamers and therefore alter the binding affinity and specificity. Such risks can be prevented by using random aptamer pools containing modified nucleotides during the SELEX selection [88, 89]. In addition, the negatively charged cell membrane may interfere with the aptamer's binding capabilities, but this can be overcome by increasing the binding affinity and specificity of aptamers toward their cell surface receptors to trigger receptor-mediated endocytosis [86, 90]. Aptamers offer an attractive alternative to antibodies as binding ligands for targeted delivery of drugs to tumors. Therefore, aptamers are being intensively investigated as targeting moieties for anticancer therapies.

Aptamers by themselves can be effective therapeutic agents solely by binding to their targets. They can act as an inhibitor by blocking catalytic sites on enzymes, or ligand recognition sites in receptors, or induce loss-of-function conformational changes [91, 92]. The first aptamer approved for clinical trial for the treatment of cancer in humans is the antinucleolin aptamer AS1411 [93]. Nucleolin is a cell surface receptor associated with cell survival, growth, and proliferation [94]. It is also associated with poor clinical prognosis for some cancer types [95, 96]. Its overexpression in cancer has been associated with the apoptosis regulator BCL2 expression in chronic lymphocytic leukemia cells [97]. In in vivo xenografts, treatment with the AS1411 aptamer resulted in cytostasis and induction of cell death. An even enhanced antitumor activity was observed when AS1411 was combined with gemcitabine [98]. AS1411 was not developed by SELEX but was developed on the basis of observations that guanosine-rich oligonucleotides (GROs) possessed antiproliferative properties against cancer cells in vitro by a nonantisense mechanism [93, 99]. On the basis of promising in vitro results of GROs and in vivo results of AS1411 in inhibiting tumor cancer cell growth and the antitumor effect, AS1411 was selected for a phase I clinical trial and was the first nucleic acid-based aptamer tested for the treatment of cancer in humans [93, 98]. It has been reported to be tested in 17 patients with advanced solid tumors [99–102]. Results from the trial showed tolerability to the drug without serious toxicity, and the treatment showed promising results in patients [102]. Half of the patients had stable disease 2 months posttreatment, and the disease in these patients was stable for 2–9 months before progression. One of the patients showed a near complete response after more than 6 months [102].

Aptamers offer versatility by allowing additional functionalization either by direct conjugation to chemotherapeutic agents, such as doxorubicin or RNA interference (RNAi)-inducing agents (small interfering RNA [siRNA], micro RNA [miRNA], small hairpin [shRNA]), or by surface conjugation onto nanoparticles loaded with specific drugs. Thus, aptamers can function as an effective drug delivery system [86] (Fig. 5.4). Aptamer-mediated drug delivery offers selective binding to the target cells, while sparing the normal cells and minimizing off-target effects. The tumor penetration ability of aptamers not only allows for delivery of their payload to the tumor mass but also ensures the payload is retained for a sufficient amount of time. Additionally, aptamers can undergo receptor-mediated endocytosis upon binding to their cell targets, which is necessary for optimal results in eradicating the target cell and reducing systemic toxicity often associated with nonspecific chemotherapeutic drugs [13–16]. RNAi can be used to target specific gene expression that

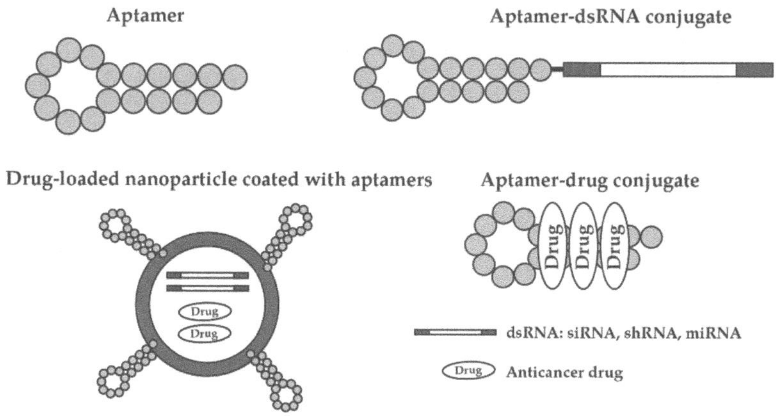

Figure 5.4 Various applications of aptamers for specific drug delivery. Aptamers can be conjugated with dsRNA for RNAi-induced gene expression knockdown, leading to increased sensitivity to anticancer therapy. Aptamers can also be used to deliver anticancer drugs by guiding drug-loaded nanoparticles or by direct conjugation with anticancer drugs.

regulates tumor growth and therapy resistance, thus increasing the sensitivity of such cells to conventional anticancer drugs. Indeed, aptamers have been conjugated to siRNA, which resulted in a reduction of gene expression similar to that observed when the siRNA was administered to cells using oligofectamine [103]. Other practical applications of aptamers involve conjugation with chemotherapeutic drugs, as well as being used as coating agents of vesicles, such as liposomes, that encapsulate drugs [104]. These applications suggest that aptamers can be used as promising drug delivery agents to target specific tumor cells with stem cell properties, which will limit therapy resistance and tumor regrowth.

5.2 Current Applications of Aptamers in Targeting CSCs

Conventional cancer treatment is effective in eliminating the majority of tumor cells in a clinical setting, but resistant CSCs are capable of surviving such treatments and pose a significant threat of relapse. Targeting the CSC population could limit the potential for tumor regrowth. However, targeting the CSC population alone does not guarantee the desired results as non-CSCs have displayed plasticity in which they de-differentiate and revert to CSCs [105, 106]. Therefore, targeting both the CSC population and the tumor bulk cells simultaneously is recommended to achieve the best outcome in cancer therapy (Fig. 5.3). The ability of aptamers to specifically target and penetrate tumors makes an exceptional diagnostic and therapeutic tool for the detection and treatment of cancer. They can be chemically modified to resist nucleases in the bloodstream and increase its circulation half-life. A variety of strategies of using aptamers for targeted therapy has been explored as a means of drug delivery and enhancing treatment efficacy, such as conjugation of drugs directly to the aptamer or by linking the aptamer to a carrier complex that encapsulates the drug [107–112]. Aptamers were also used to deliver double-stranded RNA (dsRNA) to induce RNAi-mediated gene knockdown, such as siRNA, shRNA, or miRNA, and increase sensitivity to anticancer treatments [83, 103, 113].

Developing aptamers that target CSCs can be achieved with SELEX that uses specific CSC markers in their pure form as the target molecule. Among the most commonly used CSC markers as SELEX targets are EpCAM, CD44, and CD133. These markers have all been identified to have specific roles in pluripotency, proliferations, cell–cell and cell–extracellular matrix communications, stem cell maintenance, and self-renewal [55, 114–116]. The overexpression of these surface markers in a wide variety of CSCs from different tumor types makes them attractive binding targets for CSCs. As an alternative to using pure CSC markers as the SELEX target, pools of cells enriched for CSCs can be used instead (cell-SELEX), which were either prepared by cell sorting based on known CSC markers or by enriching for cells with high self-renewal and tumorigenic capacities [85, 117–119]. In addition, generating aptamers by cell-SELEX allows for identification of markers in their natural form and the discovery of new CSC markers [120]. Next are listed aptamers generated for targeting CSCs, or potential CSCs, and their applications as drug delivery systems (Table 5.2).

Table 5.2 Aptamer specific to prospective CSC targets

SELEX method	Perspective target	Aptamer selected	Reference
	EpCAM	RNA	[90]
Standard SELEX	EpCAM	DNA	[121]
	CD44	RNA	[122]
	CD133	RNA	[85]
	Brain cancer/CD133	DNA	[119]
	Liver cancer	DNA	[123]
Cell-SELEX	Colon cancer/sphere culture	DNA	[117]
	Prostate cancer/E-cadherin and CD44	DNA	[118]
	Leukemia	DNA	[124]
	Leukemia	DNA	[125]
	Brain cancer/EGFRvIII^{high}	DNA	[126]

5.2.1 *Aptamers Generated by Standard SELEX*

5.2.1.1 Epithelial cell adhesion molecule

EpCAM is a transmembrane glycoprotein that plays a significant role in cell–cell adhesion. It has been identified as a cell surface marker for human embryonic stem cells and as a marker for pluripotency and proliferation [114–116]. It is also known to be a tumor-associated antigen due to its expression in a wide variety of cancers [127]. Its expression in colon cancer was found to be 1000-fold higher than in normal colonic tissues, and EpCAM mRNA was found to be over 100-fold overexpressed in breast and ovarian cancers [128–130]. The abundance and overexpression of EpCAM cell surface protein in a variety of cancer types make it an attractive target for diagnosis and targeted therapy.

The first-published EpCAM-specific RNA aptamer was described in 2011 [90]. The selection method used the EpCAM molecule as the target by attaching the cytoplasmic domain to a solid support in order to develop aptamers that recognize the extracellular portion of the target, which allows aptamer recognition on the cell surface. The nucleotide library was modified with 2-fluoropyrimidines and a 3′-inverted deoxythymidine cap to inhibits degradation by 3′-exonucleases, as well as the addition of a fluorophore label DY647 at the 5′-end to facilitate detection and biodistribution studies in animal models. The product of selection is a 19-nt long aptamer

(EpDT3) with a dissociation constant (K'_d) of 12 ± 6.5 nM for the human gastric carcinoma cells Kato III. The EpDT3 aptamer was tested on a panel of $EpCAM^+$ human cancer cell lines, including MCF-7, SW480, T47D, HT-29, and MDA-MB-231, and the binding results corresponded accordingly with each known EpCAM expression level for the cells. Furthermore, the aptamer did not bind to $EpCAM^-$ cells, thus verifying the aptamer's ability to specifically target $EpCAM^+$ cells only. The aptamer also exhibited successful receptor-mediated internalization by endocytosis, making the aptamer a suitable candidate for an efficient drug delivery system to $EpCAM^+$ CSCs.

A derivative of the aptamer (Ep23) was successfully intercalated with the chemotherapeutic drug doxorubicin at a ratio of 0.4:1 (aptamer:doxorubicin), which resulted in superior drug delivery to CSCs than free doxorubicin (personal communication). The aptamer–doxorubicin conjugate displayed higher tumor penetration in the colorectal adenocarcinoma cell line HT-29. CSC targeting was assessed by the reduction in sphere-forming cells and lower tumorigenic potential in xenograft models. Systemic treatment of animals with the aptamer–doxorubicin conjugate resulted in decreased stem cell markers (CD44, CD24, EpCAM) and a higher survival rate than treatment with doxorubicin alone (personal communication). The aptamer–doxorubicin conjugate was also applied to the retinoblastoma cell lines Y79 and WERI-Rb1 [107]. Though it is yet to be determined if the CSC population had been targeted in the retinoblastoma cell lines, the aptamer–doxorubicin conjugate led to an increased cytotoxic effect than free doxorubicin and the conjugate spared the noncancerous Müller glial cells.

The Ep23 aptamer was also conjugated to survivin-specific siRNA, which targets the expression of the antiapoptotic protein that has been linked to CSC resistance [131, 132]. The aptamer–siRNA chimera was successful in reducing the survivin expression in breast adenocarcinoma MCF-7 and colorectal adenocarcinoma HT-29 cells at the protein and RNA levels. The aptamer–siRNA chimera had no effect on the CSC population, but when combined with either doxorubicin or 5-fluorouracil, it resulted in a decrease in tumor sphere growth and tumor formation in xenografts compared to treatments with either of the drugs alone. Systemic treatment of tumor-bearing mice also showed a slower tumor growth rate and improved survival rates (personal communication).

The delivery of encapsulated drugs was also used with the aptamer. It was used to deliver curcumin-loaded nanoparticles (CUR-NPs) for targeted delivery to colorectal adenocarcinoma cells HT-29 [108]. The aptamer-curcumin-nanoparticle (Apt-CUR-NP) complex had a high curcumin-loading efficiency and overcame the poor solubility of curcumin. Apt-CUR-NPs preserved their binding specificity to the $EpCAM^+$ HT-29 cells and showed enhanced cellular uptake when compared to CUR-NPs without aptamer-mediated guidance and subsequently showed higher cytotoxicity to the cells in vitro. In addition, Apt-CUR-NPs resulted in increased bioavailability of curcumin in vivo compared to that of free curcumin, demonstrating that aptamer-directed CUR-NPs enhance drug delivery efficacy to targeted cells. The aptamer was also used as a binding ligand for a nanocomplex composed of polyethyleneimine (PEI) and an EpCAM-specific siRNA [113]. The nanocomplex was shown to be stable, and the electrostatic interaction between the noncore, the siRNA, and the aptamer helped improve the stability further. The resultant nanocomplex (PEI-Apt-siRNA) had EpCAM aptamers on the surface to enable EpCAM recognition on the cell targets and the EpCAM siRNA within the PEI. Although PEI has cytotoxic effects, 0.3 μg/mL of PEI had minimal cytotoxicity in vitro. The cellular uptake of the PEI-Apt-siRNA nanocomplex was successful and led to an efficient siRNA-induced silencing of EpCAM in MCF-7 and WERI-Rb1 cells compared to the native siRNA transfection with lipofectamine. Following effective EpCAM silencing, they showed a significant decrease in cell viability.

Another EpCAM-specific aptamer was selected using EpCAM-coated beads [121]. The enriched pool of DNA aptamers showed a binding affinity with a K'_d value of 8.4 ± 2.2 nM. The highest binding affinity aptamer selected from the pool had a K'_d value of 8.6 ± 2.5 nM (JYK-01). It was later modified at the 5′-end of the aptamer with an amine group and conjugated to a diacetylene monomer (ACP probe) as an imaging probe that produces fluorescence upon binding to EpCAM. The ACP probes conjugated to the JYK-01 aptamer showed specific labeling and fluorescence switching when specifically bound to $EpCAM^+$ cells, which prompts specific visualization of the target cells.

The wide variety of applications for EpCAM-specific aptamers shows promise as a therapeutic and a diagnostic tool against cancer. The ability to deliver anticancer drugs by various means of delivery and the ability to target CSCs make the aptamer an effective anti-CSC agent. Additionally, the specific visualization of $EpCAM^+$ cells by fluorescence switching shows a promising role in cancer screening and diagnosis.

5.2.1.2 CD44

CD44 has been shown to play an important role in cell–cell and cell–extracellular matrix communications, as well as an essential role in stem cell maintenance and self-renewal through the transcription of stem cell genes such as *BMI-1* and *Oct-3/4* [55]. It's been reported to be overexpressed in multiple types of cancer, including breast, prostate, pancreas, ovarian, and colon cancers, and is extensively used for the enrichment and identification of CSCs [57, 133, 134]. CD44 is the main receptor for hyaluronic acid, which was considered a promising method for targeted therapy [135]. However, hyaluronic acid lacks specificity as it can bind to several other targets, which may lead to off-target effects [135]. Developing aptamers specific to CD44 may hold promise as an effective tool against CSCs with minimal risk to the neighboring healthy cells.

A study reported the development of CD44-specific RNA aptamer using SELEX [122]. The selected aptamer (apt1) had an initial binding affinity with a K'_d of 81.3 $\pm$ 30.6 nM, and it was also modified with a 2'O-methyl group for better nuclease resistance. The binding specificity of the Apt1 aptamer was evaluated across a different panel of $CD44^+$ cells, including MDA-MB-231, MCF7, and T47D. The binding reactions of the aptamer across these cell lines were found to be at different intensities and in accordance with each of their CD44 expression levels. The aptamer alone had no effect on cell growth and proliferation, but it was later conjugated to the surface of PEGylated liposomes for specific drug delivery (apt1-lip) [109]. The conjugation did not negatively affect the binding affinity of the aptamers; in fact, it improved its binding affinity to a K_d value of 6.2 ± 1.6 nM. However, the binding of Apt1 alone showed higher saturation. The changes were likely related to the multivalent binding of the aptamers across the liposome surface that increased the binding affinity, but the steric hindrance decreased the saturation level. The cellular uptake of apt1-lip was confirmed on $CD44^+$ cell lines A549 and MDA-MB-231. The safety profile of apt1-lip was assessed by its effect on the secretion of inflammatory cytokines. They found no increase in the secretion of the inflammatory cytokines IL-12p70, TNF, IL-10, IL-6, IL-1β, and IL-8 as a result of Apt1-Lip treatment at 400 nM. The inability to induce an inflammatory response by tumor cells holds promise for a safe and efficient drug delivery system for therapeutic application.

5.2.2 *Aptamers Generated by Cell-SELEX*

Standard SELEX is usually performed with highly purified molecules as the target to avoid nonspecific binding. Using a pure and homogeneous target molecule has the advantage of high rates of selecting high-affinity aptamers. However, in the case of biological systems, using pure forms of target molecules away from their native system has a risk of stability and loss of native conformation. Structural conformity plays a significant role in aptamer binding and specificity; therefore careful consideration must be given when using standard SELEX for generating aptamers against CSC markers. Alternatively, a theoretical study has suggested that complex heterogeneous targets can also be used to generate specific aptamers, and it was later experimentally confirmed with aptamers generated against red blood cell ghosts and live African trypanosomes [136–138]. Cancer cells have been extensively used as targets for cell-SELEX and led to the development of aptamers against a wide variety of cancer cell types [139, 140]. The advantages of cell-SELEX over standard SELEX include using target molecules in their native conformation, and it can be performed without any knowledge regarding the cell surface molecules of the target cells, thus enabling de novo generation of cell-specific aptamers [120]. A successfully generated aptamer by cell-SELEX against the mouse embryonic stem cells efficiently bound to the target cells and discriminated against all examined differentiated cells makes it possible for generating aptamers that can distinguish CSCs from terminally differentiated tumor cells [141]. Using cell-SELEX may produce aptamers against a previously unrecognized cell-specific surface molecule, which can be applied for de novo identification of new novel markers of the target cell [120]. This makes cell-SELEX an effective method for the generation of aptamers specific to CSCs.

5.2.3 *CD133*

CD133 is considered to be one of the universal markers for CSCs due to its expression in a wide variety of cancer types. It is a pentaspan, highly glycosylated membrane glycoprotein that is associated with cholesterol in the plasma membrane [142, 143]. It is known to be expressed in somatic stem and progenitor cells, as well as in developing epithelial and differentiated cells. Its exact function has yet to be fully elucidated, but it has been linked to the Notch-signaling pathway, which is critical for binary cell fate, differentiation of intestinal epithelium, and lymphopoiesis [144]. CD133 has gained significant attention due to its reported role as a CSC marker in a wide variety of cancers, including breast, colon, glioma, liver, lung, melanoma, ovarian, pancreatic, and prostate cancers [57]. Its role as a CSC marker has been under heated debate, as previously mentioned, but it was shown that the AC133 epitope of CD133 served as a more reliable CSC marker [60, 61].

An RNA aptamer specific to CD133 was generated by cell-SELEX [85]. Given that CD133 is a complex pentaspan protein, the target molecule was transiently expressed on HEK293T cells, which are naturally CD133, to preserve its native conformation. The CD133 expression on the cell was confirmed by the AC133 antibody. The resultant aptamer, A15, showed high binding affinity, and its specificity to CD133 was further confirmed with $CD133^+$ HT-29 and $CD133^-$ T98G and HEK293T cells. The aptamer was later confirmed to be internalized upon binding to its target cells via receptor-mediated endocytosis, as well as being able to penetrate tumor spheres and be retained at the center for at least 24 hours. The specificity and penetrative ability of the aptamer demonstrates its ability to target $CD133^+$ CSCs of many cancer types and may hold promise for the delivery of therapeutics to glioblastoma CSCs, as similarly observed with paclitaxel-loaded nanoparticles in reducing glioblastoma proliferation [145]. The aptamer displayed therapeutic potential as a binding ligand for salinomycin-loaded nanoparticles (SAL-NPs) to target $CD133^+$ osteosarcoma CSCs [110]. The aptamer-guided nanoparticle (Apt-SAL-NP) showed specific cytotoxicity toward the $CD133^+$ Saos-2 osteosarcoma cells and showed the best therapeutic effect in Saos-2 xenograft mice when compared with either SAL-NPs or salinomycin-only treatments.

Another aptamer was selected against CD133 using the cell-SELEX method to identify glioblastoma CSCs [119]. Their SELEX targets were functionally validated $CD133^+$ CSCs derived from pooled tumors of the same parental glioblastoma xenografts. They produced DNA aptamers (A1–A6) that showed high binding affinities in the nanomolar range. The cells that bound to the aptamers displayed high levels of CSC markers such as sox2, while the cells that failed to bind with the aptamers expressed high levels of differentiation markers such as the glial fibrillary acidic protein (GFAP). The aptamer-targeted cells also displayed the CSC functionality of sphere formation. Additionally, both aptamer-targeted cells and nontarget cells were implanted into the brains on immunocompromised mice. The aptamer-targeted cells showed higher tumor volumes and greater propensity to invade into normal brain than the nontarget cells, providing further evidence of the aptamers' ability to target glioblastoma CSCs. It was concluded that the aptamers have therapeutic importance for the delivery of drugs to glioblastoma CSCs, as it was demonstrated that paclitaxel-loaded nanoparticles could reduce glioblastoma proliferation and cross the blood–brain barrier [145, 146].

5.2.4 *Liver Cancer*

Liver cancer has been shown to express a wide variety of CSC markers, including CD133, CD44, and EpCAM [31]. A DNA aptamer against liver cancer was selected using cell-SELEX with the BNL 1ME A.7R.1 (MEAR) mouse liver hepatoma cell line as the target and its normal counterpart BNL CL.2 (BNL) cells as the control [123]. The selected aptamers were capable of targeting only the cell surface molecules present on MEAR cells. Seven of the aptamers were found to bind MEAR cells with high affinities in the nanomolar range, with the aptamer TLS11a having the highest binding affinity with a K_d' value of 4.51 ± 0.39. The TLS11a aptamer also displayed binding with H23 and Huh7 liver cancer cell lines but did not bind to normal liver tissue, making it a valuable probe for liver cancer analysis. A modified version of the aptamer TLS11a-GC, which specifically targets the hepatocellular carcinoma cell line LH86, was conjugated to the antitumor drug doxorubicin [111]. The TLS11a aptamer was shown to intercalate with a maximum of two doxorubicin molecules. By adding a long GC tail to the 5′-end of the aptamer, the newly modified TLS11a-GC formed a dimer structure and increased the doxorubicin-loading capacity to 56 doxorubicin molecules according to the Nupack calculation, producing a TLS11a-GC-to-doxorubicin ratio of 1:28. The effect of TLS11a-GC-Dox on the cell viability of LH86 and showed a decrease in cell viability, but the observed toxicity was less than that of free-doxorubicin treatment. However, further experiments were conducted on normal human liver cells Hu1229 and showed that free-doxorubicin treatment had significant toxicity, while the TLS11a-GC-Dox conjugate showed lower toxicity, indicating that the TLS11a-GC-Dox conjugate induces toxicity to the target cells and decreases the uptake of doxorubicin by nontarget cells. Tumors in immunodeficient mice were induced with LH86 by subcutaneous injection and were treated with IV injection into the tail veil to measure the antitumor effect of the drugs. Both free doxorubicin and the TLS11a-GC-DOX conjugate showed obvious antitumor inhibition. The TLS11a-GC-Dox-treated group had a more efficient effect compared to the free-doxorubicin-treated group, indicating that the TLS11a-GC-Dox conjugate targeted the tumor cells and achieved higher local doxorubicin concentration in the tumor site than free doxorubicin.

5.2.5 *Colon Cancer*

A CSC-specific DNA aptamer was developed using a novel automated cell-SELEX on a microfluidic microchip [117]. Sphere culture of the ileocecal colorectal adenocarcinoma HCT-8 cell line was used as the CSC target for the selection and the adherent culture (nonsphere culture) of the cells as the control selection. The system produced four high affinity aptamers that could specifically bind to colon CSCs, and their specificity was further confirmed with other cell lines, including A549, MCF7, BxPC3 HepG2s, HeLa, and NIH3T3 cells. The selected aptamers, CSC-16, CSC-19, CSC-35, and CSC-37, showed relatively high capture rates toward the CSC population of the target cell (HCT-8) with capture rates of 51.7$\pm$ 5.3%, 52.9$\pm$ 4.7%, 45.6$\pm$ 9.1%, and 54.9$\pm$ 4.4%, respectively. Their novel microfluidic system cell-SELEX was effective in selecting aptamers against colons CSCs, which could be applied in personalized medicine for aptamer screening or drug discovery [117].

5.2.6 *Prostate Cancer*

CSC-specific aptamers were selected using cell-SELEX on the prostate carcinoma cell line DU145 [118]. Though the aptamers were not selected for CD44 specifically, the resultant DNA aptamers showed high binding affinity to cells expressing the CSC markers E-cadherin and CD44. The aptamer CSC13 was shown to have a high binding affinity with a K_d' value of 2.2 $\pm$ 1.0 nM. Its specificity was later confirmed with other cell lines, including PC3, LNCaP, and VCaP. Cell targets of the CSC13 aptamer showed higher sphere-forming capacity and higher tumorigenicity than those observed in non-CSC13 target cells. The CSC13 aptamer was later used as a delivery system for gold nanorods (AuNRs) and increased the CSC deaths by near-infrared (NIR) irradiation [112]. Direct irradiation of the cells by NIR maintained high cell viability due to low light absorption [147]. However, guiding the AuNR with the aptamer led to an increase in cell death.

5.2.7 *Leukemia*

The precursor T-cell acute lymphoblastic leukemia (ALL) cell line, CCRF-CEM, was shown to produce soluble factors that enhanced erythroid burst formation in human peripheral blood and bone marrow cells, and it was concluded that the soluble factors produced may be important for the regulation of human erythropoiesis and thus required for the proliferation of early erythroid progenitors

[148]. CSCs have been shown to produce soluble factors that promoted survival and proliferation, which may suggest a possible link to leukemic stem cells, and therefore suggested a suitable target for inhibiting the proliferation of early erythroid progenitors [149]. DNA aptamers against CCRF-CEM were developed by cell-SELEX and found to be capable of identifying cancer cells in real clinical specimens [124]. Human Burkitt's lymphoma cell line, Ramos, was used as the negative control. The selection produced 10 aptamers with high binding affinity for CCRF-CEM. Among them, the sgc8 aptamer had the highest binding affinity with a K_d' value of 0.8 $\pm$ 0.09 nM and showed no binding to the Ramos cell line. The aptamer also recognized other T-cell ALL cell lines, including Sup-T1, Molt-4, and jurkat. The sgc8 aptamer could not recognize cancer cells from B-cell lymphoma patients, but it did recognize cancer cells from T-cell ALL patients, which were closely related to the CCRF-CEM target cells used in cell-SELEX, thus outlining the clinical and diagnostic potential of the aptamer. The sgc8 aptamer demonstrated its potential for drug-loaded liposome delivery to target cells [150]. The aptamer-liposome showed high specific binding to target cells and subsequent release of a model drug (FITC-Dextran). Alternatively, the aptamer (sgc8c) was later used to conjugate with doxorubicin and showed no adverse effect on the binding specificity of CCRF-CEM cells, which showed a K_d' value of 2.0 $\pm$ 0.2 nM [151]. Doxorubicin is released when it's internalized in the acidic endosomal environment, and the potency of the drug reserved its potency. Additionally, the nonspecific uptake of drugs by nontarget cells was inhibited when the drug was conjugated to the aptamer, displaying the specificity of the conjugate to the target cells [151].

Aptamers against the CCRF-CEM cell line allowed for specific recognition of cells from complex mixtures, including whole blood samples, by using aptamer-nanoparticle conjugations for a two-particle assay [152]. Aptamer–magnetic nanoparticle conjugations were used for target cell extraction, while aptamer–fluorescent nanoparticle conjugations were simultaneously added for sensitive cell detection. This two-particle assay requires sufficient aptamer-binding sites on the target cells. Otherwise, different aptamers or other recognition elements can be used to overcome competitive binding. The assay displayed rapid detection of target cells from complex samples compared to immunophenotypic and polymerase chain reaction (PCR)-based analysis and demonstrated its application in complex samples [152]. Equal amounts of CEM and Ramos cells were mixed for testing the two-particle assay. Both magnetic and fluorescent nanoparticles were simultaneously added, which resulted in collecting the CEM cells in the sample with bright fluorescence that made them easily distinguishable. The assay also demonstrated its application in clinical samples by using whole blood spiked with CEM cells to mimic clinical samples and led to the extraction of 40% of the spiked cells. The extraction efficiency was consistent with current immunomagnetic separation [153, 154]. This aptamer-based assay identifies target cells from the complex mixture with collection efficiencies rivaling or surpassing current methods for cellular detection from clinical samples [152].

Aptamers for acute myelogenous leukemia (AML) were developed with cell-SELEX on NB4 cells derived from AML M3 to generate DNA aptamers [125]. Out of 10 candidate aptamers, 3 were found to have high binding affinities to NB4 cells with K_{d}' values of 2.77 nM for JH6, 7.57 nM for JH19, and 12.37 nM for K19. The selected aptamers (JH6, JH19, and K19) showed no binding to lymphocytes, but they did show binding on mature and immature granulocytes and monocytes. These results suggest that the aptamers may recognize myeloid-specific surface markers. Additionally, the aptamers had low, but statistically significant, levels of binding to CD34(+) early hematopoietic precursors. The binding of the aptamers was abolished when NB4 cells were treated with trypsin to remove cell surface markers, indicating that the target molecules of the aptamers may be directly or indirectly related to surface proteins on the cell membrane. It was later confirmed that the likely binding target of the aptamer K19 is siglic-5, and it was capable of competing with anti-Siglec-5 antibody for binding to NB4 cells. Siglec-5 has been reported to be expressed on granulocytes and therefore can be used as a biomarker for granulocytic maturation and AML cell detection and can potentially mediate targeted therapy against AML cells [125, 155]. In the study, the researchers conjugated a biotin-labeled K19 aptamer with streptavidin-labeled saporin, and it was found that the aptamer mediated the uptake of saporin and inhibited NB4 cell proliferation in vitro.

5.2.8 *Glioblastoma Multiforme*

RNA aptamers were developed against glioblastoma CSCs using cell-SELEX. A candidate aptamer, S-2, showed specific binding to glioblastoma CSCs and was successfully internalized, showing great promise for targeted delivery and imaging of glioblastoma [156]. DNA aptamers were developed that can identify human glioblastoma multiforme (GBM) overexpressing the epidermal growth factor receptor variant III (EGFRvIII), which is linked to radiation and chemotherapeutic resistance, as well as the CSC phenotype of self-renewal in GBM and invasive breast carcinoma [126, 157, 158]. The selection used EGFRvIII-overexpressing human glioma U87Δ cells as the binding target for cell-SELEX and produced four DNA aptamers (32, 41, 43, and 47) that showed high binding affinities with K'_d values less than 100 nM and were able to distinguish the target cells from negative control human glioma U87MG cells and HEK293 cells. Aptamer 47 was shown to be localized on the cell surface, and aptamers 41, 43, and 32 were successfully internalized into the cell through endocytosis. Interestingly, aptamer 32 demonstrated high binding affinity to the EGFRvIII protein with a K'_d value of 0.62 ± 0.04 nM. These DNA aptamers show promising potential as molecular probes for the diagnosis and treatment of GBM. Other DNA aptamers were generated by cell-SELEX using the glioblastoma cell line A172 as the target [159]. The selection produced five aptamers (GMT 3, GMT 4, GMT 5, GMT 8, and GMT 9) that showed high affinity in the nanomolar range, with the highest K'_d value of 61.82 ± 6.37 for GMT 4. The aptamers showed high binding affinity to other types of cell lines, including the glioblastoma cell line U87MG. Aptamers GMT 4 and GMT 8 also showed high binding affinity for CCRF-CEM, which is a T-lymphoblast cell line, indicating that glioblastoma and acute lymphoblastic leukemia have the same molecular recognition site for these two aptamers. The binding target of these aptamers is more likely associated with cell surface proteins as their binding affinities decreased after cell treatment with proteinase K.

5.3 Conclusion and Final Remarks

Ongoing research on CSCs continually highlights their significance to cancer growth, metastasis, and survival. Their role as cancer-initiating cells and their resistance to conventional cancer therapy make them prime targets for anticancer treatments. The over-expression of specific cell surface markers on CSCs allows for their identification and enables development of binding ligands of high affinity and specificity, such as antibodies and aptamers. With binding affinities comparable to monoclonal antibodies, ability to penetrate tumors, and relative ease of chemical modification, aptamers showed great potential for targeted therapy. Aptamer-mediated drug delivery showed improved delivery to target cells and reduced off-target effects, which successfully induced higher cytotoxicity in cancer cells and reduced the CSC population by the observed reduction of self-renewal and tumorigenicity. In addition to their therapeutic potential, selecting aptamers against CSC markers by cell-SELEX allows for the identification of new markers, which may aid in a deeper understanding of CSCs and their underlying mechanisms. The ambitious research on aptamers as effective tools for guiding cancer therapeutics to their targets holds promise for significant clinical outcomes and improved quality of life for cancer patients.

References

1. Markowitz, S. D., and Bertagnolli, M. M. (2009). Molecular basis of colorectal cancer, *New Engl. J. Med.*, **361**, pp. 2449–2460.
2. Anderson, E. C., Hessman, C., Levin, T. G., Monroe, M. M., and Wong, M. H. (2011). The role of colorectal cancer stem cells in metastatic disease and therapeutic response, *Cancers (Basel)*, **3**, pp. 319–339.
3. Cetin, I., and Topcul, M. (2012). Cancer stem cells in oncology, *J. BUON*, **17**, pp. 644–648.

4. Magee, J. A., Piskounova, E., and Morrison, S. J. (2012). Cancer stem cells: impact, heterogeneity, and uncertainty, *Cancer Cell,* **21**, pp. 283–296.
5. Nguyen, L. V., Vanner, R., Dirks, P., and Eaves, C. J. (2012). Cancer stem cells: an evolving concept, *Nat. Rev. Cancer,* **12**, pp. 133–143.
6. Al-Hajj, M., Wicha, M. S., Benito-Hernandez, A., Morrison, S. J., and Clarke, M. F. (2003). Prospective identification of tumorigenic breast cancer cells, *Proc. Natl. Acad. Sci. U S A,* **100**, pp. 3983–3988.
7. Dalerba, P. (2007). Phenotypic characterization of human colorectal cancer stem cells, *Proc. Natl. Acad. Sci. U S A,* **104**, pp. 10158–10163.
8. O'Brien, C. A., Pollett, A., Gallinger, S., and Dick, J. E. (2007). A human colon cancer cell capable of initiating tumour growth in immunodeficient mice, *Nature,* **445**, pp. 106–110.
9. Aguirre-Ghiso, J. A. (2007). Models, mechanisms and clinical evidence for cancer dormancy, *Nat. Rev. Cancer,* **7**, pp. 834–846.
10. Prabhakar, U., et al. (2013). Challenges and key considerations of the enhanced permeability and retention effect for nanomedicine drug delivery in oncology, *Cancer Res.,* **73**, pp. 2412–2417.
11. Zhong, Y., Meng, F., Deng, C., and Zhong, Z. (2014). Ligand-directed active tumor-targeting polymeric nanoparticles for cancer chemotherapy, *Biomacromolecules,* **15**, pp. 1955–1969.
12. Hollis, C. P., et al. (2013). Biodistribution and bioimaging studies of hybrid paclitaxel nanocrystals: lessons learned of the EPR effect and image-guided drug delivery, *J. Controlled Release,* **172**, pp. 12–21.
13. Bertrand, N., Wu, J., Xu, X., Kamaly, N., and Farokhzad, O. C. (2014). Cancer nanotechnology: the impact of passive and active targeting in the era of modern cancer biology, *Adv. Drug Delivery Rev.,* **66**, pp. 2–25.
14. Allen, T. M. (2002). Ligand-targeted therapeutics in anticancer therapy, *Nat. Rev. Cancer,* **2**, pp. 750–763.
15. Modery-Pawlowski, C. L., and Gupta, A. S. (2014). Heteromultivalent ligand-decoration for actively targeted nanomedicine, *Biomaterials,* **35**, pp. 2568–2579.
16. Jaracz, S., Chen, J., Kuznetsova, L. V., and Ojima, I. (2005). Recent advances in tumor-targeting anticancer drug conjugates, *Bioorg. Med. Chem.,* **13**, pp. 5043–5054.
17. Harding, F. A., Stickler, M. M., Razo, J., and DuBridge, R. B. (2010). The immunogenicity of humanized and fully human antibodies: residual immunogenicity resides in the CDR regions, *MAbs,* **2**, pp. 256–265.

18. Schrama, D., Reisfeld, R. A., and Becker, J. C. (2006). Antibody targeted drugs as cancer therapeutics, *Nat. Rev. Drug Discovery,* **5**, pp. 147–159.
19. Radom, F., Jurek, P. M., Mazurek, M. P., Otlewski, J., and Jeleń, F. (2013). Aptamers: molecules of great potential, *Biotechnol. Adv.,* **31**, pp. 1260–1274.
20. Panowksi, S., Bhakta, S., Raab, H., Polakis, P., and Junutula, J. R. (2014). Site-specific antibody drug conjugates for cancer therapy, *MAbs,* **6**, pp. 34–45.
21. Patrawala, L. (2005). Side population is enriched in tumorigenic, stem-like cancer cells, whereas ABCG2+ and ABCG2- cancer cells are similarly tumorigenic, *Cancer Res.,* **65**, pp. 6207–6219.
22. Klonisch, T., et al. (2008). Cancer stem cell markers in common cancers - therapeutic implications, *Trends Mol. Med.,* **14**, pp. 450–460.
23. Cox, C. V., Diamanti, P., Evely, R. S., Kearns, P. R., and Blair, A. (2009). Expression of CD133 on leukemia-initiating cells in childhood ALL, *Blood,* **113**, pp. 3287–3296.
24. Liu, Q., Li, J. G., Zheng, X. Y., Jin, F., and Dong, H. T. (2009). Expression of CD133, PAX2, ESA, and GPR30 in invasive ductal breast carcinomas, *Chin. Med. J. (Engl.),* **122**, pp. 2763–2769.
25. Fillmore, C. M., and Kuperwasser, C. (2008). Human breast cancer cell lines contain stem-like cells that self-renew, give rise to phenotypically diverse progeny and survive chemotherapy, *Breast Cancer Res.,* **10**, p. R25.
26. Hwang-Verslues, W. W., et al. (2009). Multiple lineages of human breast cancer stem/progenitor cells identified by profiling with stem cell markers, *PLoS One,* **4**, p. e8377.
27. Yeung, T. M., Gandhi, S. C., Wilding, J. L., Muschel, R., and Bodmer, W. F. (2010). Cancer stem cells from colorectal cancer-derived cell lines, *Proc. Natl. Acad. Sci. U S A,* **107**, pp. 3722–3727.
28. Vermeulen, L., et al. (2008). Single-cell cloning of colon cancer stem cells reveals a multi-lineage differentiation capacity, *Proc. Natl. Acad. Sci. U S A,* **105**, pp. 13427–13432.
29. Singh, S. K., et al. (2004). Identification of human brain tumour initiating cells, *Nature,* **432**, pp. 396–401.
30. Zeppernick, F., et al. (2008). Stem cell marker CD133 affects clinical outcome in glioma patients, *Clin. Cancer. Res.,* **14**, pp. 123–129.
31. Mikhail, S., and He, A. R. (2011). Liver cancer stem cells, *Int. J. Hepatol.,* **2011**, p. 486954.

32. Rivera, C., Rivera, S., Loriot, Y., Vozenin, M. C., and Deutsch, E. (2011). Lung cancer stem cell: new insights on experimental models and preclinical data, *J. Oncol.,* **2011**, p. 549181.
33. Silva, I. A., et al. (2011). Aldehyde dehydrogenase in combination with CD133 defines angiogenic ovarian cancer stem cells that portend poor patient survival, *Cancer Res.,* **71**, pp. 3991–4001.
34. Li, C., et al. (2007). Identification of pancreatic cancer stem cells, *Cancer Res.,* **67**, pp. 1030–1037.
35. Simeone, D. M. (2008). Pancreatic cancer stem cells: implications for the treatment of pancreatic cancer, *Clin. Cancer Res.,* **14**, pp. 5646–5648.
36. Sergeant, G., Vankelecom, H., Gremeaux, L., and Topal, B. (2009). Role of cancer stem cells in pancreatic ductal adenocarcinoma, *Nat. Rev. Clin. Oncol.,* **6**, pp. 580–586.
37. Hermann, P. C., Bhaskar, S., Cioffi, M., and Heeschen, C. (2010). Cancer stem cells in solid tumors, *Semin. Cancer Biol.,* **20**, pp. 77–84.
38. Griend, D. J. V., et al. (2008). The role of CD133 in normal human prostate stem cells and malignant cancer-initiating cells, *Cancer Res.,* **68**, pp. 9703–9711.
39. Wei, C., Guomin, W., Yujun, L., and Ruizhe, Q. (2007). Cancer stem-like cells in human prostate carcinoma cells DU145: the seeds of the cell line? *Cancer Biol. Ther.,* **6**, pp. 763–768.
40. Patrawala, L., et al. (2006). Highly purified CD44+ prostate cancer cells from xenograft human tumors are enriched in tumorigenic and metastatic progenitor cells, *Oncogene,* **25**, pp. 1696–1708.
41. Trerotola, M., et al. (2010). CD133, Trop-2 and alpha2beta1 integrin surface receptors as markers of putative human prostate cancer stem cells, *Am. J. Transl. Res.,* **2**, pp. 135–144.
42. Furth, J., and Kahn, M. C. (1937). The transmission of leukemia in mice with a single cell, *Am. J. Cancer,* **31**, pp. 276–282.
43. Makino, S. (1956). Further evidence favoring the concept of the stem cell in ascites tumors of rats, *Ann. N. Y. Acad. Sci.,* **63**, pp. 818–830.
44. Hewitt, H. B. (1958). Studies of the dissemination and quantitative transplantation of a lymphocytic leukaemia of CBA mice, *Br. J. Cancer,* **12**, pp. 378–401.
45. Bruce, W. R., and Van Der Gaag, H. (1963). A quantitative assay for the number of murine lymphoma cells capable of proliferation in vivo, *Nature,* **199**, pp. 79–80.

46. Pierce, G. B., and Wallace, C. (1971). Differentiation of malignant to benign cells, *Cancer Res.*, **31**, pp. 127–134.
47. Pierce, G. B., and Speers, W. C. (1988). Tumors as caricatures of the process of tissue renewal: prospects for therapy by directing differentiation, *Cancer Res.*, **48**, pp. 1996–2004.
48. Nowell, P. C. (1976). The clonal evolution of tumor cell populations, *Science*, **194**, pp. 23–28.
49. Uckun, F. M. (1995). Leukemic cell growth in SCID mice as a predictor of relapse in high-risk B-lineage acute lymphoblastic leukemia, *Blood*, **85**, pp. 873–878.
50. Lapidot, T. (1994). A cell initiating human acute myeloid leukaemia after transplantation into SCID mice, *Nature*, **367**, pp. 645–648.
51. Bonnet, D., and Dick, J. E. (1997). Human acute myeloid leukemia is organized as a hierarchy that originates from a primitive hematopoietic cell, *Nat. Med.*, **3**, pp. 730–737.
52. Ricci-Vitiani, L. (2007). Identification and expansion of human colon-cancer-initiating cells, *Nature*, **445**, pp. 111–115.
53. Singh, S. K. (2004). Identification of human brain tumour initiating cells, *Nature*, **432**, pp. 396–401.
54. McCarthy, N. (2012). Cancer stem cells: tracing clones, *Nat. Rev. Cancer*, **12**, pp. 579–579.
55. Hurt, E. M., Kawasaki, B. T., Klarmann, G. J., Thomas, S. B., and Farrar, W. L. (2008). CD44(+)CD24(−) prostate cells are early cancer progenitor/stem cells that provide a model for patients with poor prognosis, *Br. J. Cancer*, **98**, pp. 756–765.
56. Shi, M. F., et al. (2010). Identification of cancer stem cell-like cells from human epithelial ovarian carcinoma cell line, *Cell. Mol. Life Sci.*, **67**, pp. 3915–3925.
57. Medema, J. P. (2013). Cancer stem cells: the challenges ahead, *Nat. Cell Biol.*, **15**, pp. 338–344.
58. Maetzel, D., et al. (2009). Nuclear signalling by tumour-associated antigen EpCAM, *Nat. Cell Biol.*, **11**, pp. 162–171.
59. Shmelkov, S. V., et al. (2008). CD133 expression is not restricted to stem cells, and both CD133 + and CD133- metastatic colon cancer cells initiate tumors, *J. Clin. Invest.*, **118**, pp. 2111–2120.
60. Kemper, K. (2010). The AC133 epitope, but not the CD133 protein, is lost upon cancer stem cell differentiation, *Cancer Res.*, **70**, pp. 719–729.

61. Mak, A. B. (2011). CD133 protein N-glycosylation processing contributes to cell surface recognition of the primitive cell marker AC133 epitope, *J. Biol. Chem.,* **286**, pp. 41046–41056.

62. Waldron, N. N., et al. (2011). Targeting tumor-initiating cancer cells with dCD133KDEL shows impressive tumor reductions in a xenotransplant model of human head and neck cancer, *Mol. Cancer Ther.,* **10**, pp. 1829–1838.

63. Tolba, F. M., Foda, M. E., Kamal, H. M., and Elshabrawy, D. A. (2013). Expression of CD133 in acute leukemia, *Med. Oncol.,* **30**, p. 527.

64. Ma, S., Lee, T. K., Zheng, B. J., Chan, K. W., and Guan, X. Y. (2007). CD133+ HCC cancer stem cells confer chemoresistance by preferential expression of the Akt/PKB survival pathway, *Oncogene,* **27**, pp. 1749–1758.

65. Wilson, B. J. (2011). ABCB5 identifies a therapy-refractory tumor cell population in colorectal cancer patients, *Cancer Res.,* **71**, pp. 5307–5316.

66. Dylla, S. J. (2008). Colorectal cancer stem cells are enriched in xenogeneic tumors following chemotherapy, *PLoS One,* **3**, p. e2428.

67. Todaro, M. (2007). Colon cancer stem cells dictate tumor growth and resist cell death by production of interleukin-4, *Cell Stem Cell,* **1**, pp. 389–402.

68. Li, X. (2008). Intrinsic resistance of tumorigenic breast cancer cells to chemotherapy, *J. Natl. Cancer Inst.,* **100**, pp. 672–679.

69. Creighton, C. J., et al. (2009). Residual breast cancers after conventional therapy display mesenchymal as well as tumor-initiating features, *Proc. Natl. Acad. Sci. U S A,* **106**, pp. 13820–13825.

70. Tamura, K., et al. (2010). Accumulation of CD133-positive glioma cells after high-dose irradiation by gamma knife surgery plus external beam radiation, *J. Neurosurg.,* **113**, pp. 310–318.

71. Dean, M., Fojo, T., and Bates, S. (2005). Tumour stem cells and drug resistance, *Nat. Rev. Cancer,* **5**, pp. 275–284.

72. Johnstone, R. W., Ruefli, A. A., and Lowe, S. W. (2002). Apoptosis: a link between cancer genetics and chemotherapy, *Cell,* **108**, pp. 153–164.

73. Abdraboh, M. E., et al. (2011). Survivin is a novel target of CD44-promoted breast tumor invasion, *Am. J. Pathol.,* **179**, pp. 555–563.

74. Maycotte, P., Jones, K. L., Goodall, M. L., Thorburn, J., and Thorburn, A. (2015). Autophagy supports breast cancer stem cell maintenance by regulating IL6 secretion, *Mol. Cancer Res.,* **13**, pp. 651–658.

75. Wu, S., Wang, X., Chen, J., and Chen, Y. (2013). Autophagy of cancer stem cells is involved with chemoresistance of colon cancer cells, *Biochem. Biophys. Res. Commun.*, **434**, pp. 898–903.
76. Milla, L. A., González-Ramírez, C. N., and Palma, V. (2012). Sonic Hedgehog in cancer stem cells: a novel link with autophagy, *Biol. Res.*, **45**, pp. 223–230.
77. Petros, R. A., and DeSimone, J. M. (2010). Strategies in the design of nanoparticles for therapeutic applications, *Nat. Rev. Drug Discovery*, **9**, pp. 615–627.
78. Hirsjärvi, S., Passirani, C., and Benoit, J. P. (2011). Passive and active tumour targeting with nanocarriers, *Curr. Drug Discovery Technol.*, **8**, pp. 188–196.
79. Danhier, F., Feron, O., and Préat, V. (2010). To exploit the tumor microenvironment: passive and active tumor targeting of nanocarriers for anti-cancer drug delivery, *J. Controlled Release*, **148**, pp. 135–146.
80. Jayasena, S. (1999). Aptamers: an emerging class of molecules that rival antibodies in diagnostics, *Clin. Chem.*, **45**, pp. 1628–1650.
81. Chari, R. V., Miller, M. L., and Widdison, W. C. (2014). Antibody-drug conjugates: an emerging concept in cancer therapy, *Angew. Chem., Int. Ed. Engl.*, **53**, pp. 3796–3827.
82. Bouchard, P. R., Hutabarat, R. M., and Thompson, K. M. (2010). Discovery and development of therapeutic aptamers, *Annu. Rev. Pharmacol. Toxicol.*, **50**, pp. 237–257.
83. Dassie, J. P., et al. (2009). Systemic administration of optimized aptamer-siRNA chimeras promotes regression of PSMA-expressing tumors, *Nat. Biotechnol.*, **27**, pp. 839–846.
84. White, R. R., Sullenger, B. A., and Rusconi, C. P. (2000). Developing aptamers into therapeutics, *J. Clin. Invest.*, **106**, pp. 929–934.
85. Shigdar, S., et al. (2013). RNA aptamers targeting cancer stem cell marker CD133, *Cancer Lett.*, **330**, pp. 84–95.
86. Shigdar, S., et al. (2011). Clinical applications of aptamers and nucleic acid therapeutics in haematological malignancies, *Br. J. Haematol.*, **155**, pp. 3–13.
87. Shigdar, S., et al. (2013). Aptamers as theranostic agents: modifications, serum stability and functionalisation, *Sensors (Basel)*, **13**, pp. 13624–13637.
88. Keefe, A. D., Pai, S., and Ellington, A. (2010). Aptamers as therapeutics, *Nat. Rev. Drug Discovery*, **9**, pp. 537–550.

89. Keefe, A. D., and Cload, S. T. (2008). SELEX with modified nucleotides, *Curr. Opin. Chem. Biol.,* **12**, pp. 448–456.
90. Shigdar, S., et al. (2011). RNA aptamer against a cancer stem cell marker epithelial cell adhesion molecule, *Cancer Sci.,* **102**, pp. 991–998.
91. Missailidis, S., and Hardy, A. (2009). Aptamers as inhibitors of target proteins, *Expert Opin. Ther. Pat.,* **19**, pp. 1073–1082.
92. Ulrich, H., and Wrenger, C. (2009). Disease-specific biomarker discovery by aptamers, *Cytometry A,* **75A**, pp. 727–733.
93. Ireson, C. R., and Kelland, L. R. (2006). Discovery and development of anticancer aptamers, *Mol. Cancer Ther.,* **5**, pp. 2957–2962.
94. Srivastava, M., and Pollard, H. B. (1999). Molecular dissection of nucleolin's role in growth and cell proliferation: new insights, *FASEB J.,* **13**, pp. 1911–1922.
95. Derenzini, M., Sirri, V., Trerè, D., and Ochs, R. L. (1995). The quantity of nucleolar proteins nucleolin and protein B23 is related to cell doubling time in human cancer cells, *Lab. Invest.,* **73**, pp. 497–502.
96. Derenzini, M. (2000). The AgNORs, *Micron,* **31**, pp. 117–120.
97. Otake, Y., et al. (2007). Overexpression of nucleolin in chronic lymphocytic leukemia cells induces stabilization of bcl2 mRNA, *Blood,* **109**, pp. 3069–3075.
98. Vorhies, J. S., and Nemunaitis, J. J. (2007). Nucleic acid aptamers for targeting of shRNA-based cancer therapeutics, *Biologics,* **1**, pp. 367–376.
99. Bates, P. J., Kahlon, J. B., Thomas, S. D., Trent, J. O., and Miller, D. M. (1999). Antiproliferative activity of G-rich oligonucleotides correlates with protein binding, *J. Biol. Chem.,* **274**, pp. 26369–26377.
100. Xu, X., et al. (2001). Inhibition of DNA replication and induction of S phase cell cycle arrest by G-rich oligonucleotides, *J. Biol. Chem.,* **276**, pp. 43221–43230.
101. Dapić, V., et al. (2002). Antiproliferative activity of G-quartet-forming oligonucleotides with backbone and sugar modifications, *Biochemistry,* **41**, pp. 3676–3685.
102. Laber, D. A., et al. (2005). Update on the first phase I study of AGRO100 in advanced cancer, *J. Clin. Oncol.,* **23**, p. 3064.
103. Chu, T. C., Twu, K. Y., Ellington, A. D., and Levy, M. (2006). Aptamer mediated siRNA delivery, *Nucleic Acids Res.,* **34**, pp.
104. Das, M., Mohanty, C., and Sahoo, S. K. (2009). Ligand-based targeted therapy for cancer tissue, *Expert Opin. Drug. Delivery,* **6**, pp. 285–304.

105. Meacham, C. E., and Morrison, S. J. (2013). Tumour heterogeneity and cancer cell plasticity, *Nature,* **501**, pp. 328–337.
106. Cabrera, M. C., Hollingsworth, R. E., and Hurt, E. M. (2015). Cancer stem cell plasticity and tumor hierarchy, *World J. Stem Cells,* **7**, pp. 27–36.
107. Subramanian, N., et al. (2012). Target-specific delivery of doxorubicin to retinoblastoma using epithelial cell adhesion molecule aptamer, *Mol. Vis.,* **18**, pp. 2783–2795.
108. Li, L., et al. (2014). Epithelial cell adhesion molecule aptamer functionalized PLGA-lecithin-curcumin-PEG nanoparticles for targeted drug delivery to human colorectal adenocarcinoma cells, *Int. J. Nanomed.,* **9**, pp. 1083–1096.
109. Alshaer, W., Hillaireau, H., Vergnaud, J., Ismail, S., and Fattal, E. (2015). Functionalizing liposomes with anti-CD44 aptamer for selective targeting of cancer cells, *Bioconjug. Chem.,* **26**, pp. 1307–1313.
110. Ni, M., et al. (2015). Poly(lactic-co-glycolic acid) nanoparticles conjugated with CD133 aptamers for targeted salinomycin delivery to $CD133^+$ osteosarcoma cancer stem cells, *Int. J. Nanomed.,* **10**, pp. 2537–2554.
111. Meng, L., et al. (2012). Targeted delivery of chemotherapy agents using a liver cancer-specific aptamer, *PLoS One,* **7**, p. e33434.
112. Wang, J., et al. (2013). Aptamer-conjugated nanorods for targeted photothermal therapy of prostate cancer stem cells, *Chem. Asian J.,* **8**, pp. 2417–2422.
113. Subramanian, N., et al. (2015). EpCAM aptamer mediated cancer cell specific delivery of EpCAM siRNA using polymeric nanocomplex, *J. Biomed. Sci.,* **22**, p. 4.
114. Choo, A. B., et al. (2008). Selection against undifferentiated human embryonic stem cells by a cytotoxic antibody recognizing podocalyxin-like protein-1, *Stem Cells,* **26**, pp. 1454–1463.
115. González, B., Denzel, S., Mack, B., Conrad, M., and Gires, O. (2009). EpCAM is involved in maintenance of the murine embryonic stem cell phenotype, *Stem Cells,* **27**, pp. 1782–1791.
116. Ng, V. Y., Ang, S. N., Chan, J. X., and Choo, A. B. H. (2010). Characterization of epithelial cell adhesion molecule as a surface marker on undifferentiated human embryonic stem cells, *Stem Cells,* **28**, pp. 29–35.
117. Hung, L.-Y., et al. (2015). Screening of aptamers specific to colorectal cancer cells and stem cells by utilizing on-chip cell-SELEX, *Sci. Rep.,* **5**, p. 10326.

118. Sefah, K., et al. (2013). Cell-based selection provides novel molecular probes for cancer stem cells, *Int. J. Cancer,* **132**, pp. 2578–2588.

119. Kim, Y., et al. (2013). Aptamer identification of brain tumor-initiating cells, *Cancer Res.,* **73**, pp. 4923–4936.

120. Ohuchi, S. (2012). Cell-SELEX technology, *Biores. Open Access,* **1**, pp. 265–272.

121. Jung, Y. K., Woo, M. A., Soh, H. T., and Park, H. G. (2014). Aptamer-based cell imaging reagents capable of fluorescence switching, *Chem. Commun.,* **50**, pp. 12329–12332.

122. Ababneh, N., et al. (2013). In vitro selection of modified RNA aptamers against CD44 cancer stem cell marker, *Nucleic Acid Ther.,* **23**, pp. 401–407.

123. Shangguan, D., et al. (2008). Identification of liver cancer-specific aptamers using whole live cells, *Anal. Chem.,* **80**, pp. 721–728.

124. Shangguan, D., et al. (2006). Aptamers evolved from live cells as effective molecular probes for cancer study, *Proc. Natl. Acad. Sci. U S A,* **103**, pp. 11838–11843.

125. Yang, M., et al. (2014). Developing aptamer probes for acute myelogenous leukemia detection and surface protein biomarker discovery, *J. Hematol. Oncol.,* **7**, p. 5.

126. Tan, Y., et al. (2013). DNA aptamers that target human glioblastoma multiforme cells overexpressing epidermal growth factor receptor variant III in vitro, *Acta Pharmacol. Sin.,* **34**, pp. 1491–1498.

127. Gires, O., and Bauerle, P. A. (2010). EpCAM as a target in cancer therapy, *J. Clin. Oncol.,* **28**, pp. e239–240; author reply pp. e241–242.

128. Yu, G., et al. (2008). CpG island methylation status in the EpCAM promoter region and gene expression, *Oncol. Rep.,* **20**, pp. 1061–1067.

129. Went, P., et al. (2006). Frequent high-level expression of the immunotherapeutic target Ep-CAM in colon, stomach, prostate and lung cancers, *Br. J. Cancer,* **94**, pp. 128–135.

130. Sterzynska, K., Kempisty, B., Zawierucha, P., and Zabel, M. (2012). Analysis of the specificity and selectivity of anti-EpCAM antibodies in breast cancer cell lines, *Folia Histochem. Cytobiol.,* **50**, pp. 534–541.

131. Altieri, D. C. (2013). Targeting survivin in cancer, *Cancer Lett.,* **332**, pp. 225–228.

132. Di Stefano, A. B., et al. (2010). Survivin is regulated by interleukin-4 in colon cancer stem cells, *J. Cell. Physiol.,* **225**, pp. 555–561.

133. Du, L., et al. (2008). CD44 is of functional importance for colorectal cancer stem cells, *Clin. Cancer. Res.,* **14**, pp. 6751–6760.

134. Bapat, S. A. (2010). Human ovarian cancer stem cells, *Reproduction,* **140**, pp. 33–41.
135. Choi, K. Y., Saravanakumar, G., Park, J. H., and Park, K. (2012). Hyaluronic acid-based nanocarriers for intracellular targeting: interfacial interactions with proteins in cancer, *Colloids Surf. B. Biointerfaces,* **99**, pp. 82–94.
136. Vant-Hull, B., Payano-Baez, A., Davis, R. H., and Gold, L. (1998). The mathematics of SELEX against complex targets, *J. Mol. Biol.,* **278**, pp. 579–597.
137. Morris, K. N., Jensen, K. B., Julin, C. M., Weil, M., and Gold, L. (1998). High affinity ligands from in vitro selection: complex targets, *Proc. Natl. Acad. Sci. U S A,* **95**, pp. 2902–2907.
138. Homann, M., and Göringer, H. U. (1999). Combinatorial selection of high affinity RNA ligands to live African trypanosomes, *Nucleic Acids Res.,* **27**, pp. 2006–2014.
139. Phillips, J. A., Lopez-Colon, D., Zhu, Z., Xu, Y., and Tan, W. (2008). Applications of aptamers in cancer cell biology, *Anal. Chim. Acta,* **621**, pp. 101–108.
140. Zhang, Y., Chen, Y., Han, D., Ocsoy, I., and Tan, W. (2010). Aptamers selected by cell-SELEX for application in cancer studies, *Bioanalysis,* **2**, pp. 907–918.
141. Iwagawa, T., Ohuchi, S. P., Watanabe, S., and Nakamura, Y. (2012). Selection of RNA aptamers against mouse embryonic stem cells, *Biochimie,* **94**, pp. 250–257.
142. Mizrak, D., Brittan, M., and Alison, M. R. (2008). CD133: molecule of the moment, *J. Pathol.,* **214**, pp. 3–9.
143. Jászai, J., Fargeas, C. A., Florek, M., Huttner, W. B., and Corbeil, D. (2007). Focus on molecules: prominin-1 (CD133), *Exp. Eye Res.,* **85**, pp. 585–586.
144. Ulasov, I. V., Nandi, S., Dey, M., Sonabend, A. M., and Lesniak, M. S. (2011). Inhibition of sonic hedgehog and notch pathways enhances sensitivity of CD133(+) glioma stem cells to temozolomide therapy, *Mol. Med.,* **17**, pp. 103–112.
145. Gao, H., et al. (2012). Precise glioma targeting of and penetration by aptamer and peptide dual-functioned nanoparticles, *Biomaterials,* **33**, pp. 5115–5123.
146. Dilnawaz, F., et al. (2012). The transport of non-surfactant based paclitaxel loaded magnetic nanoparticles across the blood brain barrier in a rat model, *Biomaterials,* **33**, pp. 2936–2951.

147. Zharov, V. P., Mercer, K. E., Galitovskaya, E. N., and Smeltzer, M. S. (2006). Photothermal nanotherapeutics and nanodiagnostics for selective killing of bacteria targeted with gold nanoparticles, *Biophys. J.*, **90**, pp. 619–627.
148. Hamburger, A. W. (1980). Enhancement of human erythroid progenitor cell growth by media conditioned by a human t-lymphocyte line, *Blood*, **56**, pp. 633–639.
149. Bose, D., et al. (2011). Chemoresistant colorectal cancer cells and cancer stem cells mediate growth and survival of bystander cells, *Br. J. Cancer*, **105**, pp. 1759–1767.
150. Kang, H., O'Donoghue, M. B., Liu, H., and Tan, W. (2010). A liposome-based nanostructure for aptamer directed delivery, *Chem. Commun.*, **46**, pp. 249–251.
151. Huang, Y. F., et al. (2009). Molecular assembly of an aptamer-drug conjugate for targeted drug delivery to tumor cells, *ChemBioChem*, **10**, pp. 862–868.
152. Herr, J. K., Smith, J. E., Medley, C. D., Shangguan, D., and Tan, W. (2006). Aptamer-conjugated nanoparticles for selective collection and detection of cancer cells, *Anal. Chem.*, **78**, pp. 2918–2924.
153. Stanciu, L. A., Shute, J., Holgate, S. T., and Djukanović, R. (1996). Production of IL-8 and IL-4 by positively and negatively selected CD4+ and CD8+ human T cells following a four-step cell separation method including magnetic cell sorting (MACS), *J. Immunol. Methods*, **189**, pp. 107–115.
154. Benez, A., Geiselhart, A., Handgretinger, R., Schiebel, U., and Fierlbeck, G. (1999). Detection of circulating melanoma cells by immunomagnetic cell sorting, *J. Clin. Lab. Anal.*, **13**, 229–233.
155. Virgo, P., et al. (2003). Identification of the CD33-related Siglec receptor, Siglec-5 (CD170), as a useful marker in both normal myelopoiesis and acute myeloid leukaemias, *Br. J. Haematol.*, **123**, pp. 420–430.
156. Zhou, J., Cui, Q., Shi, Y., and Rossi, J. J. (2015). Evolution of cell-specific RNA aptamers against glioblastoma cancer stem cells via live cell-based SELEX, *American Society of Gene and Cell Therapy, (ASGCT) 18th Annual Meeting*.
157. Del Vecchio, C. A., et al. (2012). Epidermal growth factor receptor variant III contributes to cancer stem cell phenotypes in invasive breast carcinoma, *Cancer Res.*, **72**, pp. 2657–2671.

158. Horvath, S., et al. (2006). Analysis of oncogenic signaling networks in glioblastoma identifies ASPM as a molecular target, *Proc. Natl. Acad. Sci. U S A,* **103**, pp. 17402–17407.

159. Bayrac, A. T., et al. (2011). In vitro selection of DNA aptamers to glioblastoma multiforme, *ACS Chem. Neurosci.,* **2**, pp. 175–181.

Chapter 6

Aptamers as Therapeutic Tools in Neurological Diseases

Lukas Aaldering,[a] Shilpa Krishnan,[b] Sue Fletcher,[b,c] Stephen D. Wilton,[b,c] and Rakesh N. Veedu[b,c]

[a] *Institute of Plant Biology and Biotechnology (IBBP), University of Münster, Schloßplatz 3, Münster 48149, Germany*
[b] *Western Australian Neuroscience Research Institute, Perth 6150, Australia*
[c] *Center for Comparative Genomics, Murdoch University, Perth 6150, Australia*
r.veedu@murdoch.edu.au

6.1 Introduction

Nucleic acid aptamers have attracted considerable attention in recent years as tools for targeted molecular recognition. These synthetic nucleic acid ligands can fold into three-dimensional shapes (Fig. 6.1) for binding defined molecular targets with specific high-affinity recognition. With potential targets ranging from small chemicals to large proteins, these characteristics make aptamers an attractive platform for applications relating to drug delivery, biosensing, and theranostic. Aptamers are generally selected using systematic evolution of ligands by exponential enrichment (SELEX) [1] procedures. Ever since their introduction to the scientific

Aptamers: Tools for Nanotherapy and Molecular Imaging
Edited by Rakesh N. Veedu

ISBN 978-981-4669-83-2 (Hardcover), 978-981-4669-84-9 (eBook)
www.panstanford.com

community, aptamers have been applied in various research fields toward developing novel therapeutics or as tools for diagnostics and imaging [2, 3]. In 2004, Pegaptanib sodium (Macugen) was approved by the FDA for the treatment of age-related macular degeneration (AMD). Due to their functional similarities to antibodies, aptamers are often termed as "chemical antibodies." However, aptamers possess certain advantages over antibodies, such as stability, production scale, method and time, chemical fabrication, and costs. Furthermore, they are not immunogenic and can be synthesized in vitro in large quantities, up to kilograms. Herein, we describe recent advances in aptamer technology in the field of neurological diseases.

6.2 Blood–Brain Barrier

Most drugs (large or small molecules) are limited from entering the brain due to a highly selective permeability barrier called the blood–brain barrier (BBB), as shown in Fig. 6.1 [4]. The BBB is located at the interface of blood and the brain, and its primary function is to maintain homeostasis of the brain. In doing so, it offers significant protection against pathogens and toxins within the blood and also

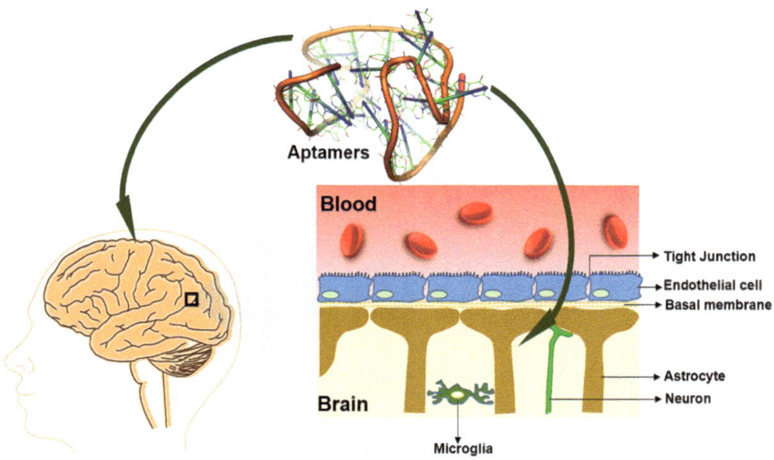

Figure 6.1 Schematic representation of the blood–brain barrier and aptamers toward targeted therapy of neurological diseases.

limits the entry and effectiveness of neurotherapeutics. Therefore, to successfully treat neurological diseases, some drugs have to be administered into the cerebrospinal fluid [5, 6].

Unlike antibodies, aptamers can be utilized for target-specific therapy of neurological diseases on the basis of their relatively small size and nearly no immunogenicity [7]. In addition to these properties, aptamers with BBB-penetrating abilities can be selected during in vivo SELEX [8].

6.3 Multiple Sclerosis

Multiple sclerosis (MS) is a chronic inflammatory and demyelinating disorder of the nervous system. Although the exact underlying mechanisms are still unclear, it is believed to be as either an autoimmune or an oligodendrogliopathy disease [9]. Etiologies for MS remain unknown, but they are attributed to a broad field of factors, ranging from infections, xenobiotics, and genetics to behavioral aspects such as smoking and environmental factors like sunlight exposure [10–12]. Currently there is no cure for MS and the available medications can only shorten the duration of attacks or relapses and help reduce the rate of the attacks to slow the progression of the disease [13].

Remyelination is a naturally occurring process in the body to restore damaged myelin sheaths after an MS attack. However, this restoration process often leads to only incomplete recovery [14]. Recently, a 40-nucleotide DNA aptamer was identified that exhibits affinity toward murine myelin [15]. It binds to multiple myelin components in vitro, and intraperitoneal (IP) injection in mice showed improved distribution and uptake in central nervous system (CNS) tissues. Furthermore, the aptamer promoted remyelination of CNS lesions in mice infected by Theiler's virus [15]. Thus, this aptamer could prove valuable in the body recovery after an MS attack and could palliate MS symptoms.

Advances in aptamers not only are limited to restoration and recovery but also are a significant tool to inhibit inflammation. Midkine (MK), a heparin-binding growth factor, is associated with MS-related inflammation. Due to its suppressing effect on regulatory

T (Treg) cell expansion, it is a crucial mediator of autoimmune tolerance [16]. Administration of an anti-MK aptamer decreased the effect of MK on Treg cells and led to increased cells populations and mitigation of experimental autoimmune encephalomyelitis-related symptoms. Ishiguro et al. used another approach by using an aptamer to reduce the pro-inflammatory activity of the cytokine interleukin-17A (IL-17A) [17]. The aptamer binds to human and mouse IL-17A and blocks the interaction between IL-17A and its receptor. This inhibited the phosphorylation of IL-17A-signaling factors IκB and JNK and the production of IL-6 in human and mouse cells. Thus the signaling cascade is disrupted.

In the field of MS diagnostics, a 2'-F modified RNA aptamer was selected to bind specifically to proteolytic autoantibodies against myelin basic protein [18]. This high-affinity aptamer was conjugated with a $Ca^{(2+)}$-regulated photoprotein and used in a bioluminescent microplate assay to detect these antibodies. The technique is not only sensitive but also simple and fast, demonstrating the potential to be used as a specific laboratory test for the diagnosis of MS.

6.4 Brain Tumor

The most common and lethal form of malignant brain tumors in adults is glioblastoma. This type of tumor is characterized by rapid growth and is highly invasive with its capacity to spread into critical neurological areas within the brain. Currently many treatments fail [19–21] due to its high resistance to apoptosis, so the tumor is surgically removed in most cases.

Since their development, the potential of aptamers in brain tumor therapy was poorly explored; however, the field has gained momentum in recent years. In 2012, Kang et al. developed two aptamers, GBM 128 and GBM 131, with high affinity and specificity against glioblastoma cells and without any nonspecific binding to normal astraglial cells or normal brain tissue [22].

Tenascin-C is an extracellular glycoprotein that is overexpressed on glioma cells and therefore acts as a marker for brain tumors. Several aptamers have been developed to either bind or facilitate imaging of glioma cells [23–26]. These aptamers are of high value for diagnostics for the identification of glioblastoma biomarkers, molecular imaging, or targeted drug delivery. In addition, Chen et al. reported an interesting approach by coupling aptamers to adenovirus vectors via biotin–avidin interaction [27]. The modification improved the transduction efficiency and further highlights the potential impact of aptamers in improving cancer gene therapy.

Like other cancers of the nervous system, glioblastoma appears to possess a molecular cell hierarchy, the apex of which are the so-called tumor-initiating cells (TICs), also called cancer stem cells. TICs are said to generate new tumor cells through the process of cell deviation and differentiation. Thus, identifying these cells is one, if not the most, important goal in cancer treatment [28, 29]. By using cell-SELEX with TICs, non-TIC, and human neural progenitor cells, Kim et al. were able to identify aptamers with the capacity to bind and internalize into glioblastoma cells that proliferate and initiate new tumors [30]. A combination of these aptamers with other therapeutics payload might improve the overall treatment efficacy of glioblastoma cells significantly.

6.5 Stroke

According to the Global Burden of Disease Study 2013, stroke is the second-most frequent cause of death in developed countries [31]. The pathology of an ischemic stroke is mainly decreased blood supply to parts of the brain due to blood clot formation. The size and associated inflammations can vary the severity of stroke and can lead to irreversible damage to brain function. To treat an ischemic stroke, the blockage of blood flow must be cleared, which can be achieved by either mechanical removal of the clot or thrombolysis. In current treatments, thrombolysis is performed with an enzyme called tissue plasminogen activator (tPA), and this approach, however, is associated with serious side effects, including thrombolysis-related symptomatic intracranial hemorrhage [32]. Thrombolysis-related hemorrhage is one of the major obstacles in the development of stroke treatments, because new therapeutics need to be immediately reversible in case of hemorrhage occurrence. Toward this, aptamers have a great advantage that their functionality can easily be altered by binding to a small complementary oligonucleotide sequence antidote.

An aptamer against the blood-clotting factor IXa has demonstrated efficient anticoagulant properties in mice, pigs, and humans. It interferes with thrombin synthesis by blocking factor IXa, which is needed for conversion of prothrombin to thrombin. The function was reversible by the administration of a short oligonucleotide antidote [33–35]. In 2012, Blake et al. evaluated this aptamer antidote in a murine model of ischemic stroke [36]. They showed that the aptamer not only reduced the thrombin formation but also was reversible using an antidote oligomer in the setting of intracranial hemorrhage, reducing the hematoma volume and thereby improving survival.

6.6 Parkinson Disease

Parkinson disease (PD) is a neurodegenerative disorder of the CNS, mainly affecting the motor neuron system. PD is typically characterized by the accumulation of α-synuclein proteins into Lewy bodies and by the loss of dopaminergic neurons within parts of the midbrain. Recently, aptamers have been developed that can bind α-synuclein monomers [37] or its oligomer [38]. These aptamers can be used as a tool for drug development or for the design of diagnostic tools for PD.

The glial cell line–derived neurotrophic factor (GDNF) promotes survival and differentiation of dopaminergic neurons in vitro [39]. GDNF is known to signal through a receptor tyrosine kinase called RET. In this process, GDNF forms a complex with GDNF family receptor α (GFRα), a glycosyl phosphatidylinositol (GPI)-linked ligand-binding subunit. This complex contains homodimers of GDNF and GFRα and initiates the dimerization of RET, leading to transphosphorylation of specific tyrosine residues and subsequently to cellular signaling [40]. A nuclease resistant 2′-fluoropyrimidine RNA aptamer (D4) was identified after a whole-cell SELEX screen using RET-expressing cells. This aptamer exhibited high affinity and specificity to the extracellular domain of RET monomers, inhibiting its dimerization and signaling [41]. While having no direct therapeutic use, these kinds of aptamer inhibitors could be used to further investigate important signaling components.

Retinoic acid induces differentiation in several neuroblastoma cell lines. This differentiation is coupled to increase expression of functional neurotrophic factor receptors, including RET. The aptamer D4 was used to investigate the functions of RET in retinoic acid–induced cell differentiation [42].

6.7 Alzheimer Disease

Alzheimer disease (AD) is the most common neurodegenerative disease and is accountable for more than 80% of all cases of dementia in the elderly. AD is characterized by the progressive loss of mental, behavioral functionality and a declining ability to learn [43]. The pathophysiology of AD is very complex, and different hypotheses have been postulated to explain the development and pathogenesis of AD. The amyloid beta (Aβ) hypotheses was prevailing for the last two decades; however, recently it was shown that this hypothesis alone does not account for the complexity of AD's pathophysiology [44].

Aptamers showed great promise and advances in imaging and diagnosis of AD. Recently, multiple aptamers have been developed to facilitate imaging of the amyloid-β peptides. In 2014, Farrar et al. developed a fluorescently tagged anti-amyloid-β-aptamer, β55, that can efficiently bind amyloid plaques in both human Alzheimer disease brain tissue ex vitro and in vivo using App/PS1 transgenic mice [45]. Similarly in 2015, Babu et al. developed a simple luminescence aptamer-ruthenium complex that binds and inhibits the formation of amyloid-β oligomers [46]. In 2012, Baird et al. used a slow-flow-rate aptamer array to investigate age-dependent changes in the cerebrospinal fluid proteome. Using this technology, over 200 proteins were identified that could be valuable biomarkers for the diagnosis and treatment of neurological diseases [47].

AD is characterized by two major abnormalities, abnormal extracellular amyloid-β protein disposition and intracellular neurofibrillary tangle (NFT) formation, leading to neuronal degeneration [48].

The amyloid-β protein is generated by the sequential cleavage of the amyloid-β precursor protein (APP) by β-site amyloid precursor protein cleaving enzyme 1 (BACE1) [49]. This enzyme contains a short cytoplasmic tail (B1-CT). Rentmeister et al. identified an RNA aptamer that binds specifically to the B1-CT tail without interfering with the B1-CTregulated BACE1 transport [50]. It is remarkable that this aptamer is able to discriminate binding regions within a 3 kDa peptide. Therefore, this aptamer could be used to further investigate and elucidate B1-CT function without affecting cellular localization or other important biological activities.

6.8 Myasthenia Gravis

Myasthenia gravis (MG) is a disease that interferes with the signal transduction between neurons and muscle cells. Pathologically, MG is characterized by antibody binding to the nicotinic acetylcholine receptor (nAChR) and thus inhibition of the excitatory effect of acetylcholine, which results in fluctuating muscle weakness and fatigue. MG is treated with acetylcholinesterase inhibitors, which increase the bioavailability of acetylcholine by inhibiting an acetylcholine-cleaving enzyme, or by administering immunosuppressants.

In 1996, an aptamer showed protective properties against AChR antibodies. The aptamer binds to a monoclonal antibody (Mab198) that recognizes the main immunogenic region of human AChR and autoimmune antibodies for MG patients. Furthermore, it is able to inhibit the antibodies from down-modulating AChR expression in vitro [51]. However, the aptamer was of limited therapeutic use due to low-affinity binding to mAb198 relative to that of AChR. Thus, attempts have been made to improve the aptamer, initially by screening 3′-end-extended libraries in vitro [52]. In a subsequent study, the secondary configuration and binding motif of the aptamer was further investigated and a minimized version of the aptamer was identified [53]. The improved aptamers showed increased binding affinities and AChR-protective properties and were more suitable as therapeutic tools in MG treatment.

6.9 Variant Creutzfeldt–Jakob Disease

The Creutzfeldt–Jakob disease (CJD) is a degenerative neurological disorder and a form of spongiform encephalopathy. It is often called the human form of mad cow disease and is invariably fatal. The accumulation of abnormally folded insoluble prion protein (PrP), especially in the brain, is a characteristic of CJD. PrP replicates within the host by promoting the misfolding of native proteins into the diseased conformation [54]. The number of misfolded proteins increases exponentially and leads to rapid neurodegeneration by disrupting cell functions and causing cell death. So far, there is no cure, and the known medication pentosan polysulfate (PPS) only helps to slow the progression of the disease [55].

Toward therapeutic treatments, different aptamers were used to stop the conversion of normal to abnormal PrPs [56–60]. In 2002, Proske et al. developed a modified RNA aptamer against PrP, targeting the amino residues between 90 and 129, a domain thought to be important in the conversion of PrP from its normal to abnormal conformation [58]. They were able to show a reduction in misfolded PrP levels with the administration of this aptamer into the medium of prion-infected neuroblastoma cells.

Murkami et al. achieved a similar result in 2008. They identified an RNA aptamer consisting of only 12 nucleotides (R12) that binds the normal cellular form of bovine PrP with high affinity [59]. And recently they showed that R12 exhibits antiprion activity in mouse neural cells by binding and stabilizing the normal form of PrP and thus blocking its conversion to abnormal PrP [60]. These advances could be the first steps to a successful therapy or prophylaxis of CJD.

6.10 Pain and Stress-Relevant Aptamers

Neuropeptides are small protein-like molecules that facilitate communication between neurons. The neuropeptide nociception/orphanin FQ (N/OFQ) is an endogenous ligand of the opioid receptor-like 1 receptor and is associated with the regulation of several biological functions, such as pain and stress. Aptamers have been developed to investigate or antagonize some neuropeptides.

Faulhammer et al. [61] developed an L-enantiomeric oligonucleotide ligand that is able to antagonize the binding between N/OFQ and the ORL1 receptor.

6.11 Stability of Aptamers

Every organism possesses series of enzymes to synthesize, modify, or hydrolyze nucleic acids. The nucleic acid–cleaving enzymes, called nucleases, play an important role as they are responsible for nucleic acid turnover and act as a defensive mechanism against pathogenic bacteria and viruses. Oligonucleotides composed of naturally occurring DNA and RNA nucleotides show poor nuclease resistance and decreased target-binding affinity with a limited in vivo half-life and bioavailability. These limitations restrict the applicability of synthetic oligonucleotides with natural nucleotides in therapeutic development.

However, aptamers can be chemically modified to increase stability [62] and the in vivo half-life. A broad range of nucleic acid modifications can be applied to improve the aptamers (Fig. 6.2). There are two different strategies to incorporate chemical modifications in aptamers: (1) using modified nucleoside triphosphates during SELEX processes to develop chemically modified aptamers [63] and (2) introducing modified nucleotides to a preselected DNA or RNA aptamer during solid-phase chemical synthesis, usually called the post-SELEX approach [64]. The first method is often challenging as most of the modified nucleotides are not good substrates of polymerases for the mandatory enzymatic synthesis protocols used in SELEX. However, some 2′-sugar-modified nucleotides, for example, 2′-O-methyl (2′-OMe), 2′-fluoro (2′-F), and locked nucleic acid (LNA), are tolerated by polymerases [65, 66]. The post-SELEX procedure has been successfully utilized to modify various preexisting aptamers, and this may still be a realistic approach for developing chemically modified aptamers.

6.12 Summary

Aptamers offer great promise toward the treatment of neurological diseases, although the technology has been underutilized in the last

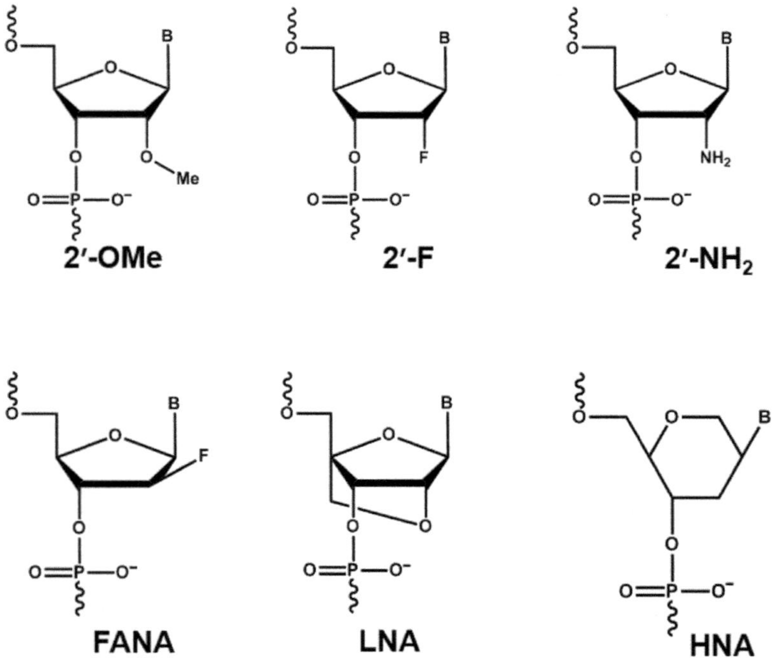

Figure 6.2 Examples of sugar-modified nucleotides for aptamer selections using SELEX-based protocols. 2′-O-methyl (2′-OMe), 2′-fluoro (2′-F), 2′-amino (2′-NH_2), 2′-fluoroarabino (2′-FANA), locked nucleic acid (LNA), and hexitol nucleic acid (HNA). See Ref. [65] for more chemical modifications and their enzymatic recognition studies.

two decades. The low cost and simplicity of aptamer development protocols, the ability of aptamers to bind against almost all targets, high stability, and low or no immunogenicity make this technology an attractive platform toward the development of promising theranostics against neurological diseases. Furthermore, it is worth mentioning that the functionality of aptamers can easily be quenched or altered by binding to a small, complementary oligonucleotide sequence antidote. With respect to their relatively small size compared to the antibody counterpart, aptamers can be easily tethered to other agents (e.g., small chemicals, peptides, and nanoparticles) for improving BBB penetration.

References

1. Tuerk, C., and Gold, L. (1990). Systematic evolution of ligands by exponential enrichment: RNA ligands to bacteriophage T4 DNA polymerase, *Science,* **249**, pp. 505–510.
2. Keefe, A. D., Pai, S., and Ellington, A. (2010). Aptamers as therapeutics, *Nat. Rev. Drug Discovery,* **9**, pp. 537–550.
3. Famulok, M., Hartig, J. S., and Mayer, G. (2007). Functional aptamers and aptazymes in biotechnology, diagnostics, and therapy, *Chem. Rev.*, **107**, pp. 3715–3743.
4. Pardridge, W. M. (2005). The blood-brain barrier: bottleneck in brain drug development, *NeuroRx*, **2**, pp. 3–14.
5. Pardridge, W. M. (2011). Drug transport in brain via the cerebrospinal fluid, *Fluids Barriers CNS,* **8**, p. 7.
6. Chen, Y., Imai, H., Ito, A., and Saito, N. (2013). Novel modified method for injection into the cerebrospinal fluid via the cerebellomedullary cistern in mice, *Acta Neurobiol. Exp. (Wars)*, **73**, pp. 304–311.
7. De Franciscis, V., Esposito, C. L., Catuogno, S., Cellai, L., and Cerchia, L. (2009). Aptamers as innovative diagnostic and therapeutic agents in the central nervous system, *CNS Neurol. Disord. Drug Targets*, **8**, pp. 393–401.
8. Cheng, C., Chen, Y. H., Lennox, K. A., Behlke, M. A., and Davidson, B. L. (2013). In vivo SELEX for identification of brain-penetrating aptamers, *Mol. Ther. Nucleic Acids*, **2**, p. e67.
9. Nakahara, J., Maeda, M., Aiso, S., and Suzuki, N. (2012). Current concepts in multiple sclerosis: autoimmunity versus oligodendrogliopathy, *Clin. Rev. Allergy Immunol.*, **42**, pp. 26–34.
10. Ascherio, A., and Munger, K. L. (2007). Environmental risk factors for multiple sclerosis. Part I: the role of infection, *Ann. Neurol.*, **61**, pp. 288–299.
11. Ascherio, A., and Munger, K. L. (2007). Environmental risk factors for multiple sclerosis. Part II: noninfectious factors, *Ann. Neurol.*, **61**, pp. 504–513.
12. Granieri, E. (2000). Exogeneous factors in the aetiology of multiple sclerosis, *J. Neurovirol.*, **6**, pp. S141–S146.
13. Compston, A., and Coles, A. (2002). Multiple sclerosis, *Lancet*, **359**, pp. 1221–1231.
14. Chang, A., Tourtellotte, W. W., Rudick, R., and Trapp, B. D. (2002). Premyelinating oligodendrocytes in chronic lesions of multiple sclerosis, *N. Engl. J. Med.*, **346**, pp. 165–173.

15. Nastasijevic, B., Wright, B. R., Smestad, J., Warrington, A. E., Rodriguez, M., and Maher, L. J. (2012). Remyelination induced by a DNA aptamer in a mouse model of multiple sclerosis, *PLoS One,* **7**, p. e39595.

16. Wang, J., Takeuchi, H., Sonobe, Y., Jin, S., Mizuno, T., Miyakawa, S., Fujiwara, M., Nakamura, Y., Kato, T., Muramatsu, H., Muramatsu, T., and Suzumura, A. (2008). Inhibition of midkine alleviates experimental autoimmune encephalomyelitis through the expansion of regulatory T cell population, *Proc. Natl. Acad. Sci. U S A,* **105**, pp. 3915–3920.

17. Ishiguro, A., Akiyama, T., Adachi, H., Inoue, J., and Nakamura, Y. (2011). Therapeutic potential of anti-interleukin-17A aptamer: suppression of interleukin-17A signaling and attenuation of autoimmunity in two mouse models, *Arthritis Rheum.*, **63**, pp. 455–466.

18. Vorobjeva, M. A., Krasitskaya, V. V., Fokina, A. A., Timoshenko, V. V., Nevinsky, G. A., Venyaminova, A. G., and Frank, L. A. (2014). RNA aptamer against autoantibodies associated with multiple sclerosis and bioluminescent detection probe on its basis, *Anal. Chem.*, **86**, pp. 2590–2594.

19. Castro, M. G., Cowen, R., Williamson, I. K., David, A., Jimenez-Dalmaroni, M. J., Yuan, X., Bigliari, A., Williams, J. C., Hu, J., and Lowenstein, P. R. (2003). Current and future strategies for the treatment of malignant brain tumors, *Pharmacol. Ther.*, **98**, pp. 71–108.

20. Laperriere, N. J., and Bernstein, M. (1994). Radiotherapy for brain tumors, *CA Cancer J. Clin.*, **44**, pp. 96–108.

21. Ng, H. K., and Lam, P. Y. (1998). The molecular genetics of central nervous system tumors, *Pathology,* **30**, pp. 196–202.

22. Kang, D., Wang, J., Zhang, W., Song, Y., Li, X., Zou, Y., Zhu, M., Zhu, Z., Chen, F., and Yang, C. J. (2012). Selection of DNA aptamers against glioblastoma cells with high affinity and specificity, *PLoS One,* **7**, p. e42731.

23. Chen, X.-C., Deng, Y.-L., Lin, Y., Pang, D.-W., Qing, H., Qu, F., and Xie, H.-Y. (2008). Quantum dot-labeled aptamer nanoprobes specifically targeting glioma cells, *Nanotechnology,* **19**, p. 235105.

24. Daniels, D. A., Chen, H., Hicke, B. J., Swiderek, K. M., and Gold, L. (2003). A tenascin-C aptamer identified by tumor cell SELEX: systematic evolution of ligands by exponential enrichment, *Proc. Natl. Acad. Sci. U S A,* **100**, pp. 15416–15421.

25. Hicke, B. J., Marion, C., Chang, Y. F., Gould, T., Lynott, C. K., Parma, D., Schmidt, P. G., and Warren, S. (2001). Tenascin-C aptamers are generated using tumor cells and purified protein, *J. Biol. Chem.*, **276**, pp. 48644–48654.

26. Förster, C., Brauer, A. B. E., Brode, S., Schmidt, K. S., Perbandt, M., Meyer, A., Rypniewski, W., Betzel, C., Kurreck, J., Fürste, J. P., and Erdmann, V. A. (2006). Comparative crystallization and preliminary X-ray diffraction studies of locked nucleic acid and RNA stems of a tenascin C-binding aptamer, *Acta Crystallogr. Sect. F Struct. Biol. Cryst. Commun.*, **62**, pp. 665–668.
27. Chen, H., Zheng, X., Di, B., Wang, D., Zhang, Y., Xia, H., and Mao, Q. (2013). Aptamer modification improves the adenoviral transduction of malignant glioma cells, *J. Biotechnol.*, **168**, pp. 362–366.
28. Hjelmeland, A. B., Lathia, J. D., Sathornsumetee, S., and Rich, J. N. (2011). Twisted tango: brain tumor neurovascular interactions, *Nat. Neurosci.*, **14**, pp. 1375–1381.
29. Reya, T., Morrison, S. J., Clarke, M. F., and Weissman, I. L. (2001). Stem cells, cancer, and cancer stem cells, *Nature*, **414**, pp. 105–11.
30. Kim, Y., Wu, Q., Hamerlik, P., Hitomi, M., Sloan, A. E., Barnett, G. H., Weil, R. J., Leahy, P., Hjelmeland, A. B., and Rich, J. N. (2013). Aptamer identification of brain tumor-initiating cells, *Cancer Res.*, **73**, pp. 4923–4936.
31. GBD *2013* (2014). Global, regional, and national age–sex specific all-cause and cause-specific mortality for 240 causes of death, 1990–2013: a systematic analysis for the Global Burden of Disease Study 2013. *Lancet*, **385**, pp. 117–171.
32. DeMers, G., Meurer, W. J., Shih, R., Rosenbaum, S., and Vilke, G. M. (2012). Tissue plasminogen activator and stroke: review of the literature for the clinician, *J. Emerg. Med.*, **43**, pp. 1149–1154.
33. Rusconi, C. P., Roberts, J. D., Pitoc, G. A., Nimjee, S. M., White, R. R., Quick, G., Scardino, E., Fay, W. P., and Sullenger, B. A. (2004). Antidote-mediated control of an anticoagulant aptamer in vivo, *Nat. Biotechnol.*, **22**, pp. 1423–1428.
34. Dyke, C. K., Steinhubl, S. R., Kleiman, N. S., Cannon, R. O., Aberle, L. G., Lin, M., Myles, S. K., Melloni, C., Harrington, R. A., Alexander, J. H., Becker, R. C., and Rusconi, C. P. (2006). First-in-human experience of an antidote-controlled anticoagulant using RNA aptamer technology: a phase 1a pharmacodynamic evaluation of a drug-antidote pair for the controlled regulation of factor IXa activity, *Circulation*, **114**, pp. 2490–2497.
35. Nimjee, S. M., Keys, J. R., Pitoc, G. A., Quick, G., Rusconi, C. P., and Sullenger, B. A. (2006). A novel antidote-controlled anticoagulant reduces thrombin generation and inflammation and improves cardiac

function in cardiopulmonary bypass surgery, *Mol. Ther.*, **14**, pp. 408–415.

36. Blake, C. M., Wang, H., Laskowitz, D. T., and Sullenger, B. A. (2011). A reversible aptamer improves outcome and safety in murine models of stroke and hemorrhage, *Oligonucleotides*, **21**, pp. 11–19.
37. Tsukakoshi, K., Harada, R., Sode, K., and Ikebukuro, K. (2010). Screening of DNA aptamer which binds to alpha-synuclein, *Biotechnol. Lett.*, **32**, pp. 643–648.
38. Tsukakoshi, K., Abe, K., Sode, K., and Ikebukuro, K. (2012). Selection of DNA aptamers that recognize α-synuclein oligomers using a competitive screening method, *Anal. Chem.*, **84**, pp. 5542–5547.
39. Lin, L. F., Doherty, D. H., Lile, J. D., Bektesh, S., and Collins, F. (1993). GDNF: a glial cell line-derived neurotrophic factor for midbrain dopaminergic neurons, *Science,* **260**, pp. 1130–1132.
40. Sariola, H., and Saarma, M. (2003). Novel functions and signalling pathways for GDNF, *J. Cell Sci.*, **116**, pp. 3855–3862.
41. Cerchia, L., Ducongé, F., Pestourie, C., Boulay, J., Aissouni, Y., Gombert, K., Tavitian, B., de Franciscis, V., and Libri, D. (2005). Neutralizing aptamers from whole-cell SELEX inhibit the RET receptor tyrosine kinase, *PLoS Biol.*, **3**, p. e123.
42. Cerchia, L., D'Alessio, A., Amabile, G., Duconge, F., Pestourie, C., Tavitian, B., Libri, D., and de Franciscis, V. (2006). An autocrine loop involving ret and glial cell-derived neurotrophic factor mediates retinoic acid-induced neuroblastoma cell differentiation, *Mol. Cancer Res.*, **4**, pp. 481–488.
43. Anand, R., Gill, K. D., and Mahdi, A. A. (2014). Therapeutics of Alzheimer's disease: past, present and future,*Neuropharmacology*, **76** Pt A, pp. 27–50.
44. Hardy, J. (2009). The amyloid hypothesis for Alzheimer's disease: a critical reappraisal, *J. Neurochem.*, **110**, pp. 1129–1134.
45. Farrar, C. T., William, C. M., Hudry, E., Hashimoto, T., and Hyman, B. T. (2014). RNA aptamer probes as optical imaging agents for the detection of amyloid plaques, *PLoS One,* **9**, p. e89901.
46. Babu, E., Muthu Mareeswaran, P., Sathish, V., Singaravadivel, S., and Rajagopal, S. (2015). Sensing and inhibition of amyloid-β based on the simple luminescent aptamer-ruthenium complex system, *Talanta*, **134**, pp. 348–353.
47. Baird, G. S., Nelson, S. K., Keeney, T. R., Stewart, A., Williams, S., Kraemer, S., Peskind, E. R., and Montine, T. J. (2012). Age-dependent changes in the

cerebrospinal fluid proteome by slow off-rate modified aptamer array, *Am. J. Pathol.,* **180**, pp. 446–456.

48. Kumar, A., and Singh, A. (2015). A review on Alzheimer's disease pathophysiology and its management: an update, *Pharmacol. Rep.*, **67**, pp. 195–203.
49. Vassar, R. (2004). BACE1: the beta-secretase enzyme in Alzheimer's disease, *J. Mol. Neurosci.*, **23**, pp. 105–114.
50. Rentmeister, A., Bill, A., Wahle, T., Walter, J., and Famulok, M. (2006). RNA aptamers selectively modulate protein recruitment to the cytoplasmic domain of beta-secretase BACE1 in vitro, *RNA*, **12**, pp. 1650–1660.
51. Lee, S. W., and Sullenger, B. A. (1997). Isolation of a nuclease-resistant decoy RNA that can protect human acetylcholine receptors from myasthenic antibodies, *Nat. Biotechnol.*, **15**, pp. 41–45.
52. Hwang, B., and Lee, S.-W. (2002). Improvement of RNA aptamer activity against myasthenic autoantibodies by extended sequence selection, *Biochem. Biophys. Res. Commun.*, **290**, pp. 656–662.
53. Cho, J.-S., and Lee, S.-W. (2009). Sequence and structural features of RNA aptamer against myasthenic autoantibodies, *Oligonucleotides*, **19**, pp. 273–280.
54. Clarke, A. R., Jackson, G. S., and Collinge, J. (2001). The molecular biology of prion propagation, *Philos. Trans. R. Soc. Lond. B Biol. Sci.*, **356**, pp. 185–195.
55. Bone, I., Belton, L., Walker, A. S., and Darbyshire, J. (2008). Intraventricular pentosan polysulphate in human prion diseases: an observational study in the UK, *Eur. J. Neurol.*, **15**, pp. 458–464.
56. Rhie, A., Kirby, L., Sayer, N., Wellesley, R., Disterer, P., Sylvester, I., Gill, A., Hope, J., James, W., and Tahiri-Alaoui, A. (2003). Characterization of 2′-fluoro-RNA aptamers that bind preferentially to disease-associated conformations of prion protein and inhibit conversion, *J. Biol. Chem.*, **278**, pp. 39697–39705.
57. Sekiya, S., Noda, K., Nishikawa, F., Yokoyama, T., Kumar, P. K. R., and Nishikawa, S. (2006). Characterization and application of a novel RNA aptamer against the mouse prion protein, *J. Biochem.*, **139**, pp. 383–390.
58. Proske, D., Gilch, S., Wopfner, F., Schätzl, H. M., Winnacker, E.-L., and Famulok, M. (2002). Prion-protein-specific aptamer reduces PrPSc formation, *ChemBioChem,* **3**, pp. 717–725.
59. Murakami, K., Nishikawa, F., Noda, K., Yokoyama, T., and Nishikawa, S. (2008). Anti-bovine prion protein RNA aptamer containing tandem GGA

repeat interacts both with recombinant bovine prion protein and its beta isoform with high affinity, *Prion*, **2**, pp. 73–80.

60. Mashima, T., Nishikawa, F., Kamatari, Y. O., Fujiwara, H., Saimura, M., Nagata, T., Kodaki, T., Nishikawa, S., Kuwata, K., and Katahira, M. (2013). Anti-prion activity of an RNA aptamer and its structural basis, *Nucleic Acids Res.*, **41**, pp. 1355–1362.
61. Faulhammer, D., Eschgfäller, B., Stark, S., Burgstaller, P., Englberger, W., Erfurth, J., Kleinjung, F., Rupp, J., Dan Vulcu, S., Schröder, W., Vonhoff, S., Nawrath, H., Gillen, C., and Klussmann, S. (2004). Biostable aptamers with antagonistic properties to the neuropeptide nociceptin/orphanin FQ, *RNA*, **10**, pp. 516–527.
62. Wang, R. E., Wu, H., Niu, Y., and Cai, J. (2011). Improving the stability of aptamers by chemical modification, *Curr. Med. Chem.*, **18**, pp. 4126–4138.
63. Darmostuk, M., Rimpelová, S., Gbelcová, H., and Ruml, T. (2015). Current approaches in SELEX: an update to aptamer selection technology, *Biotechnol. Adv.*, **33**, pp. 1141–1161.
64. Kuwahara, M., and Sugimoto, N. (2010). Molecular evolution of functional nucleic acids with chemical modifications, *Molecules*, **15**, pp. 5423–5444.
65. Lauridsen, L. H., Rothnagel, J. A., and Veedu, R. N. (2012). Enzymatic recognition of 2′-modified ribinucleoside 5′-triphosphates: towards the evolution of versatile aptamers, *ChemBioChem*, **13**, pp. 19–25.
66. Veedu, R. N., and Wengel, J. (2009). Locked nucleic acid nucleoside triphosphates and polymerases: on the way towards the evolution of LNA aptamers, *Mol. Biosyst.*, **5**, pp. 787–792.

Chapter 7

Aptamers in Bacterial, Viral, and Parasitic Diseases

Henning Ulrich,[a] Arquimedes Cheffer,[a] Flávia M. Zimbres,[b] Attila Tárnok,[c,d] and Carsten Wrenger[b]

[a] *Department of Biochemistry, Institute of Chemistry, University of São Paulo, Av. Prof. Lineu Prestes 748, 05508-00 São Paulo-SP, Brazil*
[b] *Unit for Drug Discovery, Department of Parasitology, Institute of Biomedical Sciences, University of São Paulo, Av. Prof. Lineu Prestes 1374, 05508-000 São Paulo-SP, Brazil*
[c] *Department of Pediatric Cardiology, Heart Center, Universität Leipzig, Strümpellstraße 39, 04289 Leipzig, Germany*
[d] *Translational Centre for Regenerative Medicine (TRM) Leipzig, Philipp-Rosenthal-Str. 55, D-04103 Leipzig, Germany*
henning@iq.usp.br, cwrenger@icb.usp.br

Diseases, caused by bacteria, parasitic, protozoans and viruses belong to the principal challenges of human health. Globalization and increasing resistance of pathogens against drug intervention have led to the spread of infectious diseases. Here, we discuss the use of the aptamer technique as a novel approach for developing high-affinity ligands binding to these pathogens or to biomarkers liberated by them. Aptamers are raised by using the SELEX protocol, based on screening of a combinatorial RNA or DNA library for binding a target cell or protein. Target-binding sequences are recovered and amplified by using PCR techniques for the next cycle of in vitro selection. Reiterative SELEX cycles are

Aptamers: Tools for Nanotherapy and Molecular Imaging
Edited by Rakesh N. Veedu

ISBN 978-981-4669-83-2 (Hardcover), 978-981-4669-84-9 (eBook)
www.panstanford.com

performed until the combinatorial RNA or DNA pool is purified to a homogenous population, binding with high affinity and selectivity to the target pathogen. We will discuss the current status of the SELEX applications on pathogens, focusing on diagnostic and therapeutic approaches.

7.1 Introduction

While the world population and the resistance to chemotherapeutic agents are increasing rapidly, infectious diseases are becoming an outstanding issue. Moreover, development of new chemotherapeutic agents and optimization of diagnostic methods are nowadays very expensive, which should be taken into account in the discovery of new drugs and diagnostic tools targeting pathogenic organisms such as viruses, bacteria, and parasites. In the early 1990s, Tuerk and Gold, and Ellington and Szostak developed a technique for isolating nucleic acid ligands from a combinatorial library of over 10^{15} individual sequences. This technique is called **s**ystematic **e**volution of **l**igands by **ex**ponential enrichment (SELEX), and the selected products are called aptamers [6, 7]. Since then, DNA and RNA aptamers with affinity and specificity comparable to those obtained with monoclonal antibodies have been selected against a wide variety of targets of biological and therapeutic interest, from ions and small molecules to complex targets such as membrane receptors. The libraries employed in the SELEX technique consists of single-stranded nucleic acid molecules folded into their secondary and tertiary structures, which can be easily amplified by polymerase chain reaction (PCR) [8]. Aptamer selection is initiated by incubating the library with a chosen target molecule for a period of time, and the ligand–target complex is separated from the free nucleic acid sequences. The bounded oligonucleotides are then eluted and amplified by PCR. This new enriched pool is used in the next selection and amplification cycles. In general, it takes from 6 to 20 cycles for the selection of high-affinity and specificity aptamers. In practical terms, cycles are performed until the target affinity for the library stabilizes. Negative selection should be performed in order to minimize enrichment of sequences, which bind to other molecules than the SELEX target. This is particularly important in the case of not using highly pure target molecules, when binding to different or contaminating molecules can be a problem. After the last selection cycle, the obtained aptamers are cloned and sequenced for identifying consensus motifs and secondary/tertiary structures, which are involved in interaction with the target (Fig. 7.1) [9]. Lots of different conformations can be identified, among them, stem-loop structures. Although DNA stem-loops are slightly less stable in comparison to RNA loops, DNA aptamers have greater flexibility, and consequently, they can assume a higher diversity of structural conformations. This can be explained by the lack of the ribose 2′-hydroxyl group in DNA [10]. Different from DNA aptamers, for RNA aptamer selection it is necessary to introduce into the process a reverse transcription step before amplification by PCR and to use modified nucleotides (e.g., replacement of the 2′-OH group of ribose by 2′-amino, 2′-fluoride, or O-methylene functions) in order to increase nuclease resistance and, therefore, to increase their half-life in biological fluids [11, 12]. In contrast to RNA aptamers, DNA aptamers are more stable in several applications, and they do not need to be modified previously. Additionally, DNA aptamers can be easily labeled with reporter molecules such as fluorescence dyes or biotin and employed both in the visualization of the target through fluorescence and in pull-down experiments with streptavidin-coated beads [13–16]. It should be mentioned that the binding properties of an aptamer depend on its sequence and, subsequently, the generated folded stem-loop structures [13, 17]. As aforementioned, the aptamers have affinities and specificities to their selection targets comparable to those of monoclonal antibodies and their interaction with the targets reveals dissociation constant values in the range from nanomolar to picomolar [9, 18].

Besides, the obtained specificity can be so high that the aptamer is capable to differentiate among protein isoforms as well as distinct conformations of the same protein [19, 20]. Moreover, just a temperature change can be used to denature and renature the aptamers. Similarly to animal-derived antibodies, they can also be developed by an iterative in vitro procedure [16].

Hereafter, DNA- and RNA-based aptamers are likely to replace antibodies in a variety of applications, since they have great advantages over the latter, such as their stability, nonpeptide character, and independency from animal resources. Finally, the power and versatility of aptamers have already been exploited in the development of therapeutic and diagnostic tools targeting infectious diseases, as outlined below.

7.2 Aptamers for Combating Parasitic Infections

Malaria represents one of the most serious threats to mankind with an enormous impact on the mortality and morbidity, especially in sub-Saharan Africa. Malaria leads to about 500 million clinical cases each year and more than half a million people are dying annually. Currently, a vaccine is not available and the disease treatment and control solely depends on chemotherapeutics [21]. Due to the high mutational rate of the parasite and its resulting rapid adaptation to environmental changes, drug resistance and its geographic distribution are increasing. In the past, chemotherapeutics were designed to target metabolic pathways, hemoglobin degradation and hemozoin formation. Indeed hemozoin formation was significantly inhibited by heme-binding aptamers in parasite extracts. Even further, a growth inhibitory effect could be shown in aptamer-preloaded human erythrocytes [22]. So far in vivo assays have yet not been performed but in future experiments these aptamers could be evaluated as drugs. RNA aptamers have also been generated against the major parasite-derived surface protein, *Plasmodium falciparum* erythrocyte membrane protein 1 (*Pf*EMP1) [23], which is embedded in the erythrocytic membrane. The plasmodial EMP1-protein is mediating cytoadherence to human cells, such as endothelial cells of blood vessels or in spontaneous binding of uninfected erythrocytes to *P. falciparum*-infected red blood cells (rosetting). The pathogen hosts about 60 *var*-genes and is able to switch the expression from one gene to another *var*-gene, leading to a different *Pf*EMP1 protein. By doing this the parasite can avoid interference with the host immune response (antigenic variation) [24, 25]. Barfod and coworkers were able to test the selected aptamers against the recombinantly expressed N-terminal DUFFY-binding-like domain (DBL1a) of the *Pf*EMP1 protein; afterward the isolated aptamers were tested at the cellular level and found to inhibit the rosetting progress of the malarial parasite [23]. An additional approach could also focus on aptamer application at the cellular level to selectively target all secreted surface proteins of the parasite by applying the cell-SELEX technique as illustrated in Fig 7.1.

The cell-SELEX method has already been applied to *Leishmania*. The gene expression and organization in *Leishmania* have been

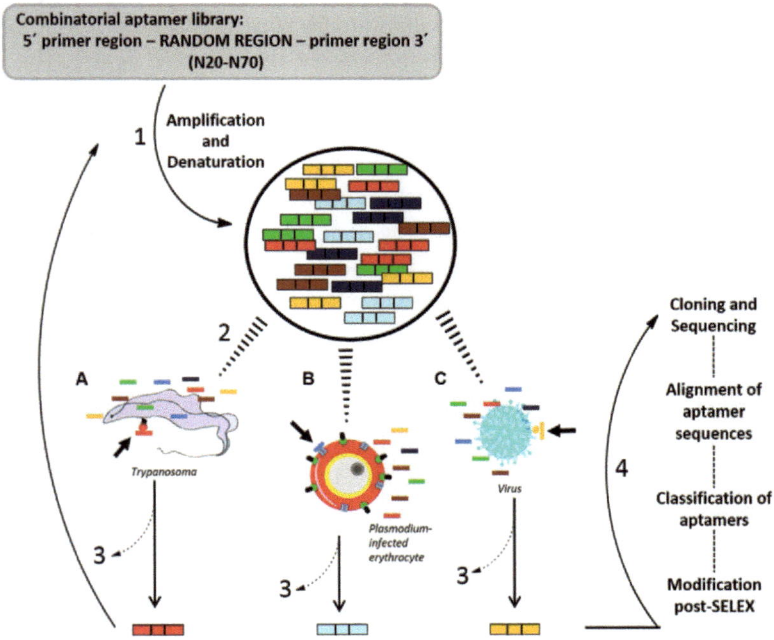

Figure 7.1 Schematic illustration of the cell-SELEX technique. The cell-SELEX method has been illustrated on (A) *Trypanaosoma*, (B) *Plasmodium*-infected erythrocyte, and (C) viruses. The combinatorial aptamer library consisting of partially randomized DNA sequences (inner random region flanked on both sites by constant sequences in the 5′ and 3′ region, N20–N70) is amplified by conventional PCR. (1) Subsequently, the derived double-stranded DNA is denatured, separated into single-stranded DNA, and applied onto the respective pathogens (2) (A–C). After several selection cycles (3), the final aptamer library is cloned and sequenced (4). By applying bioinformatics tools the aptamer consensus sequences are identified and classified (4). Post-SELEX modifications such as truncations, stabilizations, and covalent attachment of fluorescence reporters are optionally applicable (4).

analyzed recently [26, 27]. It is known that *Leishmania* does not condense its chromatin during mitosis, which might be caused by the respective histones. Normally histones are highly conserved proteins; sequence divergences have been identified in the N- and C-terminal protein regions which have been suggested as targets for diagnostic and/or therapeutic intervention. Although histone proteins reveal sequence homology Ramos et al. developed aptamers against the recombinantly expressed histone antigens H2A and H3 and the generated aptamers were highly selective solely for the parasite proteins. Aptamers were also directed against the *L. infantum* kinetoplastid membrane protein-11 (KMP-11), which is a major component of the cell membrane of kinetoplastid parasites [28]. Additionally the SELEX technique was applied for analyzing the tRNA import into the mitochondrion of *L. tropica* [29]. However, approaches toward the discovery of novel therapeutics have not yet been carried out due to missing biological activity of the selected aptamers or difficulties in accessing target proteins.

Trypanosomiasis is caused by the tropical protozoan parasite *Trypanosoma.* Due to its medical importance the causative agent of Chagas' disease, *Trypanosoma cruzi,* as well as African trypanosomes, *T. brucei ssp.*, which are responsible for sleeping sickness and Nagana in cattle, novel therapeutics and diagnostics need to be discovered. Consequently, the aptamer technique was also applied to *T. brucei,* which undergoes antigenic variation of its variant surface glycoproteins (VSGs) [30]. Three aptamer families were isolated to surface proteins of African trypanosomes with binding affinities in the nanomolar and subnanomolar range [31]. As unmodified RNA aptamers are promptly degraded in biological fluids, these cannot be used for in vivo studies. However, substitution of the 2′-OH groups of ribose residue of nucleotide by amino-fluoro-functions augments their half-life from a few seconds to several hours [9]. Such chemical modifications provided the anti–*T. brucei* aptamer with nuclease resistance [31].

As the VSG coat of *T. brucei* is variable and constantly modified for escaping the host immune response of the human host, invariant surface glycoproteins (ISGs), which comprise receptor complexes and transporters [32], were targeted for the development of RNA aptamers (Fig. 7.1). The selected RNA aptamers against live trypanosomes recognized a single 42 kDa protein 1 within the flagellar pocket of the parasite. This protein was identified as ESAG 7, a transferrin-receptor subunit [33]. The obtained results were a breakthrough, since aptamer binding of an invariant trypanosomal surface protein provided selective ligands for the non-VSG coat. Moreover, Homann and Goringer showed that with temperature increase, terminal ends of the aptamer degraded, with rapid uptake into the parasite, and were transported into the lysosome [34]. The authors concluded that the specific RNAs could provide the so-called piggy-back molecules for the transport of aptamer-coupled drugs into the lysosomal compartment of the parasite and provide a novel tool for fighting African trypanosomiasis. The SELEX technique was used as well for identifying aptamer ligands against the causative agent of Chagas' disease. *T. cruzi* needs to invade host cells for completing its life cycle. Parasite–host cell adhesion has been shown to be essential for the invasion process [35]. Host cell matrizes, such as laminin, thrombospondin, heparin sulfate, and fibronectin, have been suggested to mediate parasite–host cell contact [36, 37]. In view of that, the hypothesis was that compounds blocking parasite host cell receptor interactions would inhibit host cell invasion by *T. cruzi*. The SELEX technique was employed for the development of stable RNA aptamers blocking binding of infective *T. cruzi* trypomastigotes to epithelial monkey kidney LLC-MK(2) cells [36]. Aptamers were identified with binding affinities in the nano-molar range to parasite receptors expressed by infective trypomastigote and not by insect epimastigote forms for the host cell matrix molecules fibronectin, heparin sulfate, laminin, and thrombospondin [35, 36]. Aptamers were identified, which were effective in diminishing the cell infection rate by *T. cruzi* trypomastigotes in vitro of up to 70%. Later, an aptamer-based assay was developed for quantification of parasites in the blood of infected individuals [38]. The aptamers were coupled to biotin. Using streptavidin-paramagnetic beads, a low number of parasites, such as 5 trypanosomes in 15 ml blood, could be detected. The realized work provides a major breakthrough regarding the development of diagnostic assays for chronic Chagas' disease, where the number of circulating parasites is very low. The same group developed an enzyme-linked aptamer assay for detecting and quantifying parasite biomarkers, such as *T. cruzi* excreted secreted antigens, released into the bloodstream [39].

7.3 Aptamers and Bacterial Infections

Millions of people worldwide are affected by tuberculosis or multidrug-resistant bacterial strains such as *Staphylococcus aureus* or *Salmonella enterica*, which make these diseases a public health problem. Although there is one vaccine against tuberculosis, which consists of an attenuated strain of *Mycobacterium bovis*, bacillus Calmette–Guėrin (BCG), its efficacy is limited to tuberculosis endemic regions [40]. In 2007 Chen et al. [41] obtained an aptamer called NK2, which binds with high affinity and specificity to membrane proteins present on the surface of the virulent *M. tuberculosis* strain H37Rv, not existing on BCG. Moreover, the interaction between the NK2 aptamer and H37Rv increased the production and release of IFN-γ by CD4+T cells, which is a protective cytokine against *M. tuberculosis* infections. When the NK2 aptamers-treated mice spleen was analyzed, a reduced bacterial number as compared to the control was observed [41]. Taken together, these data demonstrate the potency of the NK2 aptamer as an antimycobacterial agent.

Salmonellae are also responsible for another common bacterial infection, and they are especially involved in food-borne illness. About 25% of all food-borne diarrhea cases require patient hospitalization. The multidrug resistant strains *Salmonella enterica* serovar Typhimurium DT104 and *S. enterica* serovar Newport cause a still more challenging salmonellosis [42]. *Salmonellae* can be found in several food (intensive livestock farming) animal species such as cows, pigs, chickens, and turkeys [43–46]. DNA aptamers were selected against outer membrane proteins (OMPs) derived from *S. enterica* serovar Typhimurium multidrug-resistant) strains. Simultaneously, negative selection steps were performed by using *Escherichia coli* OMPs and lipopolysaccharides (LPSs) as targets. Furthermore, the selected aptamers were shown to interact only with the *S. enterica* serovar Typhimurium OMPs, proving the potential of aptamers targeting *S. enterica* serovar Typhimurium multidrug resistant strains [47].

Another interesting approach would be to develop aptamers, which interfere with the formation and accessibility of the type IVB pili, since the IVB piliated *S. enterica* serovar Typhimurium A21-6 invades the human THP-1 monocytic cells more effectively than its counterpart the nonpiliated *S. enterica* serovar Typhimurium mutant strain does. Therefore, RNA aptamers were developed, which specifically interact with the pili and prevent the human THP-1 cells from being infected by the bacterium [48]. Aptamer structure analysis revealed a stem-loop, which is probably the binding site for interaction with the pili [30]. Summing up, these RNA aptamers are capable to inhibit the bacterial pathogenesis.

Other important bacteria species are *S. aureus* and *E. coli*, which naturally occur in humans. For example, around 20% of healthy adults are permanently colonized by *S. aureus*. This number rises to up to 50% in the case of transient colonization. Despite this natural occurrence, *S. aureus* can take advantage of an immune-suppressed patient and cause opportunistic infections, including life-threatening conditions such as pneumonia and endocarditis. Here, this pathogenic agent stands out as a relevant human pathogen due to methicillin-resistant *S. aureus* (MRSA) strains [49]. Besides producing potent protein toxins, these bacteria express on their surface proteins capable to bind to antibodies and, thus, to inactivate them [50]. It is well known about the existence of different sets of molecules, which precisely control bacteria proliferation at different growth states [51, 52]. It has been also demonstrated that they even undergo antigenic variations to protect themselves from the host immune system [53, 54]. Cao and coworkers have developed a whole set of specific markers for dealing with these bacterial mechanisms [55]. High-specificity and sensitivity DNA aptamers were obtained and grouped in different families regarding to their sequence homology and similarity in the predicted secondary structures. Furthermore, five of these aptamers recognized distinct molecular targets, and a more sensitive identification of different *S. aureus* strains was obtained with a combination of them rather than by using a single aptamer. These data show how a pool of specific aptamers can be employed as a highly selective diagnostic tool or to inhibit the growth of a pathogenic agent [55].

7.4 Aptamers in Viral Infections

The hepatitis C virus (HCV) is responsible for causing hepatitis C, an infectious disease that affects about 3% of the world population, leading to hepatic dysfunction conditions, like cirrhosis, in 80% of the cases. Until today there is no efficient vaccine against this infectious disease, whose treatment is based on the administration of alpha interferon (INF-α) alone or combined with ribavirin [56–58]. Nevertheless, the treatment efficacy is limited, the treatment is costly, and there is the risk of dangerous side effects [56, 57, 59]. As a strategy for the discovery of new modes of detection and to selectively interfere with virus proliferation, the alive cell surface–SELEX (CS-SELEX) technique has been developed and furthermore used to encounter single-stranded DNA aptamers targeting the HCV envelope surface glycoprotein E2 [60]. One of these aptamers has already been identified and named ZE2. One believes that the ZE2 aptamer specifically binds to the HCV-E2 envelope glycoprotein and blocks its interaction with CD81, an important HCV receptor, which, in turn, protects the human hepatocytes from the infection by HCV. Therefore, the ZE2 aptamer could be used as both a diagnostic and a therapeutic tool in HCV infections.

Another target for selection of aptamers is the avian influenza virus (AIV) H5N1, also known as an agent causing bird flu. H5N1 is a type A influenza virus, and it is responsible for major epidemics and pandemics. Using a random single-stranded DNA library and purified AIV H5N1 hemagglutinin (HA) and the whole H5N1 virus as targets, Wang et al. [61] carried out the first rounds of selection cycles, as illustrated in Fig. 7.1. After several positive target and negative selection cycles with other AIV subtypes (H5N2, H5N3, H5N9, H7N2, H2N2, and H9N2), aptamers were selected, cloned, and sequenced. Through consensus sequence analysis, secondary structures (hairpin loops and bulge loops) were predicted that are likely to play an important role in the interaction with the target.

Surface plasmon resonance identified one aptamer that binds to the H5N1 virus with a dissociation constant of 4.65 nM.

The rabies virus (RABV) causes a zoonotic disease named rabies, which is transmitted through contact with infected saliva during a bite or through direct contact with mucosal tissues. The viral infection leads to an acute fatal encephalitis, resulting in coma and death. Various warm-blooded mammals are affected by this infectious disease. To date, there is no available approved therapy against the clinical signs of the disease. As the infection is lethal in all the cases, the development of a cheap and effective drug has drawn attention. The aptamers represent, in this sense, a promising alternative. Indeed, aptamers have already been selected against RABV-infected cells by using the cell-SELEX technique [62]. These aptamers were posteriorly employed in viral titer assays, demonstrating that viral replication is inhibited in RABV-infected cells while no blockade of the canine distemper virus or canine parvovirus replication was observed, which confirms the specificity of the selected aptamers. Most importantly, the aptamers have been demonstrated to protect mice to some extent from RABV infection. Interestingly, the aptamers had a protective effect, since only circa 15% of the previously aptamer-treated animals died after inoculation with CVS-11. On the other hand, almost no mice survived when the aptamers were used for treatment [62].

Similar to other pathogens, drug resistance has also been reported in HIV infections. It is known that HIV contains various small RNA sequences that can strongly and specifically bind to viral or cellular proteins. Functional studies raised the possibility to exploit these viral RNA–protein complexes in therapeutic approaches, as it has been demonstrated for small HIV RNA regions called TAR, which might be used to block viral replication in cellular models [63]. The interaction between the envelope glycoprotein gp120 of the HIV-1 virus and its CD4 receptor in host cells plays a key role in the cellular invasion process, which makes gp120 a potential target for the selection of aptamers. In fact, RNA aptamers targeting the gp120 glycoprotein have already been generated [64, 65]. Through in vivo assays with a humanized mouse model that mimics HIV-1 replication and T-cell depletion as occurs in humans, it has been found that the anti-gp120 aptamer inhibits viral replication and protects T-cells from viral-induced decrease [66].

High-affinity and specificity aptamers have been also discovered that bind to several epitopes of the viral transcriptase reverse, an enzyme that catalyzes an essential step for viral replication [67–69]. Some of these aptamers have drawn attention due to their capability to interact with the HIV-1 reverse transcriptase at the nanomolar range, to inhibit its catalytic activity and, subsequently, to block viral replication in cell culture [68, 70–74]. Taken together, these data clearly demonstrate how aptamers, targeting HIV-1 transcriptase reverse, could be exploited as promising new therapeutic tools.

7.5 Conclusion

Aptamers have the ability to selectively identify structures in a mixture of potential targets and can also be used for ligand-mediated target purification [13, 16]. Even further aptamers can distinguish among small differences in the molecular signature of cell surface antigens [75], which underline the potency of the technique at the cellular level (Fig. 7.1). As outlined above these highly specific aptamers are capable to bind and thereby inhibit the functioning of pathogen-derived proteins in order to interfere with the proliferation of deadly human infective pathogens.

Acknowledgments

This work was financially supported by grants 2009/54325-2, 2013/10288-1, and 2011/19703-6, as well as the German-Brazilian Network project (grant 2012/50393-6) of the Fundação de Amparo à Pesquisa do Estado de São Paulo and the German Federal Ministry of Education, the Research Translational Center for Regenerative Medicine (TRM) Leipzig (BMBF, PtJ-Bio, 1315883), and the Mobilitätsprojekt Gesundheitsforschung Brasilien (grant BRA 11/081).

References

1. Shi, G., Yagyu, F., Shimizu, Y., Shimizu, K., et al. (2011). Flow cytometric assay using two fluorescent proteins for the function of the internal ribosome entry site of hepatitis C virus, *Cytometry A*, **79**, pp. 653–660.
2. Gondois-Rey, F., Granjeaud, S., Kieu Sle, T., Herrera, D., et al. (2012). Multiparametric cytometry for exploration of complex cellular dynamics, *Cytometry A*, **81**, pp. 332–342.
3. Pollara, J., Hart, L., Brewer, F., Pickeral, J., et al. (2011). High-throughput quantitative analysis of HIV-1 and SIV-specific ADCC-mediating antibody responses, *Cytometry*, **79**, pp. 603–612.
4. Mateos-Perez, J. M., Redondo, R., Nava, R., Valdiviezo, J. C., et al. (2012). Comparative evaluation of autofocus algorithms for a real-time system for automatic detection of Mycobacterium tuberculosis, *Cytometry A*, **81**, pp. 213–221.
5. Ruger, M., Bensch, G., Tungler, R., and Reichl, U. (2012). A flow cytometric method for viability assessment of Staphylococcus aureus and Burkholderia cepacia in mixed culture, *Cytometry A*, **81**, pp. 1055–1066.
6. Ellington, A. D., and Szostak, J. W. (1990). In vitro selection of RNA molecules that bind specific ligands, *Nature*, **346**, pp. 818–822.
7. Tuerk, C., and Gold, L. (1990). Systematic evolution of ligands by exponential enrichment: RNA ligands to bacteriophage T4 DNA polymerase, *Science*, **249**, pp. 505–510.
8. Ulrich, H., Martins, A. H., and Pesquero, J. B. (2005). RNA and DNA aptamers in cytomics analysis, *Curr. Protoc. Cytom.*, pp. 1–39.
9. Ulrich, H., Martins, A. H., and Pesquero, J. B. (2004). RNA and DNA aptamers in cytomics analysis, *Cytometry A*, **59**, pp. 220–231.
10. Harada, K., and Frankel, A. D. (1995). Identification of two novel arginine binding DNAs, *EMBO J.*, **14**, pp. 5798–5811.
11. Davydova, A. S., Vorobjeva, M. A., and Venyaminova, A. G. (2011). Escort aptamers: new tools for the targeted delivery of therapeutics into cells, *Acta Nat.*, **3**, pp. 12–29.
12. Ng, E. W., Shima, D. T., Calias, P., Cunningham, E. T., Jr., et al. (2011). Pegaptanib, a targeted anti-VEGF aptamer for ocular vascular disease, *Nat. Rev. Drug Discovery*, **5**, pp. 123–132.
13. Nery, A. A., Wrenger, C., and Ulrich, H. (2009). Recognition of biomarkers and cell-specific molecular signatures: aptamers as capture agents, *J. Sep. Sci.*, **32**, pp. 1523–1530.

14. Murphy, M. B., Fuller, S. T., Richardson, P. M., and Doyle, S. A. (2003). An improved method for the in vitro evolution of aptamers and applications in protein detection and purification, *Nucleic Acids Res.,* **31**, p. e110.
15. Rybak, J. N., Scheurer, S. B., Neri, D., and Elia, G. (2004). Purification of biotinylated proteins on streptavidin resin: a protocol for quantitative elution, *Proteomics*, **4**, pp. 2296–2299.
16. Ulrich, H., and Wrenger, C. (2009). Disease-specific biomarker discovery by aptamers, *Cytometry A*, **75**, pp. 727–733.
17. Famulok, M., Mayer, G., and Blind, M. (2000). Nucleic acid aptamers-from selection in vitro to applications in vivo, *Acc. Chem. Res.*, **33**, pp. 591–599.
18. Tombelli, S., Minunni, M., and Mascini, M. (2005). Analytical applications of aptamers, *Biosens. Bioelectron.,* **20**, pp. 2424–2434.
19. Conrad, R., and Ellington, A. D. (1996). Detecting immobilized protein kinase C isozymes with RNA aptamers, *Anal. Biochem.,* **242**, pp. 261–265.
20. Sekiya, S., Nishikawa, F., Noda, K., Kumar, P. K., et al. (2005). In vitro selection of RNA aptamers against cellular and abnormal isoform of mouse prion protein, *Nucleic Acids Symp. Ser.*, pp. 361–362.
21. Greenwood, B. (2009). Can malaria be eliminated? *Trans. R. Soc. Trop. Med. Hyg.*, **103**(Suppl 1), pp. S2–S5.
22. Niles, J. C., Derisi, J. L., and Marletta, M. A. (2009). Inhibiting plasmodium falciparum growth and heme detoxification pathway using heme-binding DNA aptamers, *Proc. Natl. Acad. Sci. U S A,* **106**, pp. 13266–13271.
23. Barfod, A., Persson, T., and Lindh, J. (2009). In vitro selection of RNA aptamers against a conserved region of the Plasmodium falciparum erythrocyte membrane protein 1, *Parasitol. Res.*, **105**, pp. 1557–1566.
24. Buffet, P. A., Safeukui, I., Deplaine, G., Brousse, V., et al. (2011). The pathogenesis of Plasmodium falciparum malaria in humans: insights from splenic physiology, *Blood,* **117**, pp. 381–392.
25. Flick, K., and Chen, Q. (2004). var genes, PfEMP1 and the human host, *Mol. Biochem. Parasitol.,* 134, pp. 3–9
26. Ramos, E., Piñeiro, D., Soto, M., Abanades, D. R., et al. (2007). A DNA aptamer population specifically detects Leishmania infantum H2A antigen, *Lab Invest.*, **87**, pp. 409–416
27. Ramos, E., Moreno, M., Martín, M. E., Soto, M., and Gonzalez, V. M. (2010). In vitro selection of Leishmania infantum H3-binding ssDNA aptamers, *Oligonucleotides*, **20**, pp. 207–213.

28. Moreno, M., Rincón, E., Piñeiro, D., Fernández, G., et al. (2003). Selection of aptamers against KMP-11 using colloidal gold during the SELEX process, *Biochem. Biophys. Res. Commun.*, **308**, pp. 214–218.

29. Bhattacharyya, S. N., Chatterjee, S., and Adhya, S. (2002). Mitochondrial RNA import in Leishmania tropica: aptamers homologous to multiple tRNA domains that interact cooperatively or antagonistically at the inner membrane, *Mol. Cell. Biol.*, **22**, pp. 4372–4382.

30. Graham, S. V., and Barry, J. D. (1995). Transcriptional regulation of metacyclic variant surface glycoprotein gene expression during the life cycle of Trypanosoma brucei, *Mol. Cell. Biol.*, **15**, pp. 5945–5956.

31. Lorger, M., Engstler, M., Homann, M., and Göringer, H. U. (2003). Targeting the variable surface of African trypanosomes with variant surface glycoprotein-specific, serum-stable RNA aptamers, *Eukaryot. Cell*, **2**, pp. 84–94.

32. Overath, P., Chaudhri, M., Steverding, D., and Ziegelbauer, K. (1994). Invariant surface proteins in bloodstream forms of Trypanosoma brucei, *Parasitol. Today*, **10**, pp. 53–58.

33. Homann, M., and Göringer, H. U. (1999). Combinatorial selection of high affinity RNA ligands to live African trypanosomes, *Nucleic Acids Res.*, **27**, pp. 2006–2014.

34. Göringer, H. U., Homann, M., and Lorger, M. (2003). In vitro selection of high-affinity nucleic acid ligands to parasite target molecules, *Int. J. Parasitol.*, **33**, pp. 1309–1317.

35. Alves, M. J., and Colli, W. (2008). Role of the gp85/trans-sialidase superfamily of glycoproteins in the interaction of Trypanosoma cruzi with host structures, *Subcell Biochem.*, **47**, pp. 58–69.

36. Ulrich, H., Magdesian, M. H., Alves, M. J., and Colli, W. (2002). In vitro selection of RNA aptamers that bind to cell adhesion receptors of Trypanosoma cruzi and inhibit cell invasion, *J. Biol. Chem.*, **277**, pp. 20756–20762.

37. Simmons, K. J., Nde, P. N., Kleshchenko, Y. Y., Lima, M. F., and Villalta, F. (2006). Stable RNA interference of host thrombospondin-1 blocks Trypanosoma cruzi infection, *FEBS Lett.*, **580**, pp. 2365–2370.

38. Nagarkatti, R., Bist, V., Sun, S., de Araujo, F. F., et al. (2012). Development of an aptamer-based concentration method for the detection of Trypanosoma cruzi in blood, *PLoS One,*, **7**, p. e43533.

39. Nagarkatti, R., de Araujo, F. F., Gupta, C., and Debrabant, A. (2014). Aptamer based, non-PCR, non-serological detection of Chagas disease biomarkers in Trypanosoma cruzi infected mice, *PLoS Negl. Trop. Dis.*, **8**, p. e2650.

40. Bloom, B., and Fine, P. (1994). The BCG experience: implications for future vaccines against tuberculosis, in Bloom, B. R. (Ed.), *Tuberculosis: Pathogenesis, Protection and Control* ASM Press, Washington DC, pp. 531–557.
41. Chen, F., Zhou, J., Luo, F., Mohammed, A. B., and Zhang, X. L. (2007). Aptamer from whole-bacterium SELEX as new therapeutic reagent against virulent Mycobacterium tuberculosis, *Biochem. Biophys. Res. Commun.*, **357**, pp. 743–748.
42. Gupta, A., Fontana, J., Crowe, C., Bolstorff, B., et al. (2003). Emergence of multidrug-resistant Salmonella enterica serotype Newport infections resistant to expanded-spectrum cephalosporins in the United States, *J. Infect. Dis.*, **188**, pp. 1707–1716.
43. Wells, S. J., Fedorka-Cray, P. J., Dargatz, D. A., Ferris, K., and Green, A. (2001). Fecal shedding of Salmonella spp. by dairy cows on farm and at cull cow markets, *J. Food Prot.*, **64**, pp. 3–11.
44. Malorny, B., Hoorfar, J. (2005). Toward standardization of diagnostic PCR testing of fecal samples: lessons from the detection of salmonellae in pigs, *J. Clin. Microbiol.*, **43**, pp. 3033–3037.
45. Carli, K. T., Eyigor, A., and Caner, V. (2001). Prevalence of Salmonella serovars in chickens in Turkey, *J. Food Prot.*, **64**, pp. 1832–1835.
46. Nayak, R., Kenney, P. B., Keswani, J., and Ritz, C. (2003). Isolation and characterisation of Salmonella in a turkey production facility, *Br. Poultry Sci.*, **44**, pp. 192–202.
47. Joshi, R., Janagama, H., Dwivedi, H. P., Senthil Kumar, T. M., et al. (2009). Selection, characterization, and application of DNA aptamers for the capture and detection of Salmonella enterica serovars, *Mol. Cell. Probes*, **23**, pp. 20–28.
48. Pan, Q., Zhang, X. L., Wu, H. Y., He, P. W., et al. (2005). Aptamers that preferentially bind type IVB pili and inhibit human monocytic-cell invasion by Salmonella enterica serovar typhi, *Antimicrob. Agents Chemother.*, **49**, 4052–4060.
49. Lowy, F. D. (1998). Staphylococcus aureus infections, *New Engl. J. Med.*, **339**, pp. 520–532.
50. Qtaishat, N. M., Gussin, H. A., and Pepperberg, D. R. (2013). Cysteine-terminated B-domain of Staphylococcus aureus protein A as a scaffold for targeting GABA(A) receptors, *Anal. Biochem.*, **432**, pp. 49–57.
51. Beier, D., and Gross, R. (2006). Regulation of bacterial virulence by two-component systems, *Curr. Opinion Microbiol.*, *9*, pp. 143–152.

52. Bronner, S., Monteil, H., and Prevost, G. (2004). Regulation of virulence determinants in Staphylococcus aureus: complexity and applications, *FEMS Microbiol. Rev.*, **28**, pp. 183–200.
53. Loughman, A., Sweeney, T., Keane, F. M., Pietrocola, G., et al. (2008). Sequence diversity in the A domain of Staphylococcus aureus fibronectin-binding protein A, *BMC Microbiol.*, **8**, p. 74.
54. van der Woude, M. W., and Baumler, A. J. (2004). Phase and antigenic variation in bacteria, *Clin. Microbiol. Rev.*, **17**, pp. 581–611, table of contents.
55. Cao, X., Li, S., Chen, L., Ding, H., et al. (2009). Combining use of a panel of ssDNA aptamers in the detection of Staphylococcus aureus, *Nucleic Acids Res.*, **37**, pp. 4621–4628.
56. Bellecave, P., and Moradpour, D. (2008). A fresh look at interferon-alpha signaling and treatment outcomes in chronic hepatitis C, *Hepatology*, **48**, pp. 1330–1333.
57. Boyer, N., and Marcellin, P. (2000). Pathogenesis, diagnosis and management of hepatitis C, *J. Hepatol.*, **32**, pp. 98–112.
58. Liu, C. H., Liu, C. J., Lin, C. L., Liang, C. C., et al. (2008). Pegylated interferon-alpha-2a plus ribavirin for treatment-naive Asian patients with hepatitis C virus genotype 1 infection: a multicenter, randomized controlled trial, *Clin. Infect. Dis.*, **47**, pp. 1260–1269.
59. Sarasin-Filipowicz, M., Oakeley, E. J., Duong, F. H., Christen, V., et al. (2008). Interferon signaling and treatment outcome in chronic hepatitis C, *Proc. Natl. Acad. Sci. U S A*, **105**, pp. 7034–7039.
60. Chen, F., Hu, Y., Li, D., Chen, H., and Zhang, X. L. (2009). CS-SELEX generates high-affinity ssDNA aptamers as molecular probes for hepatitis C virus envelope glycoprotein E2, *PLoS One*, **4**, p. e8142.
61. Wang, R., Zhao, J., Jiang, T., Kwon, Y. M., et al. (2013). Selection and characterization of DNA aptamers for use in detection of avian influenza virus H5N1, *J. Virol. Methods*, **189**, pp. 362–369.
62. Liang, H. R., Liu, Q., Zheng, X. X., Gai, W. W., et al. (2013). Aptamers targeting rabies virus-infected cells inhibit viral replication both in vitro and in vivo. *Virus Res.*, **173**, pp. 398–403.
63. Sullenger, B. A., Gallardo, H. F., Ungers, G. E., and Gilboa, E. (1990). Overexpression of TAR sequences renders cells resistant to human immunodeficiency virus replication, *Cell*, **63**, pp. 601–608.
64. Dey, A. K., Khati, M., Tang, M., Wyatt, R., et al. (2005). An aptamer that neutralizes R5 strains of human immunodeficiency virus type 1 blocks gp120-CCR5 interaction, *J. Virol.*, **79**, pp. 13806–13810.

65. Khati, M., Schuman, M., Ibrahim, J., Sattentau, Q., et al. (2003). Neutralization of infectivity of diverse R5 clinical isolates of human immunodeficiency virus type 1 by gp120-binding 2′F-RNA aptamers, *J. Virol.*, **77**, pp. 12692–12698.
66. Neff, C. P., Zhou, J., Remling, L., Kuruvilla, J., et al. (2011). An aptamer-siRNA chimera suppresses HIV-1 viral loads and protects from helper CD4(+) T cell decline in humanized mice, *Sci. Transl. Med.*, **3**, p. 66ra66.
67. Fisher, T. S., Joshi, P., and Prasad, V. R. (2002). Mutations that confer resistance to template-analog inhibitors of human immunodeficiency virus (HIV) type 1 reverse transcriptase lead to severe defects in HIV replication, *J. Virol.*, **76**, pp. 4068–4072.
68. Held, D. M., Kissel, J. D., Saran, D., Michalowski, D., and Burke, D. H. (2006). Differential susceptibility of HIV-1 reverse transcriptase to inhibition by RNA aptamers in enzymatic reactions monitoring specific steps during genome replication, *J. Biol. Chem.*, **281**, pp. 25712–25722.
69. Jaeger, J., Restle, T., and Steitz, T. A. (1998). The structure of HIV-1 reverse transcriptase complexed with an RNA pseudoknot inhibitor, *EMBO J.*, **17**, pp. 4535–4542.
70. Burke, D. H., Scates, L., Andrews, K., and Gold, L. (1996). Bent pseudoknots and novel RNA inhibitors of type 1 human immunodeficiency virus (HIV-1) reverse transcriptase, *J. Mol. Biol.*, **264**, pp. 650–666.
71. Joshi, P., and Prasad, V. R. (2002). Potent inhibition of human immunodeficiency virus type 1 replication by template analog reverse transcriptase inhibitors derived by SELEX (systematic evolution of ligands by exponential enrichment), *J. Virol.*, **76**, pp. 6545–6557.
72. Tuerk, C., MacDougal, S., and Gold, L. (1992). RNA pseudoknots that inhibit human immunodeficiency virus type 1 reverse transcriptase, *Proc. Natl. Acad. Sci. U S A*, **89**, pp. 6988–6992.
73. Chaloin, L., Lehmann, M. J., Sczakiel, G., and Restle, T. (2002). Endogenous expression of a high-affinity pseudoknot RNA aptamer suppresses replication of HIV-1, *Nucleic Acids Res.*, **30**, pp. 4001–4008.
74. Lange, M. J., Sharma, T. K., Whatley, A. S., Landon, L. A., et al. (2012). Robust suppression of HIV replication by intracellularly expressed reverse transcriptase aptamers is independent of ribozyme processing, *Mol. Ther.*, **20**, pp. 2304–2314.
75. Ulrich, H., and Wrenger, C. (2013). Identification of aptamers as specific binders and modulators of cell-surface receptor activity, *Methods Mol. Biol.*, **986**, pp. 17–39.

Chapter 8

Aptamers as Tools for Targeted Drug Delivery

Anila Mathew and D. Sakthi Kumar

Graduate School of Interdisciplinary New Science, Bio-Nano Electronics Research Center, Toyo University, 2100 Kujirai, Kawagoe, Saitama 350-8585, Japan

sakthi@toyo.jp

Targeted drug delivery has revolutionized therapeutic strategies in the recent years. Aptamers are small, versatile targeting moieties that find applications in various fields. Aptamers can themselves act as therapeutics or can be used as targeting ligands. Aptamers can act as targeting agents to carry drugs, nanoparticles, small RNAs, and other small ligands. Compared to other targeting moieties like antibodies, aptamers have superlative properties that make their application easy and efficient. The past three decades have seen tremendous advances in the development and use of aptamers for various applications in diverse fields.

In this chapter, we are discussing various applications of aptamers in targeted drug delivery. A brief description of aptamer selection by systematic evolution of ligands by exponential enrichment (SELEX) is also outlined. Aptamers are used as therapeutic moieties, or can be used for targeting drugs, drug-encapsulated nanoparticles,

Aptamers: Tools for Nanotherapy and Molecular Imaging
Edited by Rakesh N. Veedu

ISBN 978-981-4669-83-2 (Hardcover), 978-981-4669-84-9 (eBook)
www.panstanford.com

or other therapeutic moieties. These different approaches are discussed based on the latest research done using aptamers.

8.1 Introduction

Developments in nanoscience have revolutionized the field of nanomedicine. Drug delivery approaches have undergone dramatic transformation and advancement in recent decades. Although many potent drugs against various diseases are discovered, the use of many of these is limited due to their adverse effects. A way to overcome the adverse systemic side effects of therapeutics is by localized or targeted administration of drugs. Nanodrugs use engineered drug delivery system for controlled and targeted drug delivery. The distinct physicochemical features of nanomaterials, along with the prospect of targeting, have made them as attractive alternative to bulk materials for devising drug delivery systems. Nanodrug formulations with encapsulated, dispersed, or conjugated drugs with high drug loading capacity have been extensively reported. Targeting is facilitated by site-specific targeting ligands (aptamers, peptides, antibodies, or small molecules) or by external guidance (magnetic nanoparticles) [8]. Targeting increases the efficiency of therapeutics and decreases the unwanted side effects in nontargeted tissues. Among commonly used targeting agents, aptamers are considered versatile moieties having high selectivity and specificity.

Aptamers are short oligonucleotide or peptide molecules that can form unique three-dimensional conformations and bind to groves or clefts on specific targets [46]. Aptamers are used as versatile targeting moieties by conjugating these with nanoparticles or drugs and can also act themselves as therapeutic agents. The era of aptamers started in 1990, when two independent research groups reported these almost simultaneously. Ellington and Szostak coined the term "aptamer," while the term "SELEX," for systemic evolution of ligands by exponential enrichment, was first described by Tuerk and Gold [46, 72, 75, 77]. Aptamers are inherently superior to antibodies due to their tailored in vitro chemical synthesis. Along with low immunogenicity and small size, aptamers are also stable in ambient temperature for long periods [77].

Aptamers can be produced in large quantities by combinatorial chemistry in a controlled manner and at a cheaper cost when compared to antibodies. These can also be derivatized with detection reagents and are usually stable when stored dry at room temperature [7]. Aptamers can be reversibly denatured and refolded without any loss in their properties [52]. As aptamer selection is completely a chemical process, it can be selected against any 'aptamerogenic' targets that can even be toxins, without any batch-to-batch variations [41, 46]. The nonrecalcitrant nature and optimal pharmacokinetics of aptamers make these a preferred choice for targeting applications. The aptamer–target binding is similar to antigen–antibody interaction [65]. Structural conformation of aptamers is important for their target recognition and binding [43, 76]. Binding with other moieties like nanoparticles can enhance or decrease the binding potential of aptamers [69]. In this regard, Zeyu Xiao and Omid C. Farokhzad discussed the challenges and opportunities that crop up during aptamer interactions [91].

In nanomedicine, aptamers are used for diagnosis and therapy, as theranostic tools, and for biosensor development in various diseases [41, 43, 50–52, 65, 70, 84, 89, 98]. Aptamer–aptamer complexes, aptamer–nanoparticle complexes, or aptamer–biomacromolecule conjugates (therapeutic moiety, siRNA, proteins/enzymes, other small molecules) are used in diagnostic and therapeutic applications [9, 25, 28, 31, 40, 71, 75, 90, 104]. A number of aptamers are currently under clinical trials. The only aptamer that is currently approved by the Food and Drug Administration (FDA) for therapeutic application is Pegaptanib (Macugen) for treating age-related macular degeneration. A review by Padma Sundaram et al. [76] has listed various aptamers against various diseases that are currently under clinical studies [76].

8.2 Aptamer Selection

Since 1990 when SELEX was first reported, the method has undergone significant improvements; and currently, hundreds of aptamers can be raised in a matter of days [7, 72]. Over the years, the basic SELEX protocol has undergone evolution to select aptamers for specific objectives. In 2003, the term "cell-SELEX" was coined for the strategy where aptamers could be selected against any cells, without specific knowledge of the target molecules [91]. In a review by Guillermo Aquino-Jarquin and Julia D. Toscano-Garibay [2], a historical overview of different SELEX variants is discussed. Modifications in SELEX have helped to avoid the selection of weakly interacting ligands [2]. Aptamers can be selected under specific conditions that match the environment of its application and can be either DNA or RNA aptamers [46]. A review by Pooja Dua et al. [25] describes different SELEX strategies used for the identification of aptamers that target cell surface proteins and examine the diagnostic and therapeutic applications of the aptamers selected through the discussed methods [25]. Aptamers can be easily modified to withstand exonuclease degradation. Additionally, aptamers can be modified with other moieties like polyethylene glycol (PEG) to enhance the bioavailability and pharmacokinetic properties [107]. Zsuzsanna Szeitner et al. [77] have stressed the importance of ensuring the proper functionality of aptamers in real samples. In their opinion, consideration of aptamer application during selection and characterization has improved the performance of aptamers. They have discussed various factors that should be considered during selection of aptamers for analytical purposes [77]. Aptamer and ribozyme databases were designed by researchers to store the comprehensive sequence information that is accumulating exponentially. Detailed and structured information regarding experimental conditions for aptamer selection is also provided in these databases [22, 45, 83].

To develop highly selectable aptamers that could distinguish between two related phenotypes, differential whole cell-SELEX was coupled with the counterselection/selection approach. Apart from high selectivity, the selected aptamers were able to inhibit specific intracellular signaling pathways in the target cells [15]. To obtain aptamers having high affinity, in silico maturation by means of a genetic algorithm was applied post-SELEX process. The authors report the selection of a prostate-binding antigen targeting the DNA aptamer with high binding capability and report this genetic algorithm as a complementary process to SELEX [67]. Yuri Imaizumi et al. [38] constructed a modified DNA library with enhanced potential to select high-affinity aptamers with low dissociation constants. The modified DNA library included (E)-5-(2-(*N*-(2-(N^6-adeninyl)ethyl))carbamylvinyl)-uracil bases, which are expected to increase the chances of intermolecular and intramolecular hydrogen bond formation, along with stacking interactions and for furnishing pockets to enable target binding. The results report the affinity enhancement and superior efficiency of base-modified aptamers when compared to the aptamers selected from a natural DNA library [38].

The presence of blood–brain barrier hinders therapeutic approaches in many neurological diseases of cerebral origin. To overcome this issue, Congsheng Cheng et al. [20] have identified brain-penetrating aptamers by the in vivo SELEX method (Fig. 8.1). They developed a library of aptamers of 40 nucleotides in length,

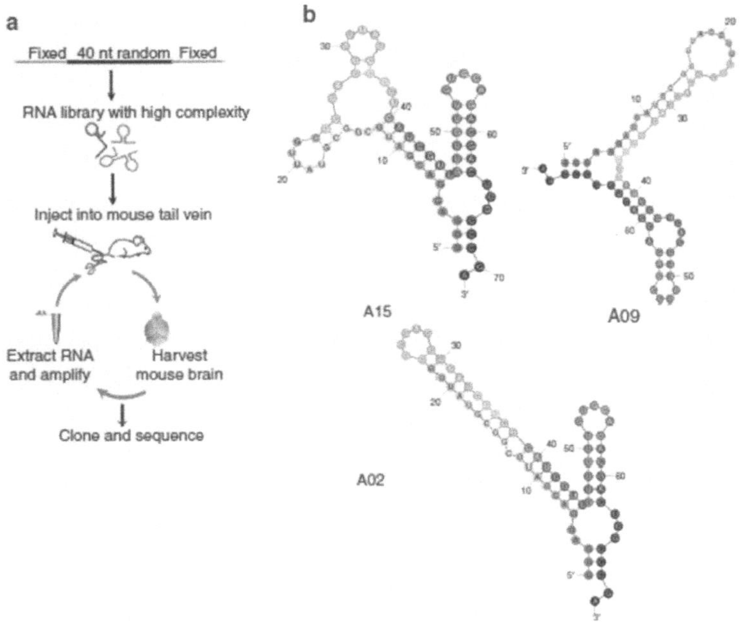

Figure 8.1 In vivo SELEX for identifying brain penetrating aptamers. Reprinted by permission from Macmillan Publishers Ltd: [*Molecular Therapy—Nucleic Acids*] (Ref. [20]), copyright (2013).

and the in vivo selection was performed in wild-type mice to identify the aptamers that penetrate brain following injection into the peripheral vasculature. The authors report that more than 15 rounds of in vivo selection were required to identify potential sequences. The resulting aptamers could bind to the brain capillary endothelia and penetrate into the parenchyma following peripheral injection. This novel study demonstrates that aptamers can be panned from the brains of living animals. Due to reticuloendothelial clearance, high levels of RNA were observed in nontarget tissues also [20].

An approach that combines cell internalization SELEX and high throughput sequencing coupled with bioinformatics analysis to select and identify cell-internalizing RNA aptamers was reported [82]. The authors suggest that apart from the classification based on sequence similarity, structural classification of the aptamers is also important for identification of the functional properties. The authors report the development of predictive tools that is valuable in target sequence identification during aptamer selection. During the experimental process, they were able to identify several RNA aptamers that targeted vascular smooth muscle cells [82]. In vitro selection and characterization of RNA aptamers that target human interleukin-6 receptor (IL-6R) were reported and developed as a potential tool for therapeutic use in different diseases, including cancer and inflammatory diseases [56]. The developed aptamer AIR-3A was only 19 nucleotides in length and retained all the necessary features required for targeting specificity. The AIR-3A aptamer could additionally serve as a delivery vehicle to transport therapeutic molecules to the cells harboring IL-6Rs [56]. Jagat R. Kanwar et al. [41] have reviewed the aptamers that are available for ocular diseases and discussed various modifications that can increase the efficiency of aptamers [41].

8.3 Aptamers as Therapeutics

Aptamers can act both as drugs and as drug carriers. Therapeutic aptamers can bind to protein targets and modulate their function in a specific manner [4, 12]. Several anticancer aptamers are capable of binding to the target cells and inhibiting the cancer cell growth. An antinucleolin aptamer is one such anticancer aptamer that is currently under phase two clinical trials [53]. Achim Boltz et al. [11] developed a chimeric DNA aptamer that can simultaneously bind to CD16α and c-Met receptors. Two each of CD16α-specific aptamers targeting natural killer cells and c-Met-specific aptamers targeting the tumor cells were coupled using linkers to form bi-specific aptamers that were used to induce antibody-dependent cell-mediated cytotoxicity (ADCC). The linkers in the bi-specific aptamers were designed specifically to replicate the distance between the complementarity-determining regions and the Fc binding domain of antibodies, thereby mimicking an environment for facilitating the ADCC mechanism [11]. A short, 19-nucleotide RNA aptamer (AIR-3A) with high affinity and specificity for human IL-6R was identified after 16 rounds of in vitro selection [56]. Interleukins play a major role in immune response and inflammatory diseases. In the reported study, the authors explain the use of streptavidin as a model cargo into the targeted cells. The aptamer binding did not interfere with the interaction of IL-6R with its natural ligands. The aptamer AIR-3A could also be used as a carrier for delivering payloads to the targeted cells. Interestingly, the authors report that the internalization of AIR-3A was temperature dependent and occurred only at 37°C but not at 4°C [56].

The ErbB-2 receptor belongs to the epidermal growth factor-related protein family of receptor tyrosine kinases and is overexpressed in many cancers. The anticancerous activity of a trimeric version of ErbB-2 receptor was tested both in vitro and in vivo (Fig. 8.2) [53]. The antiproliferative studies were conducted in human gastric cancer N87 cancer cells and in female CD-1 nude mice intradermally inoculated with N87 cells. The aptamer multimerization helped the avidity of binding to the ErbB-2 receptor and in crosslinking the receptor on the cell surface. The authors suggest that higher oligomers of the aptamer may have increased effectiveness and are not limited to trimers [53]. A nuclease-resistant therapeutic aptamer that binds specifically to the epidermal growth factor receptor (EGFR) was selected by SELEX, and the activity was studied in EGFR-positive cancer cells [29]. The results indicate that the aptamer CL4 was successful in

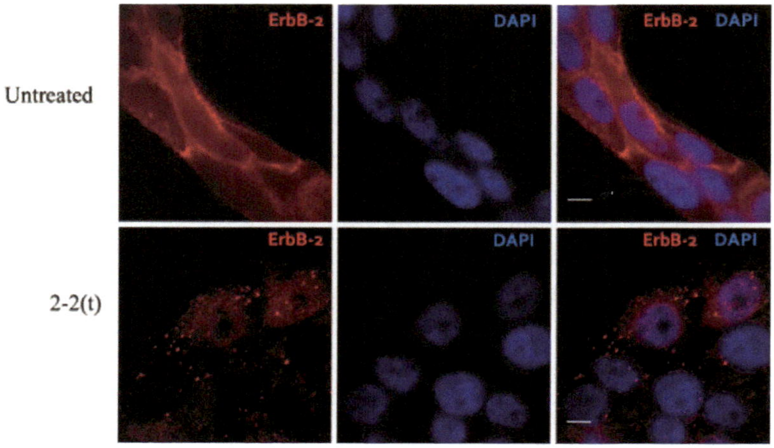

Figure 8.2 Translocation of surface ErbB-2 into intracellular puncta in N87 cells after treatment with the trimeric aptamer [53]. Adapted from Ref. [53] Copyright 2013, National Academy of Sciences, USA.

inducing apoptosis in the cells that were resistant to commonly used therapeutic EGFR inhibitors. When co-administered with cetuximab, synergistic effect in the therapeutic effect was observed, both in vitro and in vivo [29]. The chemically modified Adipo8 aptamer was tested for its selectivity in binding to white adipose tissues, the accumulation of which leads to obesity [19]. The aptamer was modified by introducing phosphorothioate linkage (PS linkage) and conjugating with PEG. The modified aptamers could selectively identify the white adipose tissue compared to other types of adipose and pre-adipose tissues. Adipo8 inhibited adipogenic differentiation of adipose tissue and holds great promise for the development of therapeutic strategies against obesity [19].

A nuclease-stabilized 33-mer RNA aptamer named Gint4.T was developed to target the platelet-derived growth factor receptor β (PDGFRβ) as a potential treatment strategy in human tumors [13]. On binding to the ectodomain of the PDGFRβ, cell migration and proliferation, as well as ligand-dependent receptor activation and downstream signaling, are inhibited. Since the heterodimerization of PDGFRβ is prevented by Gint4.T along with the transactivation of EGFR, the authors suggest that the use of an EGFR-targeting aptamer along with Gint4.T would be better at tumor therapy [13]. A highly biocompatible and biodegradable natural polysaccharide, arabinogalactan, obtained from larch trees was used to deliver antivimentin aptamer for anticancerous activity [94]. Aptamer selection was done using ascites cell organelles isolated from the cells removed from the abdominal cavity of ICR mice transplanted with tumor. Aptamers were selected after 11 rounds of SELEX performed at room temperature. Aptamer NAS-24 was selected against intracellular targets, coupled with arabinogalactan, and was delivered into the cytoplasm of cells, which led to apoptosis of cells. Free arabinogalactan or free aptamer did not result in any apoptosis [94].

To block the activity of nonstructural protein 1 (NS1) of influenza virus, a single-stranded DNA aptamer was selected by the SELEX method [87]. NS1 suppress the induction of interferon, thereby inhibiting the innate immune response of the host. The selected aptamer had a G-quadruplex structure and bound to NS1 at a low nanomolar range, without having any negative effect on cell viability. In vitro studies were conducted to determine the binding of aptamer to NS1, and the results were confirmed by confocal images and flow cytometry. Quantitative reverse transcription polymerase chain reaction (RT-PCR) and enzyme-linked immunosorbent assay (ELISA) results confirmed the effect on interferon expression [87]. Many aptamers are reported for treatment against human immunodeficiency virus (HIV). In a recent study, RNA aptamers that inhibit HIV-aspartyl protease (PR) were developed for anti-HIV gene therapy by SELEX [27]. The selected aptamers had moderate binding affinity and anti-PR inhibitory activity, which was further enhanced in the second-generation aptamer. By binding to the PR in a noncompetitive manner, the aptamers could also inhibit late events in HIV replication [27].

Tuberculosis is a serious infectious disease that is caused by *Mycobacterium tuberculosis*. A whole-bacteria SELEX strategy was applied to select aptamers against *M. tuberculosis*, and preliminary studies were conducted in a mouse model [17]. In a later report, the activity of the selected aptamer, NK2, was tested in vivo and the authors report that the aptamer had marked inhibitory effects against a virulent strain of *Mycobacterium*, the H37Rv strain [18].

Along with the NK2 aptamer, aptamers from the 10th pool of SELEX were also tested for antimycobacterial activity. Compared to NK2, the survival rate and histological assays exhibited better therapeutic effect on treatment with the aptamers from the 10th pool. The authors suggest that aptamers discarded from the early selection cycles might also prove to be significant and should be investigated thoroughly [18].

8.4 Aptamer-Conjugated Therapeutic Systems

Aptamers are versatile tools for targeted delivery of therapeutic agents. Two major strategies for the aptamer-mediated targeted strategy are (a) aptamer–drug conjugate and (b) aptamer–nanoparticle system. In an aptamer–drug conjugate system, the therapeutics are covalently or noncovalently conjugated to the aptamer, while in an aptamer–nanoparticle system, the therapeutics are attached to or encapsulated in the nanoparticle, which is targeted by the aptamer [49]. The nanoparticles can be metallic nanoparticles or nonmetallic nanoparticles. Nonmetallic nanoparticles include polymeric nanoparticles, liposomes, carbon nanotubes, and dendrimers.

8.4.1 *Aptamer–Drug Conjugates*

Aptamer-bound drug conjugates are simple yet powerful tools used for targeted therapeutic strategy. Aptamers act as carriers of a drug and deliver it in the target site. Since the binding of the drug to the aptamer is through noncovalent interaction, the amount of drug loaded and the site specificity of drug loading on the aptamer cannot be controlled. Drawing inspiration from solid-phase synthesis technology for highly controllable, automated molecular systems, a therapeutic module consisting of phosphoramidite incorporated with an anticancer drug, fluorouracil, was used to create an aptamer–drug conjugate [85]. Aptamer sgc-8 was used as the model aptamer, and a random sequence conjugated with drug was used as a control. Using this automated modular synthesis, multiple drug moieties could be coupled at specific sites on the aptamer. The spatiotemporal drug release was studied in vitro, and efficient anticancerous activity was observed in the target cells [85]. To achieve diversity in therapeutic applications, chimerization of aptamers is adopted. In chimeric aptamers, aptamers are used in combination with other aptamers, biomolecules, or functional compounds to provide a combined effect of all the components in the chimera [40, 73]. The most common example of aptamer chimerization is with drugs or therapeutic biomolecules. A chimeric aptamer–drug complex was engineered to study the delivery and therapeutic activity in retinoblastoma cells [73]. The epithelial cell adhesion molecule (EpCAM)-targeting aptamer, EpDT3, and doxorubicin were used to make the chimeric complex. The results suggest that the chimera complex was selective in destroying the cancer cells and spared the normal cells [73].

A multifunctional aptamer–polymer hybrid (APH) consisting of a targeting aptamer with a drug-loaded multifunctional block polymer was constructed on the basis of click chemistry for controlled and targeted therapy [60]. A nucleolin-binding aptamer was used as the targeting moiety, which was conjugated to the block polymer containing repeat units of ethylene glycol and ethylene glycol vinyl glycidyl ether. The drug doxorubicin was tethered to the polymer backbone via enzyme-cleavable linkers, which are designed to be cleaved by esterases inside the mammalian endosomes. It was reported that cytotoxicity was restricted to the target cells and minimal drug release was observed in nontarget cells. The authors suggest that this modular approach can be used for a variety of aptamers to target various cell surface markers [60]. Recently, a synthetic drug–DNA adduct (DDA) was developed for targeted anticancer drug delivery [106]. The DDAs were nuclease resistant and could conjugate multiple copies of anthracycline drugs with site specificity. The DDA had aptamer-mediated target recognition property, which was utilized for the targeting application. The doxorubicin–sgc8 aptamer adduct was tested in a tumor xenograft mouse model and the histopathological studies confirmed antitumor activity [106]. Targeted therapy to deliver nucleoside drug 5-fluoro-2′-deoxyuridine by SQ2 aptamer in the alkaline phosphatase placental-like 2 (APPL2) receptor expressing pancreatic cells was reported [26]. The results suggest that internalization of the aptamer was by clathrin-independent, but caveolae or dynamin-mediated, endocytic pathway. The SQ2-targeted therapy led to the inhibition of cell growth in APPL2 positive cells, while it did not show any adverse effect on normal pancreatic APPL2-negative ductal cells [26].

8.4.2 *Aptamer–Nanoparticle System*

8.4.2.1 Aptamer–nanoparticle system with metallic nanoparticles

The most commonly used metallic nanoparticles in a drug delivery system are gold and silver nanoparticles. Gold nanoparticles are versatile molecules that can be used for drug delivery and diagnostic applications, owing to their properties like low toxicity, biocompatibility, and tunable optical properties. Sang-Mi Ryou et al. [64] designed a versatile nanocomposite for in vivo delivery of functional proteins, using the gold nanoparticle–DNA aptamer conjugate. A 5′- Cys5-labeled anti-His aptamer was conjugated with gold nanoparticles, which in turn was loaded with apoptosis-inducing BIM protein to study its localization. Initial studies were done in various cell lines, followed by those in the xenograft model created by subcutaneous injection of HeLa cells in BALB/c nu/nu female mice. The nanoparticle–aptamer complex was effective for both local and systemic delivery of targeted proteins. The in vivo delivery of BIM protein was confirmed by analyzing the tumor sections from treated mice [64]. An ultra pH-sensitive chemotherapeutic system was designed using hollow gold nanospheres (HAuNS) that were loaded with doxorubicin and targeted using aptamers to selectively kill lymphoma tumor cells [97]. A 39-mer RNA aptamer specific for the CD30 tumor marker was chemically conjugated on the surface of HAuNS and subsequently surface modified using polyethylene glycol to enhance the biostability. This chemotherapeutic nanosystem was stable under normal physiological pH, but rapidly released the drug load at pH 5.0. About 80% of the drug load was released within two hours at low pH, and this system did not show any side effect on nontarget cells [97].

The in vivo biodistribution and toxicological effects of a highly anisotropic gold nanoconstruct attached to an aptamer (Apt-AuNS) was reported recently [23]. Targeting and therapeutic effects were provided by the G-quadruplex, antinucleolin DNA aptamer AS1411. The study was conducted in three different animal models, and interestingly, the nanoconstructs circulated for a longer time in female rats compared to the male rats. The authors report that Apt-AuNS is biocompatible and did not exhibit any acute toxicity in the model organisms. It was also observed that PEGylation of the nanoconstruct could reduce hepatic accumulation. The biodistribution was different in all three model organisms, and the authors stress the importance of in vivo studies to evaluate the clinical relevance and applications of nanoconstructs [23].

8.4.2.2 Aptamer–nanoparticle system with nonmetallic nanoparticles

Polymeric micelles have unique core–shell structures that allow the encapsulation of hydrophobic drugs. An anti-PSMA A10 aptamer-conjugated unimolecular micelles were loaded with doxorubicin, and the targeting and anticancerous effect of this nanosystem was studied in both in vivo and in vitro systems. This nano drug delivery system was pH-sensitive and exhibited controlled drug delivery potential [92]. An aptamer-modified liposome nanocarrier (Apt-PEG-LPs) was developed to target the mouse tumor endothelial cells under in vitro and in vivo conditions [3]. The DNA aptamer, AraHH001, was selected by the cell-SELEX method against tumor endothelial cells. The PEG-modified aptamer was added to the lipid solution while preparing the liposome. The cells took up nanocarrier by receptor-mediated endocytosis and showed considerable promise for tumor endothelial targeting under in vitro and in vivo conditions [3]. A dual therapeutic system consisting of a targeting aptamer (AS1411) and paclitaxel was developed and tested for antitumor activity [88]. Human serum albumin (HSA) was used as the drug carrier. Spherical HSA-paclitaxel nanoparticles were synthesized by self-assembly, followed by modification with aptamer AS1411. Functionalization with the aptamer significantly increased the nanoparticle uptake and cytotoxicity in target cells

(MCF-7 cells), while a decrease in the uptake and increased number of viable cells were observed when treated with noncancerous cells [88]. An aptamer-coated poly(lactic-co-glycolic acid) (PLGA) hybrid nanoparticle with core–shell lipid polymeric structure was designed for co-delivery of two anticancer drugs with different solubilities [37]. The aptamer sgc8 was used as the targeting agent in this system. Paclitaxel was encapsulated in the PLGA nanoparticle, and doxorubicin was intercalated on the aptamer. The synergistic effect of the co-delivered drugs was studied, and the results report that targeted co-delivery enhances the antitumor activity in the targeted cells [37].

A doxorubicin-loaded PEG-PLGA nanopolymerosome was covalently coupled with the EpCAM aptamer for targeted delivery into breast adenocarcinoma cells. This targeted therapeutic system was capable of pH-dependent, controlled, and sustained drug release in targeted MCF-7 cells [1]. Doxorubicin-encapsulated liposomes were conjugated to an RNA aptamer for targeted delivery to prostate-specific membrane antigen (PSMA), expressing prostate cancer cells [5]. The aptamer bound liposome was termed "aptamosome" and was about 90–100 nm in size. Drug efficacy and biodistribution of aptamosomes were tested in LNCaP prostate cancer cells and in LNCaP xenograft nude mice. Increased antitumor effect was observed in targeted liposomes containing doxorubicin [5]. Ultrasound-triggered delivery of aptamer-targeted liposomes was studied in cancer cells [59]. The liposomes were modified with thermosensitive polymer (TSP) to sensitize them to high temperature. A single-stranded aptamer targeting the platelet-derived growth factor receptor (PDGFR) was used in this study. Doxorubicin was used as the anticancer drug, and calcein was used as a model drug to study the targeting in MDA-MB-231 cells, while ultrasound triggered the drug delivery. The results indicate that TSP-modified liposomes were more sensitive to ultrasound and the drug release was induced by the nonthermal effect of ultrasound. The ultrasound-mediated cavitation also favored the internalization of drug into targeted cells [59]. Aptamer-targeted DNA nanopyramids were constructed for targeted delivery of anticancer drugs [16]. The AS1411 aptamer was incorporated into the DNA pyramid and it was resistant to nuclease degradation. The nanostructures were

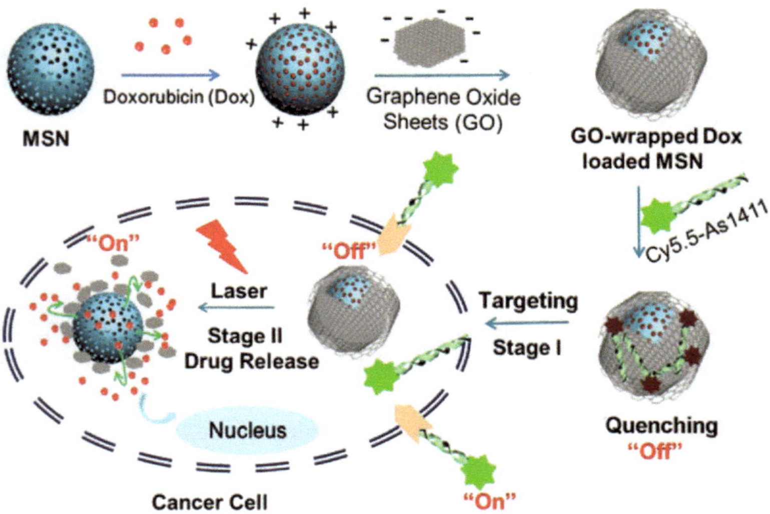

Figure 8.3 Schematic representation of aptamer-targeted graphene oxide–wrapped, doxorubicin-loaded silica nanoparticle for photoresponsive drug delivery. Reproduced from Ref. [80] with permission from The Royal Society of Chemistry.

internalized by the targeted cells and resulted in selective inhibition of the cancer cell growth. The DNA nanostructure is also capable of simultaneous delivery of multiple therapeutic moieties [16].

An aptamer-targeted photoresponsive drug delivery system (Fig. 8.3) consisting of graphene oxide wrapped and doxorubicin-loaded mesoporous silica nanoparticle was developed for cancer therapy [80]. This photoresponsive system had two "off/on" switches; the first switch was nucleolin-targeting Cy5.5 labeled-AS1411 aptamer conjugated on the surface of graphene oxide, and the second switch was laser irradiation. When irradiated with laser, graphene oxide sheets undergo expansion and vibrate, leading to release of doxorubicin from the mesoporous nanoparticle. The chemotheraputic effect and the photothermal therapeutic effect of this aptamer-targeted drug delivery system were studied in MCF-7 cells, and it was observed that the synergistic therapy exhibited higher therapeutic efficiency than when only a chemotherapeutic drug or photothermal therapy was used alone [80]. A high-drug loading, aptamer-targeted anticancer theranostic platform of calcium carbonate nanostructure (apt-CCN) was fabricated by an easy water-in-oil emulsion [99]. The negatively charged, porous CCN could incorporate a large amount of therapeutics. In this study, doxorubicin was used as the model drug. The CCN and the aptamer were linked via avidin, and a large number of aptamers could be conjugated on the CCN surface, providing a multivalent effect. This pH sensitive apt-CCN was internalized into the target cell by receptor-mediated endocytosis and the therapeutic drugs were released in response to the acidic pH in lysosome [99]. An A10 aptamer-targeted, doxorubicin-conjugated polylactide nanoconjugate was developed for controlled release of therapeutics for anticancer therapy in canine hemangiosarcoma. This targeted nanoconjugate was developed for intravenous drug administration and showed cytoreductive activity in mice bearing canine hemangiosarcoma tumors [79].

Aptamer-targeted, drug-loaded, single-walled carbon nanotubes (SWNTs) were developed to deliver anticancer drugs and the effect was studied in vitro [78]. Protein tyrosine kinase-7 binding sgc-8 aptamer was used in this study. Anticancer drug daunorubicin was loaded into the carbon nanotubes. This daunorubicin–aptamer–SWNT tertiary complex was efficiently internalized by the target cells, and drug delivery was pH dependent. When antisense oligos against the aptamer were introduced, the oligos disrupted the interaction between the aptamer and the target, thus inactivating the therapeutic effect of this system. Selective and controllable delivery is an important feature of efficient drug delivery systems [78].

8.5 Aptamer in Small RNA Deliver

Gene silencing by RNA interference (RNAi) was first reported in 1990s; and since then, it has attracted attention as a potential therapeutic approach. Although small RNA drugs are considered as efficient therapeutics, the major problem faced by small RNA therapeutics is its delivery [6, 21, 30, 46]. Small interfering RNA (siRNA) and microRNA (miRNA) are the typical activators of RNAi that have great potential in therapeutic development [30]. Jiehua

Zhou and John J. Rossi [101] have discussed the use of aptamer-targeted RNAi therapeutics in detail. In their review, covalent aptamer–siRNA chimeras, noncovalent aptamer–connector–siRNA conjugates, and aptamer-functionalized nanovectors loaded with siRNAs are discussed with examples from previous reports [101]. siRNA delivery by aptamers is facilitated by connecting the aptamer and siRNA with a linker, by forming an aptamer–siRNA chimera or by aptamer-targeted nanoparticle delivery where the nanoparticle carries the siRNA [49].

The siRNA–aptamer chimera construct is a promising approach for targeted siRNA delivery. In one of the pioneering studies, an aptamer–siRNA chimera was designed to target specific cell type and silence gene expression [54]. An anti-apoptotic gene was targeted by siRNA, and the anti-PSMA aptamer (A10) was used to target the siRNA to the target cancer cells. The siRNA activity resulted in decreased proliferation and induced apoptosis in the target cells. Modification of the binding sites on the aptamer resulted in the loss of binding activity of the aptamer to the target [54]. The anti-PSMA A9 aptamer coupled to siRNA through a modular streptavidin bridge was used to target prostate cancer cells, and the resulting gene expression was studied in LNCaP cells [21]. A 27-mer antilamin A/C siRNA was used in this study. The cells took up the siRNA–aptamer conjugate within 30 min, and efficient inhibition of gene expression was reported [21]. An siRNA–aptamer chimera was designed in a two-step process where the siRNA molecules (against eGFP) were adsorbed on to the quantum dot surface by electrostatic interaction, followed by coupling with the aptamer (anti-PSMA) [6]. This siRNA–aptamer chimera showed significantly improved gene silencing compared to nanoparticles (quantum dots) with single-step adsorbed siRNA–aptamer complexes. The authors stress the importance of preserving the orientation and three-dimensional conformation of aptamers on the nanoparticle surface for efficient aptamer activity [6]. Cell internalization SELEX was used to select an aptamer that binds to the HER2 receptor and subsequently gets internalized into the cells [81]. This cell-internalizing aptamer was conjugated with the pro-survial gene *Bcl-2*-silencing siRNA. The resulting aptamer–siRNA chimera could efficiently silence the *Bcl-2* gene and sensitize the HER2+ breast cancer cells to chemotherapy [81].

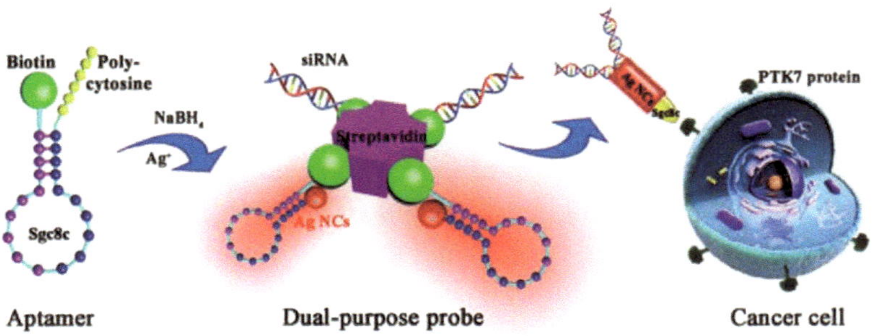

Figure 8.4 Schematic representation of sgc8c aptamer-functionalized fluorescent silver nanocluster–mediated siRNA delivery. Reproduced from Ref. [48] with permission from The Royal Society of Chemistry.

A multifunctional probe for theranostic purpose was designed using cell-specific internalization aptamer, fluorescent silver nanoclusters (Ag NCs) and therapeutic siRNA (Fig. 8.4) [48]. The Ag NCs were functionalized using the sgc8c aptamer and coupled to siRNA using streptavidin-biotin chemistry. The efficiency of vascular endothelial growth factor (VEGF)-silencing siRNA and the fluorescence produced by Ag NCs was studied in human cervical carcinoma HeLa cells. The authors report excellent fluorescence imaging potential, intracellular siRNA delivery, and gene silencing by this system [48]. SELEX was used to isolate several 2′-F-modified RNA aptamers that targeted the B-cell-activating factor (BAFF) receptor (BAFF-R) [103]. The aptamers were then conjugated with signal transducer and activator of transcription 3 (STAT3) siRNA. The BAFF-R aptamer played dual roles in this aptamer–siRNA chimera—to inhibit the proliferation and survival mediated by BAFF and to target delivery of the siRNA to the target cells. This combinatorial process is a promising therapeutic strategy against B-cell malignancies [103].

An anti-gp 120 aptamer–siRNA chimera was developed against cells expressing HIV-1 gp 120 [100]. Both the aptamer and the siRNA (anti-tat/rev siRNA) had anti-HIV activity, and the authors report that this dual functioning chimera allowed selective delivery into the target cells and inhibited HIV replication and spread [100]. A targeted approach using aptamer for combinatorial delivery of antiviral and host dicer substrate siRNA (DsiRNA) was tested for treatment in HIV infection [102]. In this facile strategy, 3′ 7-carbon linker (7C3) bound to a 16-nucleotide 2′ OMe/2′ Fl GC rich bridge was used to aid the binding of various siRNAs with the targeting aptamer. An HIV-1 gp120 aptamer targeted the DsiRNAs, and the in vivo antiviral activity was studied in a humanized mouse model. The results report the delivery of siRNA and long-term suppression of HIV-1 viral load [102]. An RNA aptamer–siRNA chimera was developed as a combination approach for the treatment of HIV infection [58]. The HIV-1 envelope protein-targeting aptamer was attached to the siRNA targeting the viral RNA through a linker. The antiviral effect was tested in the humanized Rag2-/-gc-/- (RAG-hu) mouse model, which mimics the HIV-1 replication and CD4+ T-cell depletion in humans. This nontoxic therapeutic approach reports efficient viral suppression by the dual anti-HIV activity offered by the aptamer and the siRNA [58].

A DNA aptamer against U87-EGFRvIII cells were used to deliver c-Met siRNA to the target cells and silence gene expression. The aptamer was selected by SELEX, and the results demonstrated that the aptamer was efficient in targeting the U87-EGFRvIII cells and delivering the siRNA to the cells [96]. An aptamer targeting the cytotoxic T lymphocyte-associated antigen 4 ($CTLA4_{apt}$) was covalently linked with STAT3 siRNA for tumor suppression [36]. STAT3 is a crucial nuclear transcription factor that acts as a signaling mediator in tumor cells and tumor-associated immune cells. Both local and systemic administration of the aptamer–siRNA complex significantly reduced the tumor growth and metastasis, and showed that gene silencing by this strategy is effective in both tumor-associated T-cells and tumor cells [36].

Targeted siRNA delivery into cancer cells were studied by Nithya Subramanian et al. [74]. They report a dual approach in which the EpCAM aptamer (EpApt) and EpCAM siRNA(SiEp) were used for targeted therapy in EpCAM positive cancer cells. EpCAM is a putative cancer stem cell marker and a target for cancer therapies. The aptamer and the siRNA were complexed with polyethyleneimine (PEI) to form a nanoformulation. Gene silencing was studied in two cell lines, and the targeted nanoformulation showed better gene silencing compared to the conventional silencing mechanisms [74]. Aptamer-mediated targeting was used to deliver siRNA to inhibit the mammalian target of rapamycin complex 1 (also known as mechanistic target of rapamycin complex 1, mTORC1), signaling in circulating CD8+ cells T-cells, and to enhance their differentiation into memory T-cells for enhanced protective immunity [10]. The aptamer used in this study was selected against 4-1BB, a costimulatory molecule expressed on CD8+ cells. In vivo analysis of the aptamer–siRNA complex was performed and mice treated with rapamycin, a pharmacological inhibitor of mTORC, did not induce an immunological response, while the aptamer–siRNA complex induced a potent memory response and cytotoxic effector activity in tumor-bearing mice [10]. Luiza I. Hernandez et al. [35] have described several qualitative assays to test and monitor the uptake of cell-specific internalization and subcellular localization of aptamers. The techniques included microscopy, microplate, and flow cytometry-based assays to test the cell-specific internalization and RNA-ribosome inactivating protein (RIP) assay to test subcellular localization. A novel quantitative assay called the quantitative and ultrasensitive internalization method (QUSIM) to measure the efficiency of aptamer internalization is also described in the paper. These techniques are valuable in assessing the aptamer delivery of siRNA, as translocalization of siRNA into the cytoplasm is a crucial aspect in siRNA-based therapeutics [35].

MicroRNAs (miRNAs) are short noncoding RNAs that can induce gene regulation either by inhibiting translation or by degrading complementary messenger RNA. Jan-H. Rohde et al. [62] have reported a method to deliver miRNA and miR-126 to endothelial and breast cancer cells. miR-126 is essential for vasculature growth but at the same time is implicated in the development of various cancers. The authors tested three different strategies for delivering miR-26 to endothelial cells via a transferrin receptor-targeting aptamer. One of the strategies was by conjugating pre-miR-126 to the aptamer, which resulted in the delivery of functional miR-126. On delivery to the target cells, the pre-miR-126 was endogenously processed, followed by the inhibition of the target gene vascular cell adhesion molecule-1 (VCAM-1). The aptamer–miRNA chimera resulted in endothelial cell sprouting and inhibited tumor cell proliferation and recruitment of endothelial cells [62]. Two nuclease-resistant aptamers, GL21.T and Gint4.T-targeting receptor tyrosine kinases Axl and PDGFRβ, respectively, were used to selectively target and deliver tumor suppressor antimiR-222 [14]. AntimiRs antagonize the endogenous miRNA and lead to increased levels of microRNA target proteins. A stick-annealing approach was used to form a highly stable bridge duplex between the aptamer and the antimiR. The aptamer–anti-miR complexes were rapidly internalized into the target cells by receptor-mediated pathway, and the results indicate that the cells treated with the anti-miR–aptamer complex expressed receptor-dependent selective downregulation of respective miRNA levels [14].

8.6 Aptamers in Theranostic Applications

A theranostic application refers to the simultaneous diagnostic and therapeutic effect contributed by the same moiety. Developments in nanomedicine have brought about tremendous advances in theranostic applications of nanomaterials. Aptamers have augmented theranostic applications using nanomaterials. A review by Sarah Shigdar et al. [69] discusses various types of modifications and functionalizations that help in enhancing the stability of theranostic aptamers [69]. An aptamer-tethered DNA nanotrain (aptNTr) was developed by self-assembly for targeted drug delivery in cancer therapy [105]. Consisting of only three DNA building blocks, this system was easy to synthesize and had high payload capacity. Along with the therapeutics, bioimaging agents could be coupled with this system, enabling theranostic possibilities of this complex. The aptamer sgc8 was used as the model aptamer and in vitro studies were conducted in CEM and Ramos cells. Anthracycline anticancer drugs like doxorubicin, daunorubicin, and epirubicin were used as model drugs [105].

Quantum dots are excellent imaging agents and can be used as carriers of drugs. The theranostic applications of an anticancerous, pH-responsive, aptamer-targeted quantum dot–drug conjugate were studied in ovarian cancer cells [66]. The DNA aptamer, MUC1, specific to mucin overexpressed on the cancer cells was used; and doxorubicin was used as the model anticancerous drug. The quantum dot and doxorubicin was linked via a pH-sensitive hydrazone bond, and the aptamer was attached to the quantum dot by EDC/NHS chemistry [66]. A theranostic nanocarrier system consisting of the EpCAM aptamer and quantum dot–bound PLGA nanoparticle-containing anticancer drug nutlin-3a was reported as a nanotheranostic approach in cancer therapy [24]. EDC–NHS coupling was used for simultaneous conjugation of aptamer and quantum dots on the nanoparticle. The imaging potential of this system was studied in different solid cancer cell lines and the results indicate enhanced internalization, cellular labeling, and anticancer effects by this theranostic agent [24].

A multifunctional nanoprobe consisting of a two-photon dye-doped mesoporous silica nanoparticle (TPD-MSN), MnO_2 nanosheets, and sgc8 aptamer was developed for contrast-enhanced

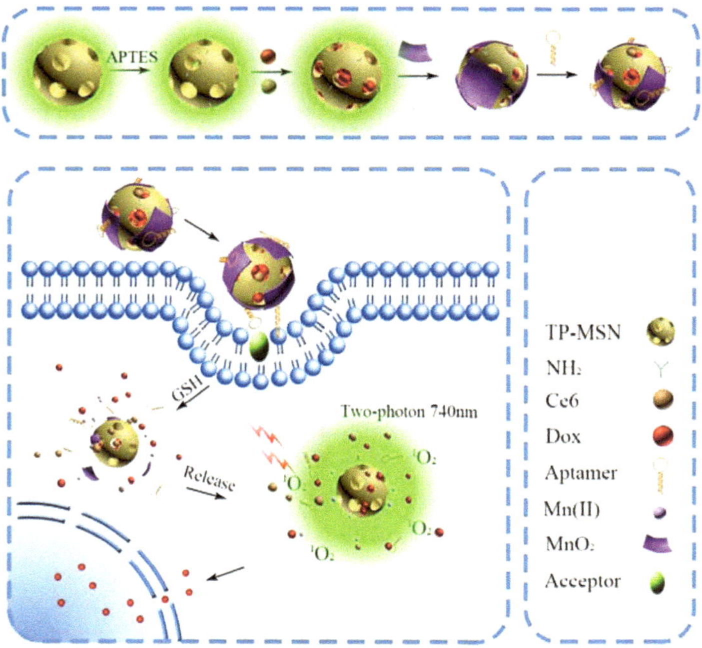

Figure 8.5 Multifunctional nanoprobe for contrast-enhanced bimodal cellular imaging and targeted therapy consisting of a two-photon dye-doped mesoporous silica nanoparticle (TPD-MSN), MnO_2 nanosheets, and the sgc8 aptamer. Reprinted with permission from Ref. [55]. Copyright 2015 American Chemical Society.

bimodal cellular imaging and targeted therapy (Fig. 8.5) [55]. Silica nanoparticles were loaded with anticancer drug doxorubicin and photosensitizer chlorin e6 (ce6). Multifunctional MnO_2 nanosheets acted as the gatekeeper for TPD-MSNs, quencher for two-photon fluorescence, and as a contrast agent for magnetic resonance imaging (MRI). After getting internalized into the target cells, intracellular glutathione reduces MnO_2 to Mn_2+ ions, which help in contrast-enhanced two-photon fluorescence imaging and magnetic resonance signal for MRI. At the same time, doxorubicin is released providing anticancerous activity and ce6 assists in photodynamic therapy (PDT) [55]. A multifunctional aptamer-targeted theranostic moiety was synthesized using silver nanoparticles and a fluorescent moiety, fluorescein isothiocyanate (FAM) [47]. Aptamer sgc8 and TDO5 were used in this study. The results report that the aptamer–silver conjugate is internalized into the cells by receptor-mediated endocytosis and results in apoptosis. The silver nanoparticles could enhance the fluorescence intensity of FAM by the metal-enhanced fluorescence effect. The silver nanoparticle-induced apoptosis was studied in different cells, and the effect was found to be cell dependent [47].

8.7 Other Therapeutic Applications Using Aptamers

In 2006, Chonglin Yang et al. [93] reported a method of aptamer-mediated programmed cell death in *Caenorhabditis elegans*, a strategy that can be used to develop therapeutics against diseases caused by abnormal apoptosis. The authors report the selection of aptamers by SELEX, which binds to a death inhibitor in *C. elegans*, Bcl-2 homolog CED-9. These aptamers are suspected to antagonize the prosurvival activity of CED-9 and lead to programmed cell death [93]. A review by Eli Gilboa et al. [33] presents the applications of aptamers in modulating the functions of immune receptors. They report that oligonucleotide aptamers can have extensive transformative impact in the medical field [33]. In a recent review, Marjan Gharagozloo and coauthors [32] discuss autoimmune disease therapy using nanomedicine. They also discuss various RNA aptamers that are studied as potential therapeutic agents in autoimmune diseases [32]. A single-stranded DNA sequence that selectively binds to the M2 antibody, which recognizes the Flag epitope, was discovered serendipitously [44]. This aptamer could directly bind to the antigen-binding pocket of the anti-Flag M2 antibody and could be used for protein purification processes, as it acted competitively with the Flag peptide in binding with the antibody. The authors suggest that this approach could be utilized in novel therapeutic strategies against autoimmune diseases caused by antinuclear antibodies [44].

To induce antitumor activity, a nucleic acid aptamer was used to deliver antigenic cargo into cancer cells to activate T-cell-mediated immunity [86]. An aptamer that recognizes the murine receptor, DEC205, was selected as the targeting agent and was conjugated with chicken antigen ovalbumin (OVA). A 42-nucleotide aptamer, 'min.2', was used for the targeting, which performed better than the larger, a round 5 aptamer. The aptamer–antigen complex resulted in cross presentation of the antigen and activation of T-cells [86]. To modulate the immune response by tumor-infiltrating immune cells, a clinically feasible, tumor stroma-targeted costimulation with a bispecific aptamer was investigated [68]. This system consisted of a 4-1BB aptamer that targets tumor stroma, conjugated to the vascular endothelial growth factor (VEGF)-targeting aptamer. On systemic administration of this VEGF-targeted 4-1BB aptamer, murine tumor models exhibited superior and potent antitumor immunity against various unrelated cancers. Since stromal products like VEGF, osteopontin, or metalloproteases are present in different cancer cells, tumor stroma–targeted costimulation is broadly applicable for therapeutic purposes than specific cell-based therapeutic strategies [68].

Two RNA aptamers, D-12 and D-26, which target the hemagglutinin protein of influenza virus, were selected to test the therapeutic activity against influenza [34]. In influenza virus, the hemagglutinin protein mediates the initial steps in viral infection– receptor (glycan) binding and membrane fusion for cell entry. Both the aptamers were found to interfere with hemagglutinin–glycan interaction. The D-26 aptamer showed higher efficiency in distinguishing viral strains, and the affinity was further improved by incorporating the aptamer with 2′-fluoropyrimidines [34]. Oncolytic viruses are tumor-selective viruses that can destroy cancer cells by oncolysis. But quick clearance from the bloodstream and the acquired immunity to repeated infections by neutralizing antibodies (nAbs) restricts the use of oncolytic viruses in therapeutic applications. In a novel and versatile approach called aptamer-facilitated virus protection (AptaVIP), two types of aptamers were raised by the modified SELEX method to increase the in vivo oncolytic viral survival as a mode of anticancer treatment [57]. DNA aptamers were selected against vesicular stomatitis virus (oncolytic virus) and against the antigen-binding fragment (Fab) of antivesicular stomatitis virus polyclonal antibodies for shielding and blocking purposes, respectively, and they were modified and bridged together to form tetrameric counterparts. The authors were successful in demonstrating the use of this dual-aptamer AptaVIP system to block the antiviral antibodies and shield the virus from antibody neutralization that resulted in 77% viral infectivity in plaque-forming assay [57].

PDT is a treatment strategy that has received attention in the recent years. A photosensitizer ce6 was conjugated with an RNA aptamer (AIR-3A) targeting human IL-6R, and the photodynamic efficiency was tested in tumor cells [42]. Compared to the nontargeted ce6, the aptamer-conjugated ce6 was internalized by the cells within 45 min and was localized in the cytosol. The PDT resulted in the reduction of cell viability to 46% [42]. An antibody-aptamer-pincer (AAP) system was developed by Sungmuk Kang and Sang Soo Hah [39] for enhanced affinity to the target by simultaneously recognizing different parts of the target. In their study, the authors developed the AAP against thrombin or human epidermal growth factor 2, and the activity of drug (doxorubicin) loaded aptamer in AAP was also studied. The results suggest that such drug-loaded AAP systems can be successfully used in drug delivery systems [39]. Patrick Ruff et al. [63] have reported a novel approach to gene correction by using an aptamer and the efficacy of this method was tested in yeast and human cells. In this aptamer-guided gene-targeting (AGT) system, a site-specific DNA-binding protein is bound to a DNA aptamer and targeted to the specific genetic loci for correction. The DNA aptamer was selected to target I-Sce1 endonucleases that recognize an 18-base-pair sequence and induce double-stranded breaks. The authors report targeted delivery of exogenous donor DNA to the double-stranded breaks in different genomic locations of both yeast and human cells. This work proves the utility of aptamers in gene targeting for potential therapeutic use [63].

Recently, aptamer-based antithrombotic drugs have captured the attention of researchers. In 2010, Thomas J. Povsic et al. [61] published a review that details the translational applications of aptamers in antithrombotics. Having reversible and highly regulatable properties, aptamers are ideal for use in thrombotic disorders [61]. The antithrombotic effect of the RA-36 aptamer was studied in a murine model of thrombosis, and prevention of thrombus formation was observed on intravenous bolus administration. The effect of RA-36 was compared with that of bivalirudin, a thrombin inhibitor [95].

8.8 Conclusion

Aptamers were first described in the 1990s and since then a lot of research has gone into its development and applications. With advances in nanomedicine, aptamers are now widely used for basic research, diagnosis, therapy, theranostic applications, prophylaxis, and biosensor development. The unique intrinsic properties of aptamers are used in designing nanovehicles for targeted therapeutic application in various diseases. Apart from targeting, therapeutic approaches using aptamers have helped in reducing the side effects in nontargeted cells/tissues. A number of aptamers developed for pharmaceutical applications are now in various stages of clinical studies. Aptamers have proved their names as multifunctional moieties with tremendous potential in pharmaceutical applications. Although the commercialization and therapeutic applications of aptamers are still underway, there is no doubt that these versatile moieties will be key players in diagnosis and treatment of several diseases.

References

1. Alibolandi, M., Ramezani, M., Sadeghi, F., Abnous, K., and Hadizadeh, F. (2015). Epithelial cell adhesion molecule aptamer conjugated PEG-PLGA nanopolymersomes for targeted delivery of doxorubicin to

human breast adenocarcinoma cell line in vitro, *Int. J. Pharm.*, **479**(1), pp. 241–251.

2. Aquino-Jarquin, G., and Toscano-Garibay, J. D. (2011). RNA aptamer evolution: two decades of SELEction, *Int. J. Mol. Sci.*, **12**(12), pp. 9155–9171.
3. Ara, M. N., Matsuda, T., Hyodo, M., Sakurai, Y., Hatakeyama, H., Ohga, N., Hida, K., and Harashima, H. (2014). An aptamer ligand based liposomal nanocarrier system that targets tumor endothelial cells, *Biomaterials*, **35**(25), pp. 7110–7120.
4. Ashrafuzzaman, M. (2014). Aptamers as both drugs and drug-carriers, *Biomed. Res. Int.*, **2014**, p. 697923.
5. Baek, S. E., Lee, K. H., Park, Y. S., Oh, D. K., Oh, S., Kim, K. S., and Kim, D. E. (2014). RNA aptamer-conjugated liposome as an efficient anticancer drug delivery vehicle targeting cancer cells in vivo, *J. Controlled Release*, **196**, pp. 234–242.
6. Bagalkot, V., and Gao, X. (2011). siRNA-aptamer chimeras on nanoparticles: preserving targeting functionality for effective gene silencing, *ACS Nano*, **5**(10), pp. 8131–8139.
7. Baird, G. S. (2010). Where are all the aptamers? *Am. J. Clin. Pathol.*, **134**(4), pp. 529–531.
8. Bamrungsap, S., Zhao, Z., Chen, T., Wang, L., Li, C., Fu, T., and Tan, W. (2012). Nanotechnology in therapeutics: a focus on nanoparticles as a drug delivery system, *Nanomedicine (Lond.)*, **7**(8), pp. 1253–1271.
9. Banerjee, J., and Nilsen-Hamilton, M. (2013). Aptamers: multifunctional molecules for biomedical research, *Mol. Med. (Berl)*, **91**(12), pp. 1333–1342.
10. Berezhnoy, A., Castro, I., Levay, A., Malek, T. R., and Gilboa, E. (2014). Aptamer-targeted inhibition of mTOR in T cells enhances antitumor immunity, *J. Clin. Invest.*, **124**(1), pp. 188–197.
11. Boltz, A., Piater, B., Toleikis, L., Guenther, R., Kolmar, H., and Hock, B. (2011). Bi-specific aptamers mediating tumor cell lysis, *J. Biol. Chem.*, **286**(24), pp. 21896–21905.
12. Bouchard, P. R., Hutabarat, R. M., and Thompson, K. M. (2010). Discovery and development of therapeutic aptamers, *Annu. Rev. Pharmacol. Toxicol.*, **50**, pp. 237–257.
13. Camorani, S., Esposito, C. L., Rienzo, A., Catuogno, S., Iaboni, M., Condorelli, G., de Franciscis, V., and Cerchia, L. (2014). Inhibition of receptor signaling and of glioblastoma-derived tumor growth by a novel PDGFRβ aptamer, *Mol. Ther.*, **22**(4), pp. 828–841.

14. Catuogno, S., Rienzo, A., Di Vito, A., Esposito, C. L., and de Franciscis, V. (2015). Selective delivery of therapeutic single strand antimiRs by aptamer-based conjugates, *J. Controlled Release*, **210**, pp. 147–159.
15. Cerchia, L., Esposito, C. L., Jacobs, A. H., Tavitian, B., and de Franciscis, V. (2009). Differential SELEX in human glioma cell lines, *PLoS One*, **4**(11), p. e7971.
16. Charoenphol, P., and Bermudez, H. (2014). Aptamer-targeted DNA nanostructures for therapeutic delivery, *Mol. Pharm.*, **11**(5), pp. 1721–1725.
17. Chen, F., Zhou, J., Luo, F., Mohammed, A. B., and Zhang, X. L. (2007). Aptamer from whole-bacterium SELEX as new therapeutic reagent against virulent *Mycobacterium tuberculosis*, *Biochem. Biophys. Res. Commun.*, **357**(3), pp. 743–748.
18. Chen, F., Zhou, J., Huang, Y. H., Huang, F. Y., Liu, Q., Fang, Z., Yang, S., Xiong, M., Lin, Y. Y., and Tan, G. H. (2013). Function of ssDNA aptamer and aptamer pool against *Mycobacterium tuberculosis* in a mouse model, *Mol. Med. Rep.*, **7**(2), pp. 669–273.
19. Chen, K., Liu, J., Tong, G., Liu, B., Wang, G., and Liu, H. (2014). Adipo8, a high-affinity DNA aptamer, can differentiate among adipocytes and inhibit intracellular lipid accumulation in vitro, *Sci. Chin. Chem.*, **58**(10), pp. 1612–1620.
20. Cheng, C., Chen, Y. H., Lennox, K. A., Behlke, M. A., and Davidson, B. L. (2013). In vivo SELEX for identification of brain-penetrating aptamers, *Mol. Ther. Nucleic Acids*, **2**, p. e67.
21. Chu, T. C., Twu, K. Y., Ellington, A. D., and Levy, M. (2006), Aptamer mediated siRNA delivery, *Nucleic Acids Res.*, **34**(10), p. e73.
22. Cruz-Toledo, J., McKeague, M., Zhang, X., Giamberardino, A., McConnell, E., Francis, T., DeRosa, M. C., and Dumontier, M. (2012). Aptamer base: a collaborative knowledge base to describe aptamers and SELEX experiments, *Database (Oxford)*, **2012**, bas006.
23. Dam, D. H., Culver, K. S., Kandela, I., Lee, R. C., Chandra, K., Lee, H., Mantis, C., Ugolkov, A., Mazar, A. P., and Odom, T. W. (2015). Biodistribution and in vivo toxicity of aptamer-loaded gold nanostars, *Nanomedicine*, **11**(3), pp. 671–679.
24. Das, M., Duan, W., and Sahoo, S. K. (2015). Multifunctional nanoparticle-EpCAM aptamer bioconjugates: a paradigm for targeted drug delivery and imaging in cancer therapy, *Nanomedicine*, **11**(2), pp. 379–389.
25. Dua, P., Kim, S., and Lee, D. K. (2011). Nucleic acid aptamers targeting cell-surface proteins, *Methods*, **54**(2), pp. 215–225.

26. Dua, P., S. S., Kim, S., and Lee, D. K. (2015). ALPPL2 aptamer-mediated targeted delivery of 5-fluoro-2′-deoxyuridine to pancreatic cancer, *Nucleic Acid Ther.*, **25**(4), pp. 180–187.
27. Duclair, S., Gautam., A., Ellington, A., and Prasad, V. R. (2015). High-affinity RNA aptamers against the HIV-1 protease inhibit both in vitro protease activity and late events of viral replication, *Mol. Ther. Nucleic Acids*, **4**, p. e228.
28. Dupont, D. M., Andersen, L. M., Botkjaer, K. A., and Andreasen, P. A. (2011). Nucleic acid aptamers against proteases, *Curr. Med. Chem.*, **18**(27), pp. 4139–4151.
29. Esposito, C. L., Passaro, D., Longobardo, I., Condorelli, G., Marotta, P., Affuso, A., de Franciscis, V., and Cerchia, L. (2011). A neutralizing RNA aptamer against EGFR causes selective apoptotic cell death, *PLoS One*, **6**(9), p. e24071.
30. Esposito, C. L., Catuogno, S., and de Franciscis, V. (2014). Aptamer-mediated selective delivery of short RNA therapeutics in cancer cells, *J. RNAi Gene Silencing*, **10**, pp. 500–506.
31. Germer, K., Leonard, M., and Zhang, X. (2013). RNA aptamers and their therapeutic and diagnostic applications, *Int. J. Biochem. Mol. Biol.*, **4**(1), pp. 27–40.
32. Gharagozloo, M., Majewski, S., and Foldvari, M. (2015). Therapeutic applications of nanomedicine in autoimmune diseases: from immuno-suppression to tolerance induction, *Nanomedicine*, **11**(4), pp. 1003–1018.
33. Gilboa, E., McNamara, J. O. 2nd, and Pastor, F. (2013). Use of oligonucleotide aptamer ligands to modulate the function of immune receptors, *Clin. Cancer Res.*, **19**(5), pp. 1054–1062.
34. Gopinath, S. C., and Kumar, P. K. (2013). Aptamers that bind to the hemagglutinin of the recent pandemic influenza virus H1N1 and efficiently inhibit agglutination, *Acta Biomater.*, **9**(11), pp. 8932–8941.
35. Hernandez, L. I., Flenker, K. S., Hernandez, F. J., Klingelhutz, A. J., Mc-Namara, J. O. 2nd, and Giangrande P. H. (2013). Methods for evaluating cell-specific, cell-internalizing RNA aptamers, *Pharmaceuticals (Basel)*, **6**(3), pp. 295–319.
36. Herrmann, A., Priceman, S. J., Kujawski, M., Xin, H., Cherryholmes, G. A., Zhang, W., Zhang, C., Lahtz, C., Kowolik, C., Forman, S. J., Kortylewski, M., and Yu, H. (2014). CTLA4 aptamer delivers STAT3 siRNA to tumor-associated and malignant T cells, *J. Clin. Invest.*, **124**(7), pp. 2977–2987.

37. Huang, F., You, M., Chen, T., Zhu, G., Liang, H., and Tan, W. (2014). Self-assembled hybrid nanoparticles for targeted co-delivery of two drugs into cancer cells, *Chem. Commun. (Camb.)*, **50**(23), pp. 3013–3015.
38. Imaizumi, Y., Kasahara, Y., Fujita, H., Kitadume, S., Ozaki, H., Endoh, T., Kuwahara, M., and Sugimoto, N. (2013). Efficacy of base-modification on target binding of small molecule DNA aptamers, *J. Am. Chem. Soc.*, **135**(25), pp. 9412–9419.
39. Kang, S., and Hah, S. S. (2014). Improved ligand binding by antibody-aptamer pincers, *Bioconjug. Chem.*, **25**(8), pp. 1421–1427.
40. Kanwar, J. R., Roy, K., and Kanwar, R. K. (2011). Chimeric aptamers in cancer cell-targeted drug delivery, *Crit. Rev. Biochem. Mol. Biol.*, **46**(6), pp. 459– 477.
41. Kanwar, J. R., Shankaranarayanan, J. S., Gurudevan, S., and Kanwar, R. K. (2014). Aptamer-based therapeutics of the past, present and future: from the perspective of eye-related diseases, *Drug Discovery Today*, **19**(9), pp. 1309–1321.
42. Kruspe, S., Meyer, C., and Hahn, U. (2014). Chlorin e6 conjugated interleukin-6 receptor aptamers selectively kill target cells upon radiation, *Mol Ther. Nucleic Acids*, **3**, p. e143.
43. Kruspe, S., Mittelberger, F., Szameit, K., and Hahn, U. (2014). Aptamers as drug delivery vehicles, *ChemMedChem*, **9**(9), pp. 1998–2011.
44. Lakamp, A. S., and Ouellette, M. M. (2011). A ssDNA aptamer that blocks the function of the anti-FLAG M2 antibody, *J. Nucleic Acids*, **2011**, Article ID 720798, 11 pages.
45. Lee, J. F., Hesselberth, J. R., Meyers, L. A., and Ellington, A. D. (2004). Aptamer database, *Nucleic Acids Res.*, **32**(Database issue), pp. D95–D100.
46. Levy-Nissenbaum, E., Radovic-Moreno, A. F., Wang, A. Z., Langer, R., and Farokhzad, O. C. (2008). Nanotechnology and aptamers: applications in drug delivery, *Trends Biotechnol.*, **26**(8), pp. 442–449.
47. Li, H., Hu, H., Zhao, Y., Chen, X., Li, W., Qiang, W., and Xu, D. (2015). Multifunctional aptamer-silver conjugates as theragnostic agents for specific cancer cell therapy and fluorescence-enhanced cell imaging, *Anal. Chem.*, **87**(7), pp. 3736–3745.
48. Li, J., Wang, W., Sun, D., Chen, J., Zhang, P., Zhang, J., Min, Q., and Zhu, J. (2013). Aptamer-functionalized silver nanoclusters-mediated cell type-specific siRNA delivery and tracking, *Chem. Sci.*, **4**, pp. 3514–3521.
49. Li, X., Zhao, Q., and Qiu, L. (2013). Smart ligand: aptamer-mediated targeted delivery of chemotherapeutic drugs and siRNA for cancer therapy, *J. Controlled Release*, **171**(2), pp. 152–162.

50. Liu, K., Lin, B., and Lan, X. (2013). Aptamers: a promising tool for cancer imaging, diagnosis, and therapy, *J. Cell. Biochem.*, **114**(2), pp. 250–255.

51. Luo, C., Wen, W., Lin, F., Zhang, X., Gu, H., and Wang, S. (2015). Simplified aptamer-based colorimetric method using unmodified gold nanoparticles for the detection of carcinoma embryonic antigen, *RSC Adv.*, **5**, pp. 10994–10999.

52. Ma, H., Liu, J., Ali, M. M., Mahmood, M. A., Labanieh, L., Lu, M., Iqbal, S. M., Zhang, Q., Zhao, W., and Wan, Y. (2015). Nucleic acid aptamers in cancer research, diagnosis and therapy, *Chem. Soc. Rev.*, **44**(5), pp. 1240–1256.

53. Mahlknecht, G., Maron, R., Mancini, M., Schechter, B., Sela, M., and Yarden, Y. (2013). Aptamer to ErbB-2/HER2 enhances degradation of the target and inhibits tumorigenic growth, *Proc. Natl. Acad. Sci U S A*, **110**(20), pp. 8170–8175.

54. McNamara, J. O. 2nd, Andrechek, E. R., Wang, Y., Viles, K. D., Rempel, R. E., Gilboa, E., Sullenger, B. A., and Giangrande, P. H. (2006). Cell type-specific delivery of siRNAs with aptamer-siRNA chimeras, *Nat. Biotechnol.*, **24**(8), pp. 1005–1015.

55. Meng, H. M., Lu, L., Zhao, X. H., Chen, Z., Zhao, Z., Yang, C., Zhang, X. B., and Tan, W. (2015). Multiple functional nanoprobe for contrast-enhanced bimodal cellular imaging and targeted therapy, *Anal. Chem.*, **87**(8), pp. 4448–4454.

56. Meyer, C., Eydeler, K., Magbanua, E., Zivkovic, T., Piganeau, N., Lorenzen, I., Grötzinger, J., Mayer, G., Rose-John, S., and Hahn, U. (2012). Interleukin-6 receptor specific RNA aptamers for cargo delivery into target cells, *RNA Biol.*, **9**(1), pp. 67–80.

57. Muharemagic, D., Zamay, A., Ghobadloo, S. M., Evgin, L., Savitskaya, A., Bell, J. C., and Berezovski, M. V. (2014). Aptamer-facilitated protection of oncolytic virus from neutralizing antibodies, *Mol. Ther. Nucleic Acids*, **3**, p. e167.

58. Neff, C. P., Zhou, J., Remling, L., Kuruvilla, J., Zhang, J., Li, H., Smith, D. D., Swiderski, P., Rossi, J. J., and Akkina, R. (2011). An aptamer-siRNA chimera suppresses HIV-1 viral loads and protects from helper CD4(+) T cell decline in humanized mice, *Sci. Transl. Med.*, **3**(66), 66ra6.

59. Ninomiya, K., Yamashita, T., Kawabata, S., and Shimizu, N. (2014). Targeted and ultrasound-triggered drug delivery using liposomes co-modified with cancer cell-targeting aptamers and a thermosensitive polymer, *Ultrason. Sonochem.*, **21**(4), pp. 1482–1488.

60. Oh, S. S., Lee, B. F., Leibfarth, F. A., Eisenstein, M., Robb, M. J., Lyna, N. A., Hawker, C. J., and Soh, H. T. (2014). Synthetic aptamer-polymer hybrid constructs for programmed drug delivery into specific target cells, *J. Am. Chem. Soc.*, **136**(42), pp. 15010–15015.
61. Povsic, T. J., Sullenger, B. A., Zelenkofske, S. L., Rusconi, C. P., and Becker, R. C. (2010). Translating nucleic acid aptamers to antithrombotic drugs in cardiovascular medicine, *J. Cardiovasc. Transl. Res.*, **3**(6), pp. 704–716.
62. Rohde, J. H., Weigand, J. E., Suess, B., and Dimmeler, S. (2015). A universal aptamer chimera for the delivery of functional microRNA-126, *Nucleic Acid Ther.*, **25**(3), pp. 141–151.
63. Ruff, P., Koh, K. D., Keskin, H., Pai, R. B., and Storici, F. (2014). Aptamer-guided gene targeting in yeast and human cells, *Nucleic Acids Res.*, **42**(7), p. e61.
64. Ryou, S. M., Yeom, J. H., Kang, H. J., Won, M., Kim, J. S., Lee, B., Seong, M. J., Ha, N. C., Bae, J., and Lee, K. (2014). Gold nanoparticle-DNA aptamer composites as a universal carrier for in vivo delivery of biologically functional proteins, *J. Controlled Release*, **196**, pp. 287–294.
65. Santosh, B., and Yadava, P. K. (2014). Nucleic acid aptamers: research tools in disease diagnostics and therapeutics, *Biomed. Res. Intl.*, **2014**, Article ID 540451, 13 pages.
66. Savla, R., Taratula, O., Garbuzenko, O., and Minko, T. (2011). Tumor targeted quantum dot-mucin 1 aptamer-doxorubicin conjugate for imaging and treatment of cancer, *J. Controlled Release*, **153**(1), pp. 16–22.
67. Savory, N., Abe, K., Sode, K., and Ikebukuro, K. (2010). Selection of DNA aptamer against prostate specific antigen using a genetic algorithm and application to sensing, *Biosens. Bioelectron.*, **26**(4), pp. 1386–1391.
68. Schrand, B., Berezhnoy, A., Brenneman, R., Williams, A., Levay, A., Kong, L. Y., Rao, G., Zhou, S., Heimberger, A. B., and Gilboa, E. (2014). Targeting 4-1BB costimulation to the tumor stroma with bispecific aptamer conjugates enhances the therapeutic index of tumor therapy, *Cancer Immunol. Res.*, **2**(9), pp. 867–877.
69. Shigdar, S., Macdonald, J., O'Connor, M., Wang, T., Xiang, D., Al Shamaileh, H., Qiao, L., Wei, M., Zhou, S. F., Zhu, Y., Kong, L., Bhattacharya, S., Li, C., and Duan, W. (2013). Aptamers as theranostic agents: modifications, serum stability and functionalisation, *Sensors (Basel)*, **13**(10), pp. 13624–13637.

70. Shiratori, I., Akitomi, J., Boltz, D. A., Horii, K., Furuichi, M., and Waga, I. (2014). Selection of DNA aptamers that bind to influenza A viruses with high affinity and broad subtype specificity, *Biochem. Biophys. Res. Commun.*, **443**(1), pp. 37–41.
71. Shum, K. T., Zhou, J., and Rossi, J. J. (2013). Nucleic acid aptamers as potential therapeutic and diagnostic agents for lymphoma, *J. Cancer Ther.*, **4**(4), pp. 872–890.
72. Stoltenburg, R., Reinemann, C., and Strehlitz, B. (2007). SELEX: a (r)evolutionary method to generate high-affinity nucleic acid ligands, *Biomol. Eng.*, **24**(4), pp. 381–403.
73. Subramanian, N., Raghunathan, V., Kanwar, J. R., Kanwar, R. K., Elchuri, S. V., Khetan, V., and Krishnakumar, S. (2012). Target-specific delivery of doxorubicin to retinoblastoma using epithelial cell adhesion molecule aptamer, *Mol. Vis.*, **18**, pp. 2783–2795.
74. Subramanian, N., Kanwar, J. R., Athalya, P. K., Janakiraman, N., Khetan, V., Kanwar, R. K., Eluchuri, S., and Krishnakumar, S. (2015). EpCAM aptamer mediated cancer cell specific delivery of EpCAM siRNA using polymeric nanocomplex, *J. Biomed. Sci.*, **22**(1), p. 4.
75. Sun, H., and Zu, Y. (2015). Aptamers and their applications in nanomedicine, *Small*, **11**(20), pp. 2352–2364.
76. Sundaram, P., Kurniawan, H., Byrne, M. E., and Wower, J. (2013). Therapeutic RNA aptamers in clinical trails, *Eur. J. Pharm. Sci.*, **48**(1–2), pp. 259–271.
77. Szeitner, Z., András, J., Gyurcsányi, R. E., and Mészáros, T. (2014). Is less more? Lessons from aptamer selection strategies, *J. Pharm. Biomed. Anal.*, **101**, pp. 58–65.
78. Taghdisi S.M., Lavaee, P., Ramezani, M., and Abnous, K. (2011). Reversible targeting and controlled release delivery of daunorubicin to cancer cells by aptamer-wrapped carbon nanotubes, *Eur. J. Pharm. Biopharm.*, **77**(2), pp. 200–206.
79. Tang, L., Tong, R., Coyle, V. J., Yin, Q., Pondenis, H., Borst, L. B., Cheng, J., and Fan, T. M. (2015). Targeting tumor vasculature with aptamer-functionalized doxorubicin-polylactide nanoconjugates for enhanced cancer therapy, *ACS Nano*, **9**(5), pp. 5072–5081.
80. Tang, Y., Hu, H., Zhang, M. G., Song, J., Nie, L., Wang, S., Niu, G., Huang, P., Lu, G., and Chen, X. (2015). An aptamer-targeting photoresponsive drug delivery system using "off-on" graphene oxide wrapped mesoporous silica nanoparticles, *Nanoscale*, **7**(14), pp. 6304–6310.

81. Thiel, K. W., Hernandez, L. I., Dassie, J. P., Thiel, W. H., Liu, X., Stockdale, K. R., Rothman, A. M., Hernandez, F. J., McNamara, J. O. 2nd, and Giangrande, P. H. (2012). Delivery of chemo-sensitizing siRNAs to HER2+-breast cancer cells using RNA aptamers, *Nucleic Acids Res.*, **40**(13), pp. 6319–6337.
82. Thiel, W. H., Bair, T., Peek, A. S., Liu, X., Dassie, J., Stockdale, K. R., Behlke, M. A., Miller, F. J. Jr., and Giangrande, P. H. (2012). Rapid identification of cell-specific, internalizing RNA aptamers with bioinformatics analyses of a cell-based aptamer selection, *PLoS One*, **7**(9), p. e43836.
83. Thodima, V., Pirooznia, M., and Deng, Y. (2006). RiboaptDB: a comprehensive database of ribozymes and aptamers, *BMC Bioinformatics*, **7**(2), p. S6.
84. Wandtke, T., Woźniak, J., and Kopiński, P. (2015). Aptamers in diagnostics and treatment of viral infections, *Viruses*, **7**(2), pp. 751–780.
85. Wang, R., Zhu, G., Mei, L., Xie, Y., Ma, H., Ye, M., Qing, F. L., and Tan, W. (2014). Automated modular synthesis of aptamer-drug conjugates for targeted drug delivery, *J. Am. Chem. Soc.*, **136**(7), pp. 2731– 2734.
86. Wengerter, B. C., Katakowski, J. A., Rosenberg, J. M., Park, C. G., Almo, S. C., Palliser, D., and Levy, M. (2014). Aptamer-targeted antigen delivery, *Mol. Ther.*, **22**(7), pp. 1375–1387.
87. Woo, H. M., Kim, K. S., Lee, J. M., Shim, H. S., Cho, S. J., Lee, W. K., Ko, H. W., Keum, Y. S., Kim, S. Y., Pathinayake, P., Kim, C. J., and Jeong, Y. J. (2013). Single-stranded DNA aptamer that specifically binds to the influenza virus NS1 protein suppresses interferon antagonism, *Antiviral Res.*, **100**(2), pp. 337–345.
88. Wu, J., Song, C., Jiang, C., Shen, X., Qiao, Q., and Hu, Y. (2013). Nucleolin targeting AS1411 modified protein nanoparticle for antitumor drugs delivery, *Mol. Pharm.*, **10**(10), pp. 3555–3563.
89. Wu, X., Chen, J., Wu, M., and Zhao, J. X. (2015). Aptamers: active targeting ligands for cancer diagnosis and therapy, *Theranostics*, **5**(4), pp. 322– 344.
90. Xiang, D., Shigdar, S., Qiao, G., Wang, T., Kouzani, A. Z., Zhou, S. F., Kong, L., Li, Y., Pu, C., and Duan, W. (2015). Nucleic acid aptamer-guided cancer therapeutics and diagnostics: the next generation of cancer medicine, *Theranostics*, **5**(1), pp. 23–42.
91. Xiao, Z., and Farokhzad, O. C. (2012). Aptamer-functionalized nanoparticles for medical applications: challenges and opportunities, *ACS Nano*, **6**(5), pp. 3670–3676.

92. Xu, W., Siddiqui, I. A., Nihal, M., Pilla, S., Rosenthal, K., Mukhtar, H., and Gong, S. (2013). Aptamer-conjugated and doxorubicin-loaded unimolecular micelles for targeted therapy of prostate cancer, *Biomaterials*, **34**(21), pp. 5244–5253.

93. Yang, C., Yan, N., Parish, J., Wang, X., Shi, Y., and Xue, D. (2006). RNA aptamers targeting the cell death inhibitor CED-9 induce cell killing in *Caenorhabiditis elegans*, *J. Biol. Chem.*, **281**(14), pp. 9137–9144.

94. Zamay, T. N., Kolovskaya, O. S., Glazyrin, Y. E., Zamay, G. S., Kuznetsova, S. A., Spivak, E. A., Wehbe, M., Savitskaya, A. G., Zubkova, O. A., Kadkina, A., Wang, X., Muharemagic, D., Dubynina, A., Sheina, Y., Salmina, A. B., Berezovski, M. V., and Zamay, A. S. (2014). DNA-aptamer targeting vimentin for tumor therapy in vivo, *Nucleic Acid Ther.*, **24**(2), pp. 160–170.

95. Zavyalova, E., Samoylenkova, N., Revishchin, A., Golovin, A., Pavlova, G., and Kopylov, A. (2014). Evaluation of antithrombotic activity of thrombin DNA aptamers by a murine thrombosis model, *PLoS One*, **9**(9), p. e107113.

96. Zhang, X., Liang, H., Tan, Y., Wu, X., Li, S., and Shi, Y. (2014). A U87-EGFRvIII cell-specific aptamer mediates small interfering RNA delivery, *Biomed. Rep.*, **2**(4), pp. 495–499.

97. Zhao, N., You, J., Zeng, Z., Li, C., and Zu, Y. (2013). An ultra pH-sensitive and aptamer-equipped nanoscale drug-delivery system for selective killing of tumor cells, *Small*, **9**(20), pp. 3477–3484.

98. Zhao, Z., Fan, H., Zhou, G., Bai, H., Liang, H., Wang, R., Zhang, X., and Tan, W. (2014). Activatable fluorescence/MRI bimodal platform for tumor cell imaging via MnO_2 nanosheet-aptamer nanoprobe, *J. Am. Chem. Soc.*, **136**(32), pp. 11220–11223.

99. Zhou, C., Chen, T., Wu, C., Zhu, G., Qiu, L., Cui, C., Hou, W., and Tan, W. (2015). Aptamer $CaCO_3$ nanostructures: a facile, pH-responsive, specific platform for targeted anticancer theranostics, *Chem. Asian J.*, **10**(1), pp. 166–171.

100. Zhou, J., Li, H., Li, S., Zaia, J., and Rossi, J. J. (2008). Novel dual inhibitory function aptamer-siRNA delivery system for HIV-1 therapy, *Mol. Ther.*, **16**(8), pp. 1481–1489.

101. Zhou, J., and Rossi, J. J. (2010). Aptamer-targeted cell-specific RNA interference, *Silence*, **1**(1), 4.

102. Zhou, J., Neff, C. P., Swiderski, P., Li, H., Smith, D. D., Aboellail, T., Remling-Mulder, L., Akkina, R., and Rossi, J. J. (2013). Functional in vivo delivery

of multiplexed anti-HIV-1 siRNAs via a chemically synthesized aptamer with a sticky bridge, *Mol. Ther.*, **21**(1), pp. 192–200.

103. Zhou, J., Tiemann, K., Chomchan, P., Alluin, J., Swiderski, P., Burnett, J., Zhang, X., Forman, S., Chen, R., and Rossi, J. (2013). Dual functional BAFF receptor aptamers inhibit ligand-induced proliferation and deliver siRNAs to NHL cells, *Nucleic Acids Res.*, **41**(7), pp. 4266–4283.
104. Zhou, J., and Rossi, J. J. (2014). Cell-type-specific, aptamer-functionalized agents for targeted disease therapy, *Mol. Ther. Nucleic Acids*, **17**, p. e169.
105. Zhu, G., Zheng, J., Song, E., Donovan, M., Zhang, K., Liu, C., and Tan, W. (2013). Self-assembled, aptamer-tethered DNA nanotrains for targeted transport of molecular drugs in cancer theranostics, *Proc. Natl. Acad. Sci. U S A*, **110**(20), pp. 7998–8003.
106. Zhu, G., Cansiz, S., You, M., Qiu, L., Han, D., Zhang, L., Mei, L., Fu, T., Chen, Z., and Tan, W. (2015). Nuclease-resistant synthetic drug-DNA adducts: programmable drug-DNA conjugation for targeted anti-cancer drug delivery, *NPG Asia Mater.*, **7**, p. e169.
107. Zhu, J., Huang, H., Dong, S., Ge, L., and Zhang, Y. (2014). Progress in aptamer-mediated drug delivery vehicles for cancer targeting and its implications in addressing chemotherapeutic challenges, *Theranostics*, **4**(9), pp. 931–944.

Chapter 9

Aptamer–Liposome Conjugates: Current Art and Future Prospects

Oliver Ries and Stefan Vogel

Department of Physics, Chemistry and Pharmacy, University of Southern Denmark, Campusvej 55, Odense 5230, Denmark

snv@sdu.dk

While liposome formulations and nucleic acid aptamer therapeutics are already in clinical use, aptamer–liposome conjugates are now getting more in the focus. The versatility of liposomes in drug delivery combined with the high target selectivity of aptamers opens up new possibilities for highly efficient and selective treatment of many diseases, most of all, many types of cancer. Furthermore, aptamer–liposome conjugates can be used as highly selective sensors in diagnostics.

9.1 Introduction

The advantages and disadvantages of aptamers compared to other targeting molecules like monoclonal antibodies, antibody fragments or peptide-based targeting molecules are described elsewhere [1–4], as well as techniques for development and evolution of aptamers

Aptamers: Tools for Nanotherapy and Molecular Imaging
Edited by Rakesh N. Veedu

ISBN 978-981-4669-83-2 (Hardcover), 978-981-4669-84-9 (eBook)
www.panstanford.com

like SELEX and cell-SELEX [5, 6]. Aptamers are promising molecules with high potential in biomedical applications (e.g., therapeutics, diagnostics, biomarker discovery, and specific targeting ligands) [4, 7, 8]. Many aptamers are in clinical trial now [9, 10], while the first clinically used aptamer has been approved by the Food and Drug Administration (FDA) back in 2006: Macugen, for treatment of age-related macular degeneration [11].

Aptamers have been combined with a wide variety of nanoparticles: metals, metal oxide and silica nanoparticles, lipid-based structures, polymers, nanorods and nanotubes, nanoshells, and dendrimers [12–15]. Liposomes are the most prominent representatives of lipid-based structures. These spherical particles composed of a lipid bilayer encapsulating an aqueous compartment offer several advantages compared to other nanoparticles. They are biocompatible, biodegradable, easy to fabricate in controllable sizes at low cost, and with low batch-to-batch variation. Surface charge and stability can be controlled by lipid composition. Coating with polyethylene glycol (PEG) leads to prolonged circulation half-life in vivo, and they can encapsulate hydrophilic substances in the interior as well as hydrophobic substances in the lipid bilayer, thereby protecting them from the external medium [13, 16]. Many liposome formulations are currently approved for drug delivery or in clinical trial [17].

Membrane anchoring is a very efficient natural process for the noncovalent conjugation (attachment) of aptamers and biomolecules, in general, to liposomes [18, 19]. Lipidated oligonucleotides have been used earlier, and the same modifications and lipid building blocks can be used for lipid-modified aptamers [20–23, 106]. Subsequent attachment to liposomes should open up new possibilities for the selective delivery of therapeutic agents and/or contrast agents encapsulated in liposomes to specific cells (e.g., cancer cells) by simultaneous reduction of possible side effects [24].

9.2 Aptamer–Liposome Conjugation Strategies

Due to commercial availability of many different lipids, natural and synthetic, a large number of liposome formulations are

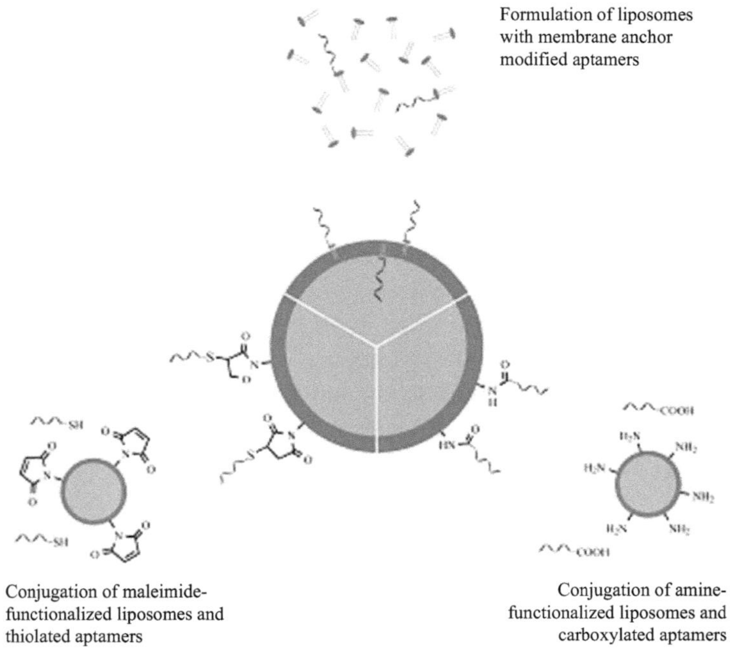

Figure 9.1 Techniques for attachment of aptamers on the liposome surface.

possible. Combined with the increasing number of aptamers, this leads to a huge amount of possible aptamer–liposome conjugate formulations. Limiting factors, on the other hand, are the (so far) used techniques for attachment of aptamers on the liposome surface: direct incorporation of a lipid membrane anchor during aptamer synthesis [25, 26], maleimide–thiol coupling, and amide bond formation between aptamer and a suitable functionalized lipid (Fig. 9.1).

A convenient approach for membrane anchor attachment is the incorporation at an early stage during aptamer synthesis. Some lipid modifications for in-house solid-supported oligonucleotide synthesis as well as lipid-modified oligonucleotides are commercially available. Other lipid modifications can be synthesized chemically and designed for special needs [27–30, 106]. Aptamers synthesized with a thiol group can be easily coupled to commercially available maleimide-functionalized lipids, while those bearing a carboxylic acid or amine functionality can be coupled to an amine- or carboxylic acid–functionalized lipid. Even though it was found that these highly amphiphilic conjugates can form micelles [31–35], it could be shown that these micelles also insert spontaneously into liposome membranes [36]. Another possibility is to incorporate lipid–aptamer conjugates during the liposome formulation process. Although the aptamer will be incorporated on the inside and outside of the liposome bilayer, the number of aptamers per liposome can be calculated. Both methods give knowledge about the exact concentration of aptamer used in each experiment and reduce the time spent for tedious determinations of coupling efficiencies.

On the other hand, the above-mentioned coupling methodologies between a thiol group and a maleimide-functionalized lipid as well as the formation of an amide bond between a lipid and an aptamer can be performed when the lipid is already incorporated in the liposome bilayer. This leads to liposomes bearing aptamers only on the outside, but successful coupling and surface coverage have to be determined by other methods, such as fluorescence, which require that the aptamer itself be fluorescently labeled or be detected otherwise.

9.3 Aptamer–Liposome Conjugates

9.3.1 *Toward Dye and Drug Delivery*

9.3.1.1 NX213 aptamer targeting VEGF

The first report of a lipid–aptamer conjugate attached to a liposome was published in 1998 [27]. Willis et al. used thereby an anchor synthesized from di(octadecyl)glycerol (DAG) ether. The lipid anchor has been attached to the 5′ end of the NX213 aptamer, identified by the same group using SELEX [37] during oligonucleotide synthesis (DAG-NX213). NX213 binds to the vascular endothelial growth factor (VEGF) with an EC_{50} of ~1 nM, which is a potent inducer of angiogenesis and is produced and secreted by most tumors and tumor cell lines [38, 39]. Incorporation of DAG-NX213 into liposomes was done by sonication of a lipid/DAG-NX213 suspension (DSPC/cholesterol 2:1, molar ratio). Liposome–aptamer conjugates (NX213-L) were approx. 50–65 nm in diameter with 34–60 aptamers per liposome, whereby one-third of the aptamers were located inside the liposomes.

Plasma pharmacokinetic studies in Sprague–Dawley rat plasma showed a slight increase in the overall clearance rate for DAG-NX213, while conjugation to liposomes led to a 3.6-fold decrease in clearance. When tested for binding to the target VEGF, NX213 and DAG-NX213 had similar binding affinities, while the binding affinity for NX213-L was around threefold lower, but still retaining most of the binding affinity, when tested against a scrambled version of the aptamer (scNX213) as negative control. Inhibition experiments with [^{125}I] VEGF and human umbilical vein endothelial cells (HUVECs) expressing the VEGF receptor, DAG-NX213 and NX213-L showed a fivefold lower potency than the free NX213 aptamer. HUVEC proliferation, measured by [^{3}H]thymidine uptake on the other hand, showed that NX213 (up to 1 μM) was unable to inhibit proliferation, while DAG-NX213 and NX213-L could inhibit HUVEC proliferation in a dose-dependent manner with EC_{50} values of $\sim$3 nM. Furthermore, Willis and coworkers performed an improved Miles assay in guinea pigs [40, 41]. Intradermal coinjection of VEGF (20 nM) and NX213 showed no inhibition, while coinjection with DAG-NX213 and NX213-L, respectively, led to up to 60% and 90% inhibition of VEGF-induced vascular permeability at an aptamer concentration of 10 μM (highest tested) with IC_{50} values of 0.3–0.5 μM. Inhibition of VEGF-induced angiogenesis was further tested with a chicken chorioallantoic membrane (CAM) assay [42]. Thereby only NX213-L (1 μM) was potent enough to suppress VEGF-induced angiogenesis, while NX213 and DAG-NX213 had little or no effect.

9.3.1.2 The sgc8 aptamer targeting tyrosine kinase 7

Tan and coworkers described in 2010 the construction of liposome–aptamer conjugates, which were loaded with a fluorescein-dextran derivate (FD) as a drug mimic [43]. They conjugated the liposomes to a 3′-thiol-modified sgc8 aptamer, which was further tetramethylrhodamine (TMR) labeled on the 5′ end by using maleimide-functionalized lipids. The sgc8 aptamer itself is selective toward leukemia CEM-CCRF cells with a K_d of 0.8 nM. The aptamer binds to the protein tyrosin kinase 7 (PTK7) [44–46].

Formulation of dye-encapsulating liposomes (HSPC/cholesterol/mPEG-DSPE/MalPEG-DSPE 2:1:0.08:0.02, molar ratio) has been performed through extrusion followed by incubation with a pre-activated aptamer, leading to aptamer–liposome conjugates (sgc8-L_{FD}) with an average size of 100 nm and approx. 250 aptamer molecules per liposome. Dialysis showed a loss of 15%–20% of the encapsulated fluorescein derivative as well as some TMR-sgc8 aptamer and lipid-linked TMR-sgc8 aptamer.

Liposome-assisted delivery of the drug mimic FD was investigated with human acute lymphoblastic leukemia cells (CEM) and human acute promyelocytic leukemia cells (NB4), the latter lacking the target membrane protein receptor PTK7 for sgc8 as negative control [45, 46]. Flow cytometry showed a significant increase for the CEM cell line in the fluorescence signal with increasing liposome concentration, while the control group showed no increase. Aptamer-free liposomes led only to a slight increase in fluorescence, indicating that the aptamer is necessary for selective and efficient uptake of the liposomes into target cells. Extended incubation times (3 and 6 h) led to the conclusion that binding of the liposome–aptamer conjugate to the target cell surface was already completed after the initially tested short incubation period (20 min), while longer time only favored nonspecific binding.

9.3.1.3 FKN-S2 aptamer targeting fractalkine

Kokkoli and coworkers focused in 2014 on the effect of different spacers between the fractalkine aptamer FKN-S2 and a lipid membrane anchor, bearing two C_{16} chains [47]. FKN-S2 was found to bind to fractalkine with a K_d of 3.4 nM. The secondary structure of the aptamer forms a stem-loop region as well as a G-quadruplex [48]. Fractalkine itself is a cell surface protein, which is expressed by a number of cancer types (e.g., epithelial ovarian cancer, adenocarcinoma) and inflamed endothelial cells [49–51]. It is thereby mainly involved through recruitment and adhesion of leukocytes [52, 53].

The effect of eight different spacers with 4, 8, or 24 PEG units, a C_{12}- or C_{24}-carbon chain, T_5- or T_{10}-oligomer, and an A_{10}-oligomer were investigated to link the 3′ end of the aptamer and the lipid membrane anchor, compared to aptamer–lipid conjugates with no spacer in between. As a lipid membrane anchor a di-cetyl glutamic acid ester derivative was used, which was coupled to the appropriate spacer or the aptamer itself via amide bonds.

Initially, the lipid–aptamer conjugates were investigated without liposomes in order to reveal the influence of various spacer units and the lipid moiety. In a radioactive competition binding assay with ^{32}P-labeled aptamer, an unlabeled competitor aptamer and fractalkine, all lipid-linked aptamers showed a significant increase in the IC_{50} value from 8.6 nM for the free aptamer to 21.8–75.6 nM. Within the spacer-containing aptamers, nucleic acid–derived linkers showed the best values, followed by those comprising polyethylene glycol spacers, while hydrocarbon chains and attachment of the lipid residue without a spacer unit resulted in the highest values. Attachment of spacer and lipid moiety at the 5′ end of the aptamer led to even higher values. Comparison with two unlipidated aptamers bearing a T_5- and T_{10}-spacer, respectively, at the 3′ end led to stronger binding and IC_{50} values of approx. 3.5 nM. This clearly indicates that lipid moiety leads to reduced binding affinity, when only amphiphilic conjugates are used in the absence of a liposome.

For a potential application lipidated 3′-T_{10}-FKN-S2-aptamer (Lip-T_{10}Apt), the conjugate with the lowest IC_{50} value against fractalkine was attached to stealth liposomes (composition not given) encapsulating calcein as drug mimic. After incubation with Lip-T_{10}Apt, the authors state an incorporation ratio of 79% for the liposome–aptamer conjugate (T_{10}Apt-L), even though it remains unclear how this value was determined. Subsequently, mouse colon adenocarcinoma cells transfected to express fractalkine (MCA-38.FKN) on their surface were treated for 1 h with T_{10}Apt-L conjugate at different aptamer concentrations (0–0.53 mol%), and calcein fluorescence was determined. Thereby with increasing aptamer concentration, binding increased and reached a plateau at 0.28 mol%, resulting presumably from sterical hindrance of liposomes on the cell surface.

9.3.1.4 ESTA aptamer targeting E-selectin

In a study toward the targeting of multiple cancer types Mann et al. in 2011 linked a thiolated aptamer against E-selectin to fluorescein-encapsulating liposomes [54]. Since E-selectin is selectively expressed by inflamed endothelial cells (e.g., in advanced tumors), this would enable a site-specific delivery of encapsulated therapeutics or contrast agents [55–59]. Earlier the same group described the development of a partly thiolated aptamer (ESTA) which was specific against E-selectin with a binding affinity of $K_d = 47$ nM [60]. ESTA was further 5′-modified with a Cy3 dye and a C10-linker bearing a carboxylic acid functionality (ESTA-Cy3) for coupling to prior extruded liposomes (DPPC/cholesterol/DSPE-PEG2000-NH_2 58.9:40:1.1, molar ratio) modified with an amino poly(ethylene glycol) lipid. The liposomes were empty (L) or encapsulating a fluorescein dye (L_{FITC}) for fluorescence studies. Aptamer–liposome conjugates (ESTA-L_{FITC}) had a diameter of 119 nm, while bearing approx. 485 aptamers per liposome on the surface.

Mann et al. used TNF-α-treated human umbilical cord vein endothelial cells (HUVECs). TNF-α induces thereby an approx. 20-fold increase in E-selectin expression on the cell membrane. When incubated with 10 nM ESTA-L_{FITC}, an increase in fluorescence, both for FITC and Cy3, could be detected inside TNF-α-treated cells, while untreated cells showed no increase after 24 h. Negative controls with FITC-encapsulating liposomes lacking the aptamer showed no fluorescence increase with TNF-α-treated or untreated HUVECs. In further in vivo studies, mice bearing MDA-MB-231-derived breast xenograft tumors were treated with ESTA-L further labeled with Rhodamine-DHPE (ESTA-LR). Five hours after injection of 3 mg of liposomes, the tumor and the major organs were isolated. Thereby an increase in red fluorescence could be observed on the tumor vascular, while the organs were not affected. When treated for 48 h, ESTA-LR could also be found accumulated in the tumor parenchyma, indicating that an elevated tumor targeting could lead to extravasation of the aptamer–liposome conjugate. For pharmacokinetic investigations of the aptamer–liposome conjugates, FBV mice were intravenously treated with ESTA-LR and liposomes without ESTA bearing the amino-PEG lipid only on the surface (aPEG-LR). Rhodamine fluorescence was thereby monitored in the plasma and compared to an untreated control group. While the volume of distribution was nearly the same in both cases, ESTA-LR conjugates showed a decrease in clearance and a slight increase in circulation half-life.

9.3.1.5 RNA aptamer Apt1 targeting the CD44 receptor protein

Fattal and coworkers presented in 2015 a work on targeting cancer stem cells (CSCs) [61]. CSCs are a small population of tumor cells, which show a high resistance to traditional cancer therapeutics. They are able to maintain and proliferate [62, 63]. Therefore, a specific treatment of CSCs in addition to normal tumor cells is crucial for eliminating tumors and decreasing tumor relapse [64, 65] since these cells are able to differentiate into other tumor cells, thereby initiating and proliferating new tumors and self-renewing themselves [66–68].

An RNA aptamer (Apt1) against the standard isoform of human CD44 receptor protein, further modified with 2′-F-pyrimidine to increase nuclease stability was used [69]. CD44 displays the most common surface marker of CSCs and is overexpressed by many tumors (e.g., colon, breast, pancreatic, head, and neck cancers) [70–73]. This multistructural and functional cell surface glycoprotein is involved in the regulation of cell adhesion, proliferation, homing, migration, motility, survival, angiogenesis, and differentiation [74].

Aptamer–liposome conjugates (Apt1-L) have been prepared by extrusion of liposomes (DPPC/cholesterol/MalPEG-DSPE 62:35:3, molar ratio) followed by incubation with 3′-thiol-5′-FITC-modified Apt1 leading to conjugates with a diameter of 140 nm and ca. 422 aptamers per liposome.

Apt1-L showed an increased binding affinity to magnetic bead immobilized CD44 (K_d 6.2 $\pm$ 1.6 nM) compared to free Apt1 (K_d 21.5 $\pm$ 3.3 nM). On the other hand, the maximum binding saturation decreased from 374 $\pm$ 45 fmol for free Apt1 to 106 $\pm$ 13 fmol for Apt1-L. Multiple aptamers on the surface increase the binding affinity to cell surface expressed CD44, while the liposomes self-block locally free binding sites due to steric hindrance. Binding specificity and selectivity of the aptamer–liposome conjugate have been carried out with human lung cancer cells (A549) and human breast cancer cells (MDA-MB-231), both known to overexpress CD44 at a high level, and mouse embryonic fibroblast cells (NIH/3T3) as negative control cells. For fluorescence assays, aptamer–liposome conjugates further modified with a Rhodamine lipid (Apt1-LR) have been used. CD44-expressing cells showed an increase in fluorescence, when treated with Apt1-LR, compared to CD44-negative cells, which showed no increase in fluorescence. Moreover, fluorescence microscopy demonstrated a higher accumulation of Apt1-LR in the cytoplasm of CD44-positive cell lines compared to CD44-negative mouse fibroblasts, both indicating the selective and cell-specific cellular uptake.

For further investigation of Apt1-L as a potential drug delivery system, the inflammatory response of the two CD44-expressing cell lines A549 and MDA-MB-231 were investigated. The absence of an inflammatory response would thereby strengthen the usefulness of this system, since cytokines associated with chronic inflammation are known to give positive feedback loops and activate signaling pathways involved in tumor maintenance, invasion, metastasis, and drug resistance [75]. The amount of inflammatory cytokines IL-12p70, TNF, IL-10, IL-6, IL-1β, and IL-8 were monitored after treatment with Apt1-L conjugate, liposomes lacking the aptamer, and lipopolysaccharide (LPS), the latter as positive control, since LPS induces strong immune responses in animals [76]. While LPS treatment showed a significant increase in the secretion of some of the cytokines (different cytokines for both cell lines), neither Apt1-L nor aptamer-free liposomes showed an increase in any of the tested cytokines.

9.3.1.6 AraHH001 aptamer targeting the cardiac protein troponin T

Harashima and coworkers used in 2014 in their approach to establish a liposomal drug delivery system an aptamer, AraHH001, derived from cell-SELEX [77]. AraHH001, targeting the cardiac protein troponin T, showed high selectivity for tumor vascular cells (K_d = 43 nm) without binding to healthy cells. Furthermore, the aptamer had a high internalization capacity, important for selective drug delivery [78].

Thiolated aptamer AraHH001 was coupled to maleimide-functionalized DSPE-PEG(2000) lipid (DSPE-AraHH001), which was further incorporated in liposomes (EPC/cholesterol/Rhodamine-DOPE/DSPE-AraHH001 70:30:1:5, molar ratio) by a short sonication protocol leading to aptamer–liposome conjugates (AraHH001-L; size not given). When incubated with mouse tumor endothelial cells (mTEC), a nearly 3.8-fold higher fluorescence was detected compared to aptamer-free control liposomes. Additionally, it could be shown that some of the conjugates were localized in the lysosomal compartment. Further investigation of the internalization process revealed that the aptamer–liposome conjugate uptake showed a dependency on the clathrin-mediated pathway. In vivo studies in mice bearing human renal cell carcinoma (OS-RC-2) cells validated a co-localization of AraHH001-L and mTECs in the tumor.

9.3.1.7 xPSM-A9 targeting prostate-specific membrane antigen

Prostate-specific membrane antigen (PSMA) expressed on prostate cancer cells has been a target for a drug delivery system designed by Baek et al. in 2014 [79]. A PSMA-specific RNA aptamer xPSM-A9 was used [80] further modified with 2′-fluoro-pyrimidines and a DNA tether for hybridization with the actual linker sequence, a short thiol- and FITC-labeled DNA strand. First, the latter was coupled to DSPE-PEG(2000)/DSPE-PEG(2000)-maleimide (1:4, molar ratio) micelles, followed by annealing of the actual aptamer to the DNA-tether, and finally, incubation with prior extruded liposomes (POPC/Cholesterol/DSPE-PEG/DSPE-Rhodamine 2:1:0.1:0.03, molar ratio) leading to aptamer–liposome conjugates (xPSM-A9-L) with a diameter of approx. 98 nm. For drug delivery experiments doxorubicin (Dox) was encapsulated during liposome extrusion (xPSM-A9-L_{Dox}).

For in vitro studies, PSMA-expressing LNCaP human prostate cancer cells and PSMA-negative PC3 human prostate cancer cells were used. When incubated with Apt-L an increase in fluorescence at the cell surface and inside the cells in the cytoplasm could be detected, while liposomes lacking the aptamer showed no increase in fluorescence. This effect could already be seen after 15 min of incubation and increased with time. A cell viability assay comparing both cell lines incubated with either free Dox, Dox-encapsulating liposomes with (xPSM-A9-L_{Dox}), or without (L_{Dox}) aptamer and nonencapsulating xPSM-A9-L conjugates demonstrated that in the case of PSMA-negative PC3 cells, only free Dox led to a reduced cell viability (ca. 50%). When incubated with LNCaP cells xPSM-A9-L_{Dox} as well as free Dox showed a cell viability of ca. 50%, while that for L_{Dox} remained at approx. 100%. This demonstrates the selective delivery of Dox and the specific uptake by PSMA-positive LNCaP cells.

In further in vivo studies, BALB/c nude mice transfected with LNCaP human prostate cancer xenografts were treated with either xPSM-A9-L_{Dox} or plain L_{Dox} (0.27 mg Dox/kg) lacking the aptamer. Fluorescence images of major inner organs and the tumor showed accumulation in liver and kidney in both cases, while only xPSM-A9-L_{Dox} accumulated in the tumor tissue as well. Mice transfected with PC3 cancer cells as control showed accumulation in both cases only in liver and kidney as expected due to the lack of PSMA on the tumor cells. For in vivo efficacy evaluation of Dox-encapsulating aptamer–liposome conjugates, LNCaP-transfected mice were treated over 4 weeks with a saline solution, free Dox, L_{Dox}, or xPSM-A9-L_{Dox} every seventh day (Dox concentration 0.3 mg/kg). Thereby xPSM-A9-L_{Dox} and L_{Dox} showed the best results. Both liposome formulations could clearly decelerate tumor growth over time compared to free Dox, while xPSM-A9-L_{Dox}-treated tumors were still significantly smaller than the ones treated with L_{Dox}.

9.3.1.8 AS1411 aptamer targeting nucleolin

A frequently used aptamer toward drug delivery by aptamer–liposome conjugation is AS1411 (also termed GRO29A or AGRO100), which was found to bind selectively to nucleolin (NCL) [81–83]. Selective and effective binding to the target could be further supported by molecular dynamics (MD) simulations, resulting in a strong negative binding energy for the AS1411–nucleoline complex [90]. NCL displays the *bcl-2* mRNA-binding protein and is the first aptamer to enter clinical cancer studies [83, 85]. Inhibition of NCL leads to increased *bcl-2* mRNA instability and cytotoxicity in a number of cancer cell lines (e.g., MCF-7, MDA-MB-231, A549, C6, and HeLa). AS1411 was found to inhibit proliferation of these cancer cell lines at a concentration of 5 μM [86].

In 2009 Lu and coworkers showed the example of liposome–aptamer-conjugate-based drug delivery of cisplatin utilizing AS1411, where they additionally investigated a complementary RNA strand as a potential antidote [87]. The IC_{50} values for free cisplatin to human breast ardenocarcinoma cells (MCF-7) and human prostate cancer cells (LNCaP) has already been determined to be 28 μM and 5.95 μM, respectively [88, 89]. The AS1411 aptamer has been modified with a $3'$-T_{12}-spacer coupled to a cholesterol-tetra(ethylene glycol) modification. The T_{12}-spacer was chosen to achieve a larger distance between the aptamer and the liposome surface in order to prevent a decrease in aptamer binding to the target. When incubated in vitro with MCF-7 cells, the aptamer alone could already show higher internalization compared to a scrambled control sequence.

Formulation of liposome–aptamer conjugates (AS1411-L) was done by extrusion of a lipid film (HSPC/cholesterol/ DSPE-mPEG2000 2:1:0.16, molar ratio) rehydrated with an aptamer solution and either calcein as fluorescent dye (AS1411-L_{calc}) or cisplatin (AS1411-L_{CP}) for drug delivery yielding conjugates of approx. 200 nm diameter and aptamers on the inside and outside (lipid/aptamer ratio $\approx$ 1680:1).

Treatment of NCL-expressing MCF-7 cells and NCL-negative LNCaP cells as control with AS1411-L_{calc} showed strong fluorescence in MCF-7 cells only. Antiproliferation of cancer cells was investigated by treatment of MCF-7 cells with AS1411-L_{CP}, nonencapsulating AS1411-L, and cisplatin-containing liposomes with a random DNA sequence, as well as of LNCaP cells with AS1411-L_{CP}. Thereby only AS1411-L_{CP}-treated MCF-7 cells showed significant cell death after 4 days (59.5%), while other formulations had only little or no effect.

The problem of overdosage and possible side effects has been addressed. Since most of the receptors and biomarkers targeted in cancer therapy are not exclusively expressed on cancer cells, only overexpressed, normal tissue could be damaged as well. When the complementary 2′-O-methyl-modified RNA strand of AS1411 (cRNA) was used with AS1411-L_{CP} and MCT-7 cells, cell death could be successfully prevented when a cRNA/aptamer ratio of 5:1 was used. Furthermore, cRNA was added at different time points after addition of AS1411-L_{CP} to the MCT-7 cells. Simultaneous addition of the liposome–aptamer conjugate and cRNA led to approx. 100% cell viability, while the viability was reduced to 43%, when cRNA was added 5 h after AS1411-L_{CP} treatment.

The identical aptamer–liposome conjugate, now encapsulating Dox, has been used by the same group in a later study [90]. A cell viability assay showed thereby for aptamer-functionalized Dox-liposomes (AS1411-L_{Dox}) a significant decrease in cell viability, equipotent as treatment with free Dox, when compared to liposomes lacking the aptamer. An in vivo study utilizing MCF-7 tumor cell transfected athymic nude mice treated with AS1411-L_{Dox} or Dox-liposomes lacking the aptamer showed that treatment with AS1411-L_{Dox} inhibited tumor growth significantly better than aptamer-free liposomes. Toxicity of the liposomes could be ruled out, since the overall health of mice was not affected during the experiment. In vitro and in vivo experiments regarding the penetration behavior of AS1411-L_{Dox} and aptamer-free liposomes showed better penetration for aptamer–liposome conjugates, most likely due to the targeting of nucleolin on the tumor cell surface. Further investigation of in vivo treated tumors showed a high level of internalization of AS1411-L_{Dox}, proving that the conjugate is actually taken up by the tumor cells.

To overcome slow drug release from aptamer-liposomes, Sung and coworkers reported on an AS1411-functionalized liposome containing Dox and ammonium bicarbonate (ABC) as thermoresponsive release trigger [84]. The corresponding ABC-triggered liposome formulation has been described earlier by the same group [91].

Aptamer–liposome conjugates containing Dox and ABC (AS1411-$L_{Dox/ABC}$) have been prepared by extrusion of a rehydrated lipid film (DPPC/cholesterol/PEG(2000)-DSPE 60:40:5, molar ratio) followed by incubation with prior prepared aptamer–lipid conjugate micelles. The aptamer–lipid conjugate was synthesized by thiol–maleimide coupling of thiol-modified AS1411 with Mal-PEG(2000)-DSPE. Stability experiments showed that AS1411-$L_{Dox/ABC}$ and aptamer-free liposomes ($L_{Dox/ABC}$) were stable at 4°C over weeks. At normal human body temperature (37°C), only a few CO_2 bubbles with some Dox release could be observed, while at 42°C a significantly increased amount of CO_2 was detected leading to a strong Dox release at the same time.

When multidrug-resistant MCF-7/ADR[a] cells were incubated in vitro with AS1411-$L_{Dox/ABC}$ followed by hyperthermia treatment, a significant increase in fluorescence within the cells could be detected compared to plain $L_{Dox/ABC}$ liposomes. Pretreatment of cells with a free AS1411 aptamer in order to block nucleolin completely inhibited Dox delivery to cells by AS1411-$L_{Dox/ABC}$ (same intensity as for plain liposomes). Furthermore, the internalization process could be followed. After internalization, AS1411-$L_{Dox/ABC}$ was located in early endosomes followed by lysosomes. Cells analyzed without hyperthermia treatment showed a slow increase in Dox fluorescence in the nucleus whereas cells treated with hyperthermia had a fast and strong fluorescence signal. The same effect could be seen in a cell viability assay. While free Dox and plain $L_{Dox/ABC}$ at 37°C and 42°C as well as AS1411-$L_{Dox/ABC}$ at 37°C did not show any effect on cell viability, AS1411-$L_{Dox/ABC}$ at 42°C led to a significant decrease (approx. 50%).

For in vivo studies, athymic nude mice were transfected with MCF-7/ADR tumor cells. After treatment with free Dox, plain $L_{Dox/ABC}$ and AS1411-$L_{Dox/ABC}$, the change in fluorescence in the tumor tissue and major organs was monitored. Free Dox-treated mice showed a generally weaker fluorescence than the liposome-treated mice, except for an accumulation of fluorescence in the heart, indicating unfavorable cardiotoxicity of Dox [95]. Regarding the tumor tissue, only a low fluorescence signal for $L_{Dox/ABC}$ before hyperthermia treatment, which increased at 42°C, was observed. However, the tumor tissue of AS1411-$L_{Dox/ABC}$-treated mice showed a significantly higher fluorescence signal at 37°C, which further increased after hyperthermia treatment showing successful delivery

---

[a]MCF-7/ADR (now NCI/ADR-RES) was developed by Batist et al. in 1986 derived supposedly from MCF-7 breast cancer cell line [92]. In 1998, Scudiero et al. reported on inconsistencies when comparing DNA-fingerprints of MCF-7 and MCF-7/ADR cell lines [93]. After debates in literature, it has to be concluded that the MCF-7/ADR cell line was originally derived from the ovarian cancer cell line OVCAR-8 [94].

of Dox into the tumor cells. Further investigations toward antitumor activity confirmed the aforementioned results. While free Dox had nearly no effect, $L_{Dox/ABC}$ could stop tumor growth at 37°C and 42°C. Only AS1411-$L_{Dox/ABC}$ treatment led to a significant decrease in tumor size, while treatment at 42°C was even more effective than at 37°C (25% and 50%, respectively, of the relative tumor volume).

In 2014 Yang and coworkers reported on the use of AS1411-aptamer–liposome conjugates for the delivery of siRNA in the treatment of melanoma [96]. The siRNA used, siBRAF, targets the mutant BRAF gene *BRAFV599E*, which seems to be crucial for cell viability and transformation [97].

Aptamer–liposome conjugates (AS1411-L) were prepared in a two-step procedure. In the first step, siRNA was added to cationic liposomes (DC-cholesterol/DOPE/Mal-PEG(3400)-DOPE 1:1:0.1, molar ratio) after extrusion, followed by coupling of the thiol-C_6-modified AS1411 aptamer by thiol–maleimide chemistry. In this case the siRNA is adsorbed on the surface of the liposomes instead of being encapsulated inside. A cytotoxicity study could show that the aptamer–liposome conjugates were, in contrast to cationic liposomes, not cytotoxic.

First internalization experiments were conducted with fluorescently labeled siRNA-carrying AS1411-L and compared to a conjugate bearing a nontargeting DNA sequence in place of the aptamer. Treatment of human melanoma cancer cells (A375), human breast cancer cells (MCF-7), normal healthy mouse hepatocytes (AML12), and mouse myoblast cells (C2C12), the latter two as negative control, showed an increase in fluorescence for A375 and MCF-7 cells treated with AS1411-L. Since the fluorescence in A375 cells was higher compared to MCF-7 cells, subsequent experiments were performed with this cell line only. Further investigation revealed that prolonged incubation time and higher siRNA concentration led to an increase in siRNA uptake. In vitro gene silencing activity analysis showed a clear reduction of activity for AS1411-L and nontargeting conjugates, increasing with higher siRNA loading, whereby the remaining activity for AS1411-L was significantly lower compared to the nontargeting one. At the same time, only AS1411-L was able to reduce BRAF protein expression in A375 cells and to reduce cell viability significantly.

In vivo studies with A375 melanoma tumor transfected nude mice intravenously treated with free siRNA, siRNA on plain liposomes, the nontargeting conjugate and AS1411-L showed significantly increased fluorescence in the tumor tissue after 48 h only for AS1411-L. Kidneys showed a high fluorescence signal indicating renal excretion of siRNA. Similar to the in vitro results, only AS1411-L was able to reduce the BRAF mRNA expression level in tumor tissue significantly. Furthermore, AS1411-L-treated tumor tissue became loose with a reduced number of tumor cells and signs of tumor necrosis.

9.3.2 *Toward a Colorimetric Diagnostic Tool*

A significantly different approach in terms of aptamer–liposome conjugate applications compared to the ones used for drug delivery has been published by Soh and coworkers in 2010 [98]. Two different aptamers, BOCK and TASSET [99, 100], both selective for thrombin, but at different binding sides, have been used as part of a colorimetric sensor system. Thrombin is a key enzyme in blood clotting and hemostasis. It converts soluble fibrinogen into insoluble fibrin, leading to blood coagulation. On the other hand, thrombin is also a major factor in thromboembolic diseases. The BOCK aptamer, selective for the fibrinogen-binding site, binds with a K_d of 2.5 nM, whereas the TASSET aptamer, selective for the heparin-binding site, showed a K_d of 1.5 nM. [101]. Both aptamers have been functionalized with diacetylene lipid membrane anchors, leading to a colorimetric sensor system after incorporation into liposomes and polymerization of the diacetylene units. Polymerization of the diacetylene moieties leads to a system with alternating triple and double bonds in the backbone and shows a blue color. External distortions in the headgroups lead to a change in the backbone length and angle, which can be seen by a characteristic blue to a red shift [102–104].

The aptamers were functionalized with an amino group for coupling to the 10,12-tricosadiynoic acid (TCDA) membrane anchor. BOCK and TASSET aptamers, as well as a random DNA sequence as a negative control were used leading to the corresponding lipidated BOCK (TCDA-BOCK) and TASSET aptamers (TCDA-TASSET) and a random DNA sequence (TCDA-RDM). Liposomes (10,12-pentacosadiynoic acid (PCDA)/DMPE/TCDA conjugate 53.3:40:6.7, molar ratio) were prepared by sonication and subsequently irradiated with UV light in order to polymerize the diacetylene moieties. Differently decorated liposomes have been formulated with a diameter of approx. 329 nm. Liposomes bearing one of the aptamers (BOCK: BOCK-L or TASSET: TASSET-L) or the random sequence (RDM-L) on the surface as well as a 1:1 mixture of both aptamers (BOCK/TASSET-L).

When tested against thrombin, the authors quantified the color change and the efficiency of binding to the target by calculation of the colorimetric response (CR). The CR displays thereby the percentage change in the ratio of absorbance for the red and blue emission. When tested with 3 μM thrombin, BOCK-L and TASSET-L bearing only one type of aptamer on the surface showed only a minor color shift (CR: 9% and 10%), while the color change for BOCK/TASSET-L, bearing both aptamers, was visible to the unaided eye (CR: 21%). In the following test the authors challenged the BOCK/TASSET-L conjugate with different thrombin concentrations (0–10 μM) and could thereby determine the "limit of visual detection" (LOD) at 0.5 μM. At this concentration the colorimetric response was found to be 10%, a value which can be seen already with the unaided eye, as proposed by Su et al. [105].

Further investigations were performed to validate that unspecific binding did not occur. Therefore, BOCK/TASSET-L was treated with bovine serum albumin (BSA), immunoglobulin G (IgG), streptavidin (STA), and fetal bovine serum (FBS, 10%). Comparable concentrations led to negligible color changes. Even 15-fold higher concentrations of BSA, IgG, and STA did not result in a significant sensor response, while BOCK/TASSET-L in FBS still resulted in a significant color change. Unspecific DNA binding could be ruled out when RDM-L was treated with thrombin. No significant color change occurred, leading to the conclusion that the change in color is caused by thrombin.

9.4 Future Prospects

Studies on the use of aptamer–liposome conjugates for drug delivery or as diagnostic tools have shown promising results. The early drug delivery studies focus mainly on the formulation of aptamer–liposome conjugates, their properties, and in vitro experiments. Later studies also include first in vivo experiments in mice.

Liposomes have a number of advantages over other nanoparticles, for example, biocompatibility, biodegradability, the possibility to encapsulate hydrophilic as well as hydrophobic molecules at high concentrations, low to no inflammatory response, and easy modification on their surface. Encapsulation of therapeutics in high dosage while reducing or eliminating the possible side effects and toxicity to healthy tissue is a clear advantage.

Regarding the conjugation process, different approaches have been shown. Aptamers have been attached to liposomes by addition of lipid–aptamer conjugates during liposome preparation, leading to aptamers located on the inside and outside of the bilayer, incubation of plain liposomes with lipid–aptamer conjugates, or coupling of thiol-modified aptamers to liposomes bearing reactive maleimide functionalities on the surface. Different spacer units connecting the lipid membrane anchor and the actual aptamer have been investigated and seem to play a role in target binding, even though this effect might also be aptamer sequence-dependent [47]. The actual structure of the membrane anchor should not significantly alter the binding efficiency, as long as it provides a strong binding within the lipid bilayer. On the other hand, attachment of a membrane anchor may result in different binding efficiencies depending on the point of attachment (5′ vs. 3′) [47]. A further advantage of liposomes is the possibility to attach more than one aptamer (or other recognition unit) on the surface in the same manner. Applying more than one aptamer for the same target at different binding sides or different aptamers specific for different targets on a cell surface may further increase binding selectivity and binding efficiency [98].

Additionally a first investigation toward an antidote to aptamer–liposome conjugates could show in vitro the feasibility of using counterstrand RNA to regulate the activity of aptamer–liposome-conjugate-mediated drug delivery [87]. This would open up an easy and cheap way to highly selective antidotes.

The problem of slow drug release once the drug-encapsulating liposomes are internalized in tumor cells was addressed as well [84].

Coencapsulation of ammonium bicarbonate together with the drug led to a thermolabile system which evolved CO_2 at temperatures above 42°C, leading to leakage and thereby drug release from the liposomal carrier.

Furthermore, the possibility to use aptamer–liposome conjugates in colorimetric assays for diagnostic applications could be successfully shown [98]. This further broadens the spectrum of the use of these highly versatile conjugates.

So far, research in aptamer–liposome conjugates for diagnostics and especially drug delivery is still in an early stage. An increasing number of studies focus now on in vivo studies, but without taking results, improvements, and advantages from other studies into account, thereby combining the positive effects. More in vivo experiments have to show the usefulness of aptamers to direct liposomes as drug delivery systems, specifically to cancer cells without harming healthy cells. Nevertheless this field of research seems to come more and more in focus and will therefore also find its way into clinical cancer treatment.

References

1. Wang, A. Z., Gu, F., Zhang, L., Chan, J. M., Radovic-Moreno, A., Shaikh, M. R., and Farokhzad, O. C. (2008). Biofunctionalized targeted nanoparticles for therapeutic applications, *Expert Opin. Biol. Ther.*, **8**, pp. 1063–1070.
2. Radom, F., Jurek, P. M., Mazurek, M. P., Otlewski, J., and Jelén, F. (2013). Aptamers: molecules of great potential, *Biotechnol. Adv.*, **31**, pp. 1260–1274.
3. Zhou, J., and Rossi, J. J. (2014). Cell-type-specific, Aptamer-functionalized agents for targeted disease therapy, *Mol. Ther. Nucleic Acids*, **3**, p. e169.
4. Nimjee, S. M., Rusconi, C. P., and Sullenger, B. A. (2005). Aptamers: an emerging class of therapeutics, *Annu. Rev. Med.*, **56**, pp. 555–583.
5. Darmostuk, M., Rimpelová, S., Gbelcová, H., and Ruml, T. (2015). Current approaches in SELEX: an update to aptamer selection technology, *Biotechnol. Adv.*, **33**, pp. 1141–1161.

6. Kanwar, J. R., Roy, K., Maremanda, N. G., Subramanian, K., Veedu, R. N., Bawa, R., and Kanwar, R. K. (2015). Nucleic acid-based aptamers: applications, development and clinical trials, *Curr. Med. Chem.*, **22**, pp. 2539–2557.
7. Keefe, A. D., Pai, S., and Ellington, A. (2010). Aptamers as therapeutics, *Nat. Rev. Drug Discovery*, **9**, pp. 537–550.
8. Betat, H., Vogel, S., Struhalla, M., Förster, H., Famulok, M., Welzel, P., and Hahn, U. (2003). Aptamers that recognize the lipid moiety of the antibiotic moenomycin A, *Biol. Chem.*, **384**, pp. 1497–1500.
9. Foy, J. W., Rittenhouse, K., Modi, M., and Patel, M. (2007). Local tolerance and systemic safety of pegaptanib sodium in the dog and rabbit, *J. Ocul. Pharmacol. Ther.*, **23**, pp. 452–466.
10. Ni, X., Castanares, M. Mukherjee, A., and Lupoid, S. E. (2011). Nucleic acid aptamers: clinical applications and promising new horizons, *Curr. Med. Chem.*, **18**, pp. 4206–4214.
11. Gragoudas, E. S., Adamis, A. P., Cunningham, E. T., Feinsod, M., and Guyer, D. R. (2004). Pegaptanib for neovascular age-related macular degeneration, *N. Engl. J. Med.*, **351**, pp. 2805–2816.
12. Ma, H., Liu, J., Ali, M. M., Mahmood, M. A. I., Labanieh, L., Lu, M., Iqbal, S. M., Zhang, Q., Zhao, W., and Wan, Y. (2015). Nucleic acid aptamers in cancer research, diagnosis and therapy, *Chem. Soc. Rev.*, **44**, pp. 1240–1256.
13. Zhang, L., Gu, F. X., Chan, J. M., Wang, A. Z., Langer, R. S., and Farokhzad, O. C. (2008). Nanoparticles in medicine: therapeutic applications and developments, *Clin. Pharmacol. Ther.*, **83**, pp. 761–769.
14. Chen, T., Shukoor, M. I., Chen, Y., Yuan, Q., Zhu, Z., Zhao, Z., Gulbakan, B., and Tan, W. (2011). Aptamer-conjugated nanomaterials for bioanalysis and biotechnology applications, *Nanoscale*, **3**, pp. 546–556.
15. Yang, L., Zhang, X., Ye, M., Jiang, J., Yang, R., Fu, T., Chen, Y., Wang, K., Liu, C., and Tan, W. (2011). Aptamer-conjugated nanomaterials and their applications, *Adv. Drug Delivery Rev.*, **63**, pp. 1361–1370.
16. Noble, G. T., Stefanick, J. F., Ashley, J. D., Kiziltepe, T., and Bilgicer, B. (2014). Ligand-targeted liposome design: challenges and fundamental considerations, *Trends Biotechnol.*, **32**, pp. 32–45.
17. Wagner, V., Dullaart, A., Bock, A., and Zweck, A. (2006). The emerging nanomedicine landscape, *Nat. Biotechnol.*, **24**, pp. 1211–1217.
18. Vogel, S., Stembera, K., Hennig, L., Findeisen, M., Giesa, S., Welzel, P., and Lampilas, M. (2001). Moenomycin analogues with modified lipid side

chains from indium-mediated Barbier-type reactions, *Tetrahedron*, **57**, pp. 4139–4146.

19. Vogel, S., Stembera, K., Hennig, L., Findeisen, M., Giesa, S., Welzel, P., Tillier, C., Bonhomme, C., and Lampilas, M. (2001). Moenomycin analogues with long-chain amine lipid parts from reductive aminations, *Tetrahedron*, **57**, pp. 4147–4160.
20. Letsinger, R. L., Zhang, G., Sun, D. K., Ikeuchi, T., and Sarin, P. S. (1989). Cholesteryl-conjugated oligonucleotides: synthesis, properties, and activity as inhibitors of replication of human immunodeficiency virus in cell cultures, *Proc. Natl. Acad. Sci. U S A*, **86**, pp. 6553–6556.
21. Saison-Behmoaras, T., Tocqué, B., Rey, I., Chassignol, M., Thuong, N. T., and Hélène, C. (1991). Short modified antisense oligonucleotides directed against Ha-ras point mutations induce selective cleavage of the mRNA and inhibit T24 cells proliferation, *EMBO J.*, **10**, pp. 1111–1118.
22. Shea, R. G., Marsters, J. C., and Bischofberger, N. (1990). Synthesis, hybridization properties and antiviral activity of lipid-oligonucleotide conjugates, *Nucleic Acids Res.*, **18**, pp. 3777–3783.
23. MacKellar, C., Graham, D., Will, D. W., Burgess, S., and Brown, T. (1992). Synthesis and physical properties of anti-HIV antisense oligonucleotides bearing terminal lipophilic groups, *Nucleic Acids Res.*, **20**, pp. 3411–3417.
24. Mann, A. P., Bhavane, R. C., Somasunderam, A., Montalvo-Ortiz, B. L., Ghaghada, K. B., Volk, D., Nieves-Alicea, R., Suh, K. S., Ferrari, M., Annapragada, A., Gorenstein, D. G., and Tanaka, T. (2011). Thioaptamer conjugated liposomes for tumor vasculature targeting, *Oncotarget*, **2**, pp. 298–304.
25. Jakobsen, U., Simonsen, A. C., and Vogel, S. (2008). DNA-controlled assembly of soft nanoparticles, *J. Am. Chem. Soc.*, **130**, pp. 10462–10463.
26. Jakobsen, U., and Vogel. S. (2009). DNA-controlled assembly of liposomes in diagnostics, *Meth. Enzym.*, **464**, pp. 233–248.
27. Willis, M. C., Collins, B., Zhang, T., Green, L. S., Sebesta, D. P., Bell, C., Kellogg, E., Gill, S. C., Magallanez, A., Knauer, S., Bendele, R. A., Gill, P. S., and Janjić, N. (1998). Liposome-anchored vascular endothelial growth factor aptamers, *Bioconjug. Chem.*, **9**, pp. 573–582.
28. Rohr, K., and Vogel, S. (2006). Polyaza crown ethers as non-nucleosidic building blocks in DNA conjugates: synthesis and remarkable stabilization of dsDNA, *ChemBioChem*, **7**, pp. 463–470.

29. Letsinger, R. L., and Chaturvedi, S. (1998). Tailored hydrophobic cavities in oligonucleotide-steroid conjugates, *Bioconjug. Chem.,* **9**, pp. 826–830.

30. Kurz, A., Bunge, A., Windeck, A., Rost, M., Flasche, W., Arbuzova, A., Strohbach, D., Müller, S., Liebscher, J. Huster, D., and Herrmann, A. (2006). Lipid-anchored oligonucleotides for stable double-helix formation in distinct membrane domains, *Angew. Chem., Int. Ed.*, **45**, pp. 4440–4444.

31. Pearce, T. R., Waybrant, B., and Kokkoli, E. (2014). The role of spacers on the self-assembly of DNA aptamer-amphiphiles into micelles and nanotapes, *Chem. Commun.*, **50**, pp. 210–212.

32. Jeong, J. H., and Park, T. G. (2001). Novel polymer-DNA hybrid polymeric micelles composed of hydrophobic poly(D,L-lactic-co-glycolic acid) and hydrophilic oligonucleotides, *Bioconjug. Chem.,* **12**, pp. 917–923.

33. Kwak, M., Minten, I. J., Anaya, D., Musser, A. J., Brasch, M., Nolte, R. J. M., Müllen, K., Cornelissen, J. J. L. M., and Hermann, A. (2010). Virus-like particles templated by DNA micelles: a general method for loading virus nanocarriers, *J. Am. Chem. Soc.*, **132**, pp. 7834–7835.

34. Wu, Y., Sefah, K., Liu, H., Wang, R., and Tan, W. (2010). DNA aptamer-micelle as an efficient detection/delivery vehicle towards cancer cells, *Proc. Natl. Acad. Sci. U S A*, **107**, pp. 5–10.

35. Tan, W., Wang, H., Chen, Y., Zhang, X., Zhu, H., Yang, C., Yang, R., and Liu, C. (2011). Molecular aptamers for drug delivery, *Trends Biotechnol.*, **29**, pp. 634–640.

36. Liu, H., Zhu, Z., Kang, H., Wu, Y., Sefan, K., and Tan, W. (2010). DNA-based micelles: synthesis, micellar properties and size-dependent cell permeability, *Chem. Eur. J.*, **16**, pp. 3791–3797.

37. Green, L. S., Jellinek, D., Bell, C., Beebe, L. A., Feistner, B. D., Gill, S. C., Jucker, F. M., and Janjić, N. (1995). Nuclease-resistent nucleic acid ligands to vascular permeability factor/vascular endothelial growth factor, *Chem. Biol.*, **2**, pp. 683–695.

38. Thomas, K. A. (1996). Vascular endothelial growth factor, a potent and selective angiogenic agent, *J. Biol. Chem.*, **271**, pp. 603–606.

39. Brown, L. F., Detmar, M., Claffey, K., Nagy, J. A., Feng, D., Dvorak, A. M., and Dvorak, H. F. (1997). Vascular permeability factor/vascular endothelial growth factor: a multifunctional angiogenic cytokine, in *Regulation of Angiogenesis* (Goldberg, I. D., and Rosen, E. M., eds.), pp. 233–269, Birkhauer, Basel, Switzerland.

40. Miles, A. A., and Miles, E. M. (1952). Vascular reactions to histamine, histamine-liberator, and leukotaxine in the skin of guinea pigs, *J. Physiol. (Lond.)*, **118**, pp. 228–257.
41. Senger, D. R., Galli, S. J., Dvorak, A. M., Peruzzi, C. A., Harvey, V. S., and Dvorak, H. F. (1983). Tumor cells secrete a vascular permeability factor that promotes accumulation of ascites fluid, *Science*, **219**, pp. 983–985.
42. Auerbach, R., Auerbach, W., and Polakowski, I. (1991). Assays for angiogenisis: a review, *Pharmacol. Ther.*, **51**, pp. 1–11.
43. Kang, H., O'Donoghue, M. B., Liu, H., and Tan, W. (2010). A liposome-based nanostructure for aptamer directed delivery, *Chem. Commun.*, **46**, pp. 249–251.
44. Shangguan, D., Li, Y., Tang, Z., Cao, Z. C., Chen, H. W., Mallikaratchy, P., Sefah, K., Yang, C. J., and Tan, W. (2006). Aptamers evolved from live cells as effective molecular probes for cancer study, *Proc. Natl. Acad. Sci. U S A*, **103**, pp. 11838–11843.
45. Shangguan, D., Cao, Z., Meng, L., Mallikaratchy, P., Sefah, K., Wang, H., Li, Y., and Tan, W. (2008). Cell-specific aptamer probes for membrane protein elucidation in cancer cells, *Proteome Res.*, **7**, pp. 2133–2139.
46. Xiao, Z., Shangguan, D., Cao, Z., Fang, X., and Tan, W. (2008). Cell-specific internalization study of an aptamer from whole cell selection, *Chem. Eur. J.*, **14**, pp. 1769–1775.
47. Waybrant, B., Pearce, T. R., and Kokkoli, E. (2014). Effect of polyethylene glycol, alkyl, and oligonucleotide spacers on the binding, secondary structure, and self-assembly of fractalkine binding FKN-S2 aptamer-amphiphiles, *Langmuir*, **30**, pp. 7465–7474.
48. Waybrant, B., Pearce, T. R., Wang, P., Sreevatsan, S., and Kokkoli, E. (2012). Development and characterization of an aptamer binding ligand of fractalkine using domain targeted SELEX, *Chem. Commun.*, **48**, pp. 10043–10045.
49. Erreni, M., Solinas, G., Brescia, P., Osti, D., Zunino, F., Colombo, P., Destro, A., Roncalli, M., Mantovani, A., Draghi, R., Levi, D., Rodriguez y Baena, R., Gaetani, P., Pelicci, G., and Allavena, P. (2010). Human glioblastoma tumours and neural cancer stem cells express the chemokine CX3CL1 and its receptor CX3CR1, *Eur. J. Cancer*, **46**, pp. 3383–3392.
50. Gaudin, F., Nasreddine, S., Donnadieu, A., Emilie, D., Combadière, C., Prévot, S., Machelon, V., and Balabanian, K. (2011). Identification of the chemokine CX3CL1 as a new regulator of malignant cell proliferation in epithelial ovarian cancer, *PLoS One*, **6**, p. e21546.

51. Hyakudomi, M., Matsubara, T., Hyakudomi, R., Yamamoto, T., Kinugasa, S., Yamanoi, A., Maruyama, R., and Tanaka, T. (2008). Increased expression of fractalkine is correlated with a better prognosis and an increased number of both CD8+ T cells and natural killer cells in gastric adenocarcinoma, *Ann. Surg. Oncol.*, **15**, pp. 1775–1782.
52. D'Haese, J. G., Demir, I. E., Friess, H., and Ceyhan, G. (2010). Fractalkine/CX3CR1: why a single chemokine-receptor duo bears a major and unique therapeutic potential, *Expert Opin. Ther. Targets*, **14**, pp. 207–219.
53. Jones, B. A., Beamer, M., and Ahmed, S. (2010). Fractalkine/CX3CL1: a potential new target for inflammatory diseases, *Mol. Interv.*, **10**, pp. 263–270.
54. Mann, A. P., Bhavane, R. C., Somasunderam, A., Montalvo-Ortiz, B. L., Ghaghada, K. B., Volk, D., Nieves-Alicea, R., Suh, K. S., Ferrari, M., Annapragada, A., Gorenstein, D. G., and Tanaka, T. (2011). Thioaptamer conjugated liposomes for tumor vasculature targeting, *Oncotarget*, **2**, pp. 298–304.
55. Burdick, M. M., McCarty, O. J., Jadhav, S., and Konstantopoulos, K. (2001). Cell-cell interactions in inflammation and cancer metastasis, *IEEE Eng. Med. Biol. Mag.*, **20**, pp. 86–91.
56. Siegel, G., and Malmsten, M. (1997). The role of the endothelium in inflammation and tumor metastasis, *Int. J. Microcirc. Clin. Exp.*, **17**, pp. 257–272.
57. Barthel, S. R., Gavino, J. D., Descheny, L., and Dimitroff, C. J. (2007). Targeting selectins and selectin ligands in inflammation and cancer, *Expert Opin. Ther. Targets*, **11**, pp. 1473–1491.
58. Bhaskar, V., Law, D. A., Ibsen, E., Breinberg, D., Cass, K. M., DuBridge, R. B., Evangelista, F., Henshall, S. M., Hevezi, P., Miller, J. C., Pong, M., Powers, R., Senter, P., Stockett, D., Sutherland, R. L., von Freeden-Jeffrey, U., Willhite, D., Murray, R., Afar, D. E., and Ramakrishnan, V. (2003). E-selectin up-regulation allows for targeted drug delivery in prostate cancer, *Cancer Res.*, **63**, pp. 6387–6394.
59. Kneuer, C., Ehrhardt, C., Radomski, M. W., and Bakowsky, U. (2006). Selectins: potential pharmacological targets?, *Drug Discovery Today*, **11**, pp. 1034–1040.
60. Mann, A. P., Somasunderam, A., Nieves-Alicea, R., Li, X., Hu, A., Sood, A. K., Ferrari, M., Gorenstein, D. G., and Tanaka, T. (2010). Identification of thioaptamer ligand against E-selectin: potential application for inflamed vascular targeting, *PLoS One*, **5**, p. e13050.

61. Alshaer, W., Hillaireau, H., Vergnaud, J., Ismail, S., and Fattal, E. (2015). Functionalizing liposomes with anti-CD44 aptamer for selective targeting of cancer cells, *Bioconjug. Chem.*, **26**, pp. 1307–1313.
62. Vermeulen, L., Sprick, M. R., Kemper, K., Stassi, G., and Medema, J. (2008). Cancer stem cells: old concepts, new insights, *Cell Death Differ.*, **15**, pp. 947–958.
63. Vermeulen, L., de Sousa e Melo, F. Richel, D. J., and Medema, J. (2012). The developing cancer stem-cell model: clinical challenges and opportunities, *Lancet Oncol.*, **13**, pp. e83–89.
64. Jones, R. J., Matsui, W. H., and Smith, B. D. (2004). Cancer stem cells: are we missing the target?, *J. Natl. Cancer Inst.*, **96**, pp. 583–585.
65. Park, C. Y., Tseng, D., and Weissman, I. L. (2009). Cancer stem cell-directed therapies: recent data from the laboratory and clinic, *Mol. Ther.*, **17**, pp. 219–230.
66. Clevers, H. (2011). The cancer stem cell: premises, promises and challenges, *Nat. Med.*, **17**, pp. 313–319.
67. Magee, J. A., Piskounova, E., and Melendez-Zajgla, J. (2012). Cancer stem cells: impact, heterogeneity, and uncertainty, *Cancer Cell*, **21**, pp. 283–296.
68. Schwarz-Cruz-y-Celis, A., and Melendez-Zajgla, J., (2011). Cancer stem cells, *Rev. Invest. Clin.*, **63**, pp. 179–186.
69. Ababneh, N., Alshaer, W., Allozi, O., Mahafzah, A., El-Khateeb, M., Hillaireau, H., Noirey, M., Fattal, E., and Ismail, S. (2013). In vitro selection of modified RNA aptamers against CD44 cancer stem cell marker, *Nucleic Acid Ther.*, **23**, pp. 401–407.
70. Al-Hajj, M., Wicha, M. S., Benito-Hernandez, A., Morrison, S. J., and Clarke, M. F. (2003). Prospective identification of tumorigenic breast cancer cells, *Proc. Natl. Acad. Sci. U S A*, **100**, pp. 3983–3988.
71. Collins, A. T., Berry, P. A., Hyde, C., Stower, M. J., and Maitland, N. J. (2005). Prospective identification of tumorigenic prostate cancer stem cells, *Cancer Res.*, **65**, pp. 10946–10951.
72. Dalerba, P., Dylla, S. J., Park, I. K., Liu, R., Wang, X., Cho, R. W., Hoey, T., Gurney, A., Huang, E. H., Simeone, D. M., Shelton, A. A., Parmiani, G., Castelli, C., and Clarke, M. F. (2007). Phenotypic charactization of human colorectal cancer stem cells, *Proc. Natl. Acad. Sci. U S A*, **104**, pp. 10158–10163.
73. Li, C., Heidt, D. G., Dalerba, P., Burant, C. F., Zhang, L., Adsay, V., Wicha, M., Clarke, M. F., and Simeone, D. M. (2007). Identification of pancreatic cancer stem cells, *Cancer Res.*, **67**, pp. 1030–1037.

74. Zoller, M. (2011). CD44: can a cancer-initiating cell profit from an abundantly expressed molecule?, *Nat. Rev. Cancer*, **11**, pp. 254–267.

75. Korkaya, H., Liu, S., and Wicha, M. (2011). Regulation of cancer stem cells by cytokine networks: attacking cancers inflammatory roots, *Clin. Cancer Res.*, **17**, pp. 6125–6129.

76. Moran, A. P. (1995). Structure-bioactivity relationships of bacterial endotoxins, *J. Toxicol. Toxin Rev.*, **14**, pp. 47–83.

77. Ara, M. N., Matsuda, T., Hyodo, M., Sakurai, Y., Hatakeyama, H., Ohga, N., Hida, K., and Harashima, H. (2014). An aptamer ligand based liposomal nanocarrier system that targets tumor endothelial cells, *Biomaterials*, **35**, pp. 7110–7120.

78. Ara, M. N., Hyodo, M., Ohga, N., Hida, K., and Harashima, H. (2012). Development of a novel DNA aptamer ligand targeting to primary cultured tumor endothelial cells by a cell-bases SELEX method, *PLoS One*, **7**, p. e50174.

79. Baek, S. E., Lee, K. H., Park, Y. S., Oh, D., Oh, S., Kim, K., and Kim, D. (2014). RNA aptamer-conjugated liposome as an efficient anticancer drug delivery vehicle targeting cancer cells in vivo, *J. Controlled Release*, **196**, pp. 234–242.

80. Lupold, S. E., Hicke, B. J., Lin, Y., and Coffey, D. S. (2002). Identification and characterization of nuclease-stabilized RNA molecules that bind human prostate cancer cells *via* the prostate-specific membrane antigen, *Cancer Res.*, **62**, pp. 4029–4033.

81. Bates, P. J., Kahlon, J. B., Thomas, S. D., Trent, J. O., and Miller, D. M. (1999). Antiproliferative activity of G-rich oligonucleotides correlates with protein binding, *J. Biol. Chem.*, **274**, pp. 26369–26377.

82. Dapić, V., Bates, P. J., Trent, J. O., Rodger, A., Thomas, S. D., and Miller, D. M. (2002). Antiproliferative activity of G-quartet-forming oligonucleotides with backbone and sugar modifications, *Biochemistry*, **41**, pp. 3676–3685.

83. Ireson, C. R., and Kelland, L. R. (2006). Discovery and development of anticancer aptamers, *Mol. Cancer Ther.*, **5**, pp. 2957–2962.

84. Liao, Z., Chuang, E., Lin, C., Ho, Y., Lin, K., Cheng, P., Chen, K., Wei, H., and Sung, H. (2015). An AS1411 aptamer-conjugated liposomal system containing a bubble-generating agent for tumor-specific chemotherapy that overcomes multidrug resistance, *J. Controlled Release*, **208**, pp. 42–51.

85. Srivastava, M., and Pollard, H. B. (1999). Molecular dissection of nucleolin's role in growth and cell proliferation: new insights, *FASEB J.*, **13**, pp. 1911–1922.

86. Soundararajan, S., Chen, W., Spicer, E. K., Courtenay-Luck, N., and Fernandes, D. J. (2008). The nucleolin targeting aptamer AS1411 destabilizes *Bcl*-2 messenger RNA in human breast cancer cells, *Cancer Res.*, **68**, pp. 2358–2365.
87. Cao, Z., Tong, R., Mishra, A., Xu, W., Wong, G. C. L., Cheng, J., and Lu, Y. (2009). Reversible cell-specific drug delivery with aptamer-functionalized liposomes, *Angew. Chem., Int. Ed.*, **48**, pp. 6494–6498.
88. Ghezzi, A., Aceto, M., Cassino, C., Gabano, E., and Osella, D. (2004). Uptake of antitumor platinum(II)-complexes by cancer cells, assayed by inductively coupled plasma mass spectrometry (ICP-MS), *J. Inorg. Biochem.*, **98**, pp. 73–78.
89. Budman, D. R., Calabro, A., and Kreis, W. (2002). Synergistic and antagonistic combinations of drugs in human prostate cancer cell lines in vitro, *Anti-Cancer Drugs*, **13**, pp. 1011–1016.
90. Xing, H., Tang, L., Yang, X., Hwang, K., Wang, W., Yin, Q., Wong, N. Y., Dobrucki, L. W., Yasui, N., Katzenellenbogen, J. A., Helferich, W. G., Cheng, J., and Lu, Y. (2013). Selective delivery of an anticancer drug with aptamer-functionalized liposomes to breast cancer cells in vitro and in vivo, *J. Mater. Chem. B*, **1**, pp. 5288–5297.
91. Chen, K. J., Liang, H. F., Chen, H. I., Wang, Y., Cheng, P. Y., Liu, H. I., Xia, Y., and Sung, H. W. (2013). A thermoresponsive bubble-generating liposomal system for triggering localized extracellular drug release, *ACS Nano*, **7**, pp. 438–446.
92. Batist, G., Tulpule, A., Sinha, B. K., Katki, A. G., Myers, C. E., and Cowan, K. H. (1986). Overexpression of a novel anionic glutathione transferase in multidrug-resistant human breast cancer cells, *J. Biol. Chem.*, **261**, pp. 15544–15549.
93. Scudiero, D. A., Monks, A., and Sausville, E. A. (1998). Cell line designation change: multidrug-resistant cell line in the NCI anticancer screen, *J. Natl. Cancer Inst.*, **90**, pp. 862.
94. Ke, W., Yu, P., Wang, J., Wang, R., Guo, C., Zhou, L., Li, C., and Li, K. (2011). MCF-7/ADR cells (re-designated NCI/ADR-RES) are not derived from MCF-7 breast cancer cells: a loss for breast cancer multidrug-resistant research, *Med. Oncol.*, **28**, pp. S135–S141.
95. Suter, T. M., and Ewer, M. S. (2013). Cancer drugs and the heart: importance and management, *Eur. Heart J.*, **34**, pp. 1102–1111.
96. Li, L., Hou, J., Liu, X., Guo, Y., Wu, Y., Zhang, L., and Yang, Z. (2014). Nucleolin-targeting liposomes guided by aptamer AS1411 for the delivery of siRNA for the treatment of malignant melanomas, *Biomaterials*, **35**, pp. 3840–3850.

97. Hingorani, S. R., Jacobetz, M. A., Robertson, G. P., Herlyn, M., and Tuveson, D. A. (2003). Suppression of $BRAF^{V599E}$ in human melanoma abrogates transformation, *Cancer Res.*, **63**, pp. 5198–5202.

98. Jung, B. Y. K., Kim, T. W., Park, H. G., and Soh, H. T. (2010). specific colorimetric detection of proteins using bidentate aptamer-conjugated polydiacetylene (PDA) liposomes, *Adv. Funct. Mater.*, **20**, pp. 3092–3097.

99. Bock, L. C., Griffin, C., Latham, J. A., Vermaas, E. H., and Toole, J. J. (1992). Selection of single-stranded DNA molecules that bind and inhibit human thrombin, *Nature*, **355**, pp. 564–566.

100. Tasset, D. M., Kubik, M. F., and Steiner, W. (1997). Oligonucleotide inhibitors of human thrombin that bind distinct epitopes, *J. Mol. Biol.*, **272**, pp. 688–698.

101. Ahmad, K. M., Xiao, Y., and Soh, H. T. (2012). Selection is more intelligent than design: improving the affinity of a bilvalent ligand through directed evolution, *Nucleic Acids Res.*, **40**, pp. 11777–11783.

102. Charych, D. H., Nagy, J. O., Spevak, W., and Bednarski, M. D. (1993). Direct colorimetric detection of a receptor-ligand interaction by a polymerized bilayer assembly, *Science*, **261**, pp. 585–588.

103. Reichert, A., Nagy, J. O., Spevak, W., and Charych, D. (1995). Polydiacetylene liposomes functionalized with sialic acid bind and colorimetrically detect influenza virus, *J. Am. Chem. Soc.*, **117**, pp. 829–830.

104. Charych, D., Cheng, Q., Reichert, A., Kuziemko, G., Stroh, M., Nagy, J. O., Spevak, W., and Stevens, R. C. (1996). A 'litmus test' for molecular recognition using artificial membranes, *Chem. Biol.*, **3**, pp. 113–120.

105. Su, Y., Li, J., and Jiang, L. (2004). Chromatic immunoassay based on polydiacetylene vesicles, *Colloids Surf. B*, **38**, pp. 29–33.

106. Ries, O., Löffler, P. M. G., and Vogel, S. (2015). Convenient synthesis and application of versatile nucleic acid lipid membrane anchors in the assembly and fusion of liposomes, *Org. Biomol. Chem.*, **13**, pp. 9673–9680.

Chapter 10

Aptamers in Medical Diagnosis

Veli Cengiz Ozalp,[a] Murat Kavruk,[b] Ozlem Dilek,[a] and Abdullah Tahir Bayrac[c]

[a] *School of Medicine, Istanbul Kemerburgaz University, Istanbul, Turkey*
[b] *Test and Calibration Center, Turkish Standards Institution (TSE), Kocaeli, Turkey*
[c] *Karamanoglu Mehmetbey University, Karaman, Turkey*
cengiz.ozalp@kemerburgaz.edu.tr

Aptamers have been used in the biomedical field as a class of biorecognition elements with many advantages. Many diseases exhibit cellular aberrations at morphological and molecular levels. Medical diagnosis via detecting molecular features is highly specific and extremely sensitive when proper recognition molecule and an efficient signal transduction system are employed, although bioanalysis of human diseases at the molecular level is an extremely challenging field. Thus, effective probes to identify and recognize biomarkers of diseases are in high demand. DNA aptamers are short single-stranded oligonucleotides, which are obtained from random combinatorial library by an in vitro method. This relatively new biorecognition agent has superior intrinsic characteristics for biosensor development. In this chapter, we first present recent advances in aptamer selection technologies and the main formats of biosensors, which were frequently employed in aptasensor development. Then, the current state of aptamers, as applied to

Aptamers: Tools for Nanotherapy and Molecular Imaging
Edited by Rakesh N. Veedu

ISBN 978-981-4669-83-2 (Hardcover), 978-981-4669-84-9 (eBook)
www.panstanford.com

medical diagnosis, is discussed for specifically cancer and diagnosis or pathogenic microorganisms. Finally, an overview of aptamer–nanomaterial conjugates is presented in many applications such as diagnosis, bioimaging, and theranostics.

10.1 Introduction

Diseases like cancer originate from genomic mutations that can cause changes of cells at morphological and molecular levels. These markers on the diseased cells are valuable targets in diagnosis, but investigating human diseases at the molecular level has been extremely challenging due to the lack of effective probes to identify and recognize distinct molecular features of diseases. Traditional diagnosis methods are mostly based on morphological examination of diseased cell samples. However, these morphological features are mostly inefficient in early diagnosis and in evaluating complex molecular alterations. Therefore, molecular characteristics at the proteomic level are best biomarkers to identify cancers because of the direct connection between genetic features and protein expression. Medical diagnosis based on molecular features can be highly specific and extremely sensitive when incorporated with proper recognition molecule and an efficient signal transduction systems. Nonetheless, identification of molecular signatures remained a challenging task, which is reflected by the fact that new biomarkers have been at the center of research efforts for effective cancer diagnosis.

Exploring nucleic acids for ligand-binding molecules resulted in the discovery of aptamers in 1990s, where their novel properties were exploited for the development of new analytical formats. Aptamers have been used for developing various kinds of biosensors such as optical, fluorometric, electrochemical, etc. Aptamers offer a very interesting alternative to previous widely used bioaffinity materials in diagnostics and also as a therapeutic tool for drug delivery systems [51]. Since aptamers have affinity properties, sensor platforms involving antibodies can potentially be replaced by aptamers in a variety of new configurations. Nucleic acid aptamers are single-stranded, short, and synthetic oligonucleotide sequences with high affinity to specific targets. Naturally, nucleic acid–based regulatory elements, called riboswitches, contain target-binding sequences. Aptamers are often named *nucleic acid antibodies*, with additional advantageous properties for biosensor development. Small size, artificial selection, in vitro chemical synthesis for economic production, relatively easy modification, and physical stability are the major superior properties of aptamers over antibodies. Aptamers can be selected in a test tube by a combinatorial procedure using whole cells, extracted membranes, or soluble proteins. The specific aptamer sequences can be converted to biosensors by combining various nanomaterials and transduction formats. The final aptasensor can be used in the identification of diseased cells, organelles, or tissues in the body. Aptasensor development is achieved in four main steps (Fig. 10.1). In vitro selection of aptamers is the major advantage compared to alternative biorecognition agents. Systematic evolution of ligands by exponential enrichment (SELEX) is a combinatorial chemistry method to identify aptamer sequences in vitro [25, 87]. The first step is to determine the target molecule that can represent the diseased state (Step I). SELEX consists of combining two major tasks, affinity separation of binding sequences through repeated cycles and amplification of binder

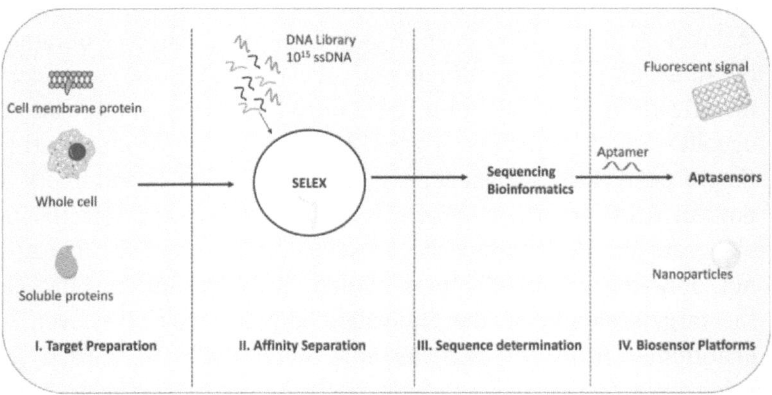

Figure 10.1 Selection of aptamers for aptasensor development in four step: (I) Biomarker target molecules are determined, (II) SELEX cycles enrich affinity sequences, (III) aptamer candidates are obtained, and finally (IV) aptamer sequence are integrated with signal transduction mechanisms.

sequence pools by polymerase chain reaction (PCR) (Step II). Step II is a relatively fast procedure that could be completed in a few weeks and completely performed in vitro. In Step III, the enriched library components are sequenced through traditional methods or next-generation methods and processed with bioinformatics techniques in order to determine the candidate aptamer sequences. Finally, aptamer sequences verified for affinity to the diseased target are integrated into various signal transduction platforms.

10.2 Aptamer Selection

Several inherent characteristics make aptamers excellent active binding elements for sensor development. Artificial selection is a powerful technique in producing aptamers for any kind of target and moreover to a specific region of the target. Aptamers are physically stable, have the ability to sustain or regain proper folding and thus binding capability after exposure to extreme conditions like high temperature and extreme pH values [20, 48]. On the other hand, there are aptamer properties that should be treated with caution. The application of aptamers in biological environments should consider the nucleases ubiquitously found in blood and intracellular environments. Various approaches to stabilize aptamers have been explored and most success obtained by spiegelmers (L-enantiomers of nucleic acids) and secondary modifications (e.g., 2′-amino-, 2′-fluoro-) [8]. The modifications usually should be applied a priori to the SELEX procedure rather than a posteriori. Spiegelmer oligonucleotides are enantiomers of natural RNA and DNA molecules. Spiegelmers were selected by elaborate methods compared to selection of aptamers, because not only the oligonucleotides should be enantiomers, but also the targets should be the equivalent mirror image. However, D-enantiomer forms of proteins have been the main hurdle for spiegelmer selection due to limitations in peptide synthesis. Thus, spiegelmers were mostly selected for cytokines and short peptide hormones. Protein spiegelmer aptamers were selected for bacterial enterotoxins and cardiac troponin I by exploiting the fact that aptamers recognize only short domains of amino acid motifs [83].

Another promising modified aptamer type has been locked nucleic acids (LNAs) as a strategy for increasing thermal and nuclease stability of nucleic acids. An LNA is initially developed as a derivative of RNA with a methylene bridge, which is composed of a connection between the 2′-oxygen of ribose and the 4′-carbon and results in a locked 3′-endo conformation. Shi et al. constructed an LNA-based aptamer diagnosis probe to improve the detection in biological fluids and to enhance the imaging properties [73]. A posteriori modification of Ramos cell–binding aptamer sequences (TD05) was prepared by adding 7 LNA and 3′-3′-T capping. The imaging ability of the DNA/LNA chimera probe was demonstrated in cell culture, serum, and in vivo on Ramos cell–implanted mouse models. The LNA incorporation increased the stability of probes about 10 times [73]. The strategy of increasing stability of aptamers by adding LNAs could be applied with greater efficiency via selecting aptamer sequences consisting of solely LNA nucleotides. The major obstacle for achieving an LNA aptamer is the availability of a polymerase that can amplify LNA sequences. In fact, a recent discovery of such polymerases that can amplify LNA triphosphates was reported and also a scheme for amplification and regeneration of LNA-containing oligonucleotide libraries were proposed [21, 24].

Traditional SELEX procedures depend on the presence of a soluble target antigen. This has been a significant drawback in that many of the cell surface receptors are extremely difficult to purify due to their hydrophobic properties. Even soluble targets may sometimes not retain their native conformation when immobilized for affinity separation, leading to nonfunctional aptamers that cannot recognize the natural structure of proteins, as expressed on living cells. A promising approach in obtaining cell specific aptamers has been the recent adaptation of traditional selection methodology, called cell-SELEX, which identifies aptamers by using whole cells as target during the selection procedure and by a negative subtraction to eliminate the common binders. This procedure has become widespread to obtain cell-specific aptamers that present opportunities in developing drug targeting systems since pathology modifies the normal cell to have marker properties that can easily be used for selecting aptamers [99]. Cell-SELEX is commonly used to obtain cancer cell–specific or bacteria-specific aptamer sequences, but the specific targets usually remain undefined due to the nature of the procedure [56].

The recent advances in aptamer technology have caused the emergence of new approaches for their study as well as their application in the treatment of human diseases. Adaptation of the method for automated procedures for high-throughput production has provided more and better performing aptamers. However, the number of aptamers available has always been limited and commercial exploitation of aptamers is still an awaited potential. Despite the large number of reports involving aptamer-incorporated sensor formats, only a limited number of products have been marketed for food analysis (mycotoxin and vascular endothelial growth factor selective aptamers) and treatment of age-related macular degeneration. The automation procedures with new technologies resulted in enhanced SELEX procedures. A two-step method for DNA aptamer selection used the semi-automated procedure of magnetic separation and solid-phase emulsion PCR [35]. A shortcoming in conventional SELEX is inherent in the method because efficiency of the whole selection process is often biased due to the high degree of diverse template sequences, which influences the efficiency of the amplification. This is considered to originate from the preferred amplification of short DNA fragments, PCR artifacts that are generated by recombination of homologous template regions and highly diverse templates, negatively influencing the formation of homoduplexes after denaturation. These adverse effects have been partly minimized with the employment of emulsion PCR in SELEX. The utilization of a water-in-oil emulsion offers the benefit of compartmentalization for highly improved efficiency with reduced bias in PCR products.

Slow off-rate modified aptamers (SOMAmers) are specific protein-binding reagents synthesized by chemical modification of aptamers. Incorporating benzyl, naphthyl, tryptamino, and isobutyl that are "mostly hydrophobic and aromatic groups added to a uracil ring," resulted in the generation of higher affinity somamer reagents to proteins by mimicking amino acid side chains [75]. Demonstration of its capabilities as a biorecognition reagent was first reported with multiplexed readout of serum proteins in 2008 [100]. The potential of SOMAmers in prorteomic applications has been demonstrated by Gold et al. in 2010 when about 800 proteins were assayed at pM detection levels. The team also identified 2 chronic kidney disease (CKD) biomarkers and proposed 58 potential CKD biomarkers [28]. Following the optimization of the procedure, Kraemer et al. utilized SOMAmers on a plate format for multiplex assay by incorporating biotin–streptavidin interaction and photocleaveble linker to the system. By using Cy3 fluorophore moeity, photocleavable SOMAmers to 9 specific proteins were immobilized on a plate for assay-based identification [44]. For example, detection of envelope protein E2 of hepatitis C virus was achieved using SOMAmer sequences. Since reverse transcription polymerase chain reaction (RT-PCR) cannot provide a healthy correlation between the amount of RNA and the virus, the envelope protein E2 has been considered a relevant indicator, which was quantified via a SOMAmer integrated in enzyme-linked aptasorbent assays (ELASAs) with a dynamic linearity from 3.91×10^2 FFU/mL to 1.00×10^5 FFU/mL, indicating that the quantification of an infectious virus titer is possible within this range of infectious virus titer [68]. SOMAmers binding interleukin-6 blocked the interaction with its receptor and thus inhibited the proliferation of tumor cells in vitro as effectively as the anticancer drug tocilizumab [31]. This study proved the potential of SOMAmers to overcome diagnostic challenges on plate-based sandwich assay platforms. The assay takes advantage of the dual nature of SOMAmers as both protein affinity-binding reagents with defined three-dimensional structures, and unique nucleotide sequences recognizable by specific DNA hybridization probes [76].

SOMAmers have also been applied for the detection of antigens of cancer cells. Early detection of malignant pleural mesothelioma (MM) in asbestos-exposed individuals was conducted in serum of 117 MM cases and 142 asbestos-exposed control over 1000 proteins by using SOMAmers [63]. These serum proteins were shown to be capable of detecting MM with an overall accuracy of 92% using the somascan proteomic assay. Another proteomic application of somascan was designed for small-cell lung cancer in order to develop novel histochemical probes [55]. By comparing the expression of 820 proteins, about 36 biomarkers were correlated with lung cancer, one-third of which had been identified by previous studies. This study was extended with more clinical samples and more biomarker protein candidates in order to strengthen the detection capability of somascan [54]. One of the latest SOMAmer study with cancer has been a proteomics analysis of cancer exosomes. By using Du145 prostate cancer cell line exosomes, researchers measured over 1000 proteins via SOMAmers and identified 300 previously unknown proteins having association with prostate cancer [93].

SOMAmers were also applied for the detection of microorganism-based diseases. For example, identifying the serum biomarkers associated with pulmonary tuberculosis treatment was one of the early applications [19]. The researchers compared 116 serum protein levels of tuberculosis (TB) patients before and after the treatment and identified 60 promising markers, some of which are related with inflammation and antimicrobial defense. A similar study was also reported for identification of proteins as candidates of biomarkers for coagulation cascade, neutrophil activity, and immunity [58]. A diagnostic assay for *Clostridium difficile* toxins (A, B, and binary toxin) was reported with a detection limit of 1 pmol/L. Binary toxin capability, which is absent in commercial antibody-based assays, is an important achievement in specific detection capability of aptamers [62]. Subnanomolar K_d for the detection of disease-related microbes via SOMAmers can also be achieved in cell surface–associated proteins. Baumstummler et al. selected a SOMAmer for the detection and capture of *Staphylococcus aureus* via its cell surface proteins. As a proof of concept for the capability of SOMAmers, the detection of *S. aureus* cells was enhanced by several orders of magnitude when the bacterial cells were captured with SOMAmers followed by qPCR analysis [6].

The promising potential of a SOMAmer can be deduced from a recent cooperation between Somalogic and New England Biolabs (NEB) when they announced to develop PCR reagents incorporated with SOMAmers for the reverse transcriptase (RTx) enzyme. By coupling a reversibly bound aptamer that inhibits RTx activity below 40°C, NEB has marketed WarmStart® RTx Reverse Transcriptase in 2014 for RNA detection, and specifically for loop-mediated isothermal amplification (LAMP) reactions with high throughput, room temperature, setup capability, and increased specificity [59]. This invention would open new possibilities to early detection of analytes related with diseases. Potential applications of SOMAmers in protein assays, disease detection, and other biotechnoloy areas have recently been reviewed in detail elsewhere [71, 100].

Finally, a recent progress in SELEX technology has been the development of immobilization free aptamer selection techniques. Especially for the targets that cannot be separated from unbounded sequences by simple washings as in cell-SELEX, immobilization to a convenient matrix is an inevitable step for affinity separation. Protein and small-molecule targets have to be immobilized on suitable platforms to be able to separate unbounding sequences from the bounding ones throughout the SELEX. Park et al. reported a technique based on graphene oxide (GO) to select aptamers without target immobilization [67]. In the GO-SELEX process, single-stranded DNA (ssDNA) is nonspecifically adsorbed by the GO, resulting in the separation of unbound DNA. Aptamers have been developed for nampt protein [67], T2 toxin [17], C-reactive protein [96], and many pesticides [84] in recent years by using GO. Also single-walled carbon nanotubes are used for their ssDNA binding affinity to enrich aptamers without immobilization [84].

10.3 Diagnostic Aptasensor Platforms

10.3.1 *Optical Aptasensors*

Fluorescent signals have been the most common transduction mechanisms in biosensor development. The delivery of fluorescent probes into tissues has always been of interest since targeting molecules coupled to nanomaterials can conveniently allow the development of smart tools to monitor cellular processes with high specificity and sensitivity [27]. Oligonucleotide optical switches are among the most popular type of optical nanosensors [4, 23, 89]. They are capable of modifying their light emission upon molecular interaction with well-defined molecular targets. For example, a noninvasive aptamer imaging method based on stem-loop- or linear-structured DNA molecular beacon (MB) was developed to determine endogenous adenosine triphosphate (ATP) concentrations at high spatiotemporal resolution [66].

10.3.2 *Electrochemical Aptasensors*

Clinical detection assays are most efficient when high sensitivity and miniaturized formats are combined with biorecognition elements and novel microfabrication technologies. Among the sensing formats, electrochemical aptasensors provide high sensitivity at low cost. Several techniques have been used for fabricating electrochemical aptasensors, such as electrochemical impedance spectroscopy, potentiometry, and amperometry. Potentiometric sensors are based on measuring the potential difference between a working and a reference electrode when there is no significant current flowing between them. An ion-selective electrode (ISE) is the most common choice as a transducer to take the advantage of detecting a wide range of ion concentrations, up to five digits of difference with a lower ppm level of detection limit values. Also, continuous measurement property is an important advantage in the fabrication of sensors. Another attractive feature of potentiometric sensors is the high signal levels independent of a sample volume. Aptamer-grafted electrodes can decrease currents due to the interference by the binding events on aptamer molecules. Kim et al. selected DNA aptamers binding to estradiol, and the aptamer was used to develop an electrochemical detection method for chemical sensing by immobilizing it on the gold electrode on the basis of avidin–biotin interaction [38]. In this study, estradiol was detected using a redox mediator where the sensitivity and selectivity of the detection was drastically increased.

Zhu et al. developed a new type of electrochemically highly sensitive, selective, and label-free sensor for the detection of anticancer drug daunomycin using a cancer cell surface lipid molecule phosphatidylserine and an aptamer co-immobilized onto the nanocomposite-conducting polymer film [39, 91]. This sensor was more sensitive than the then reported any other daunomycin detection system. Their strategy has many attractive features, such as simplicity, rapidity, no requirement for a specific label (i.e., a fluorescent or reactive moiety), and low cost, and hence could be a useful method in medical diagnosis.

10.3.3 *Nanoparticle–Aptamer Conjugates*

The development of imaging probes aims to improve specificity and sensitivity of diagnostic tools through noninvasive and quantitative detection of specific biomolecules in tissues. A promising class of molecular imaging probes consists of nanoparticles functionalized with a certain targeting agent. The aptamer–nanomaterial conjugates are of particular interest due to the unique properties of nanomaterials such as small size, increased surface-to-volume ratio, and a wide range of sensing formats, including metal and semiconductor core materials. Nanomaterials can be easily grafted with aptamers via direct covalent linkage or by noncovalent interactions. Gold nanoparticles (AuNPs) have widely been used in medical diagnosis due to their simple synthesis and unique spectral properties (e.g., surface plasmon production). AuNPs can be synthesized from the chemical reduction of $AuCl_4$ ions by agents like citric acid, and their size can be precisely controlled to have homogeneous particles with diameters in nanometer scales. AuNPs conjugated with aptamers have been extensively utilized in the medical diagnostics and imaging. The distance-dependent fluorescence quenching behavior of AuNPs provides a tool of detection of biomolecular interactions in close proximity. In addition to a strong extinction peak in the visible and near-infrared regions [18], the ability of gold nanoparticles to penetrate through tissue and easy surface modification makes them a convenient component of in vitro and in vivo systems [3]. However, these favorable properties can be efficiently exploited if a targeting mechanism can be used in conjugation with gold nanoparticles. Aptamers can spontaneously be conjugated to AuNPs via covalent binding through thiol moiety. For in vitro imaging and detection, Au particle aggregation was triggered with aptamer target interaction in order to develop a simple method for the detection of platelet-derived growth factors (PDGFs) [13]. With the presence of PDGFs, aptamers having specificity and affinity to PDGFs bind them and change conformation, allowing the partial complementary probe to release from gold nanoparticles and conjugate with aptamers. Gold nanoparticles bind to DNA probes and aggregate in the solution, resulting in a change in their absorption spectrum that can be visually detected in order to develop a colorimetric assay. Another example of aggregation-based approach of AuNPs was carried out for the p53 tumor marker with its specific aptamer [74]. The sensitivity of detection was increased tenfold by coupling aptamer affinity with the catalytic potential of AuNPs with a chemiluminescence reaction. For in vivo imaging purposes, polyethylene glycol (PEG) is generally incorporated to nanocomposites in order to increase stability and circulation time of aptamer-conjugated gold particles [57]. Gold nanoparticles were also coupled with aptamers noncovalently for the detection of analytes. For example, electrochemical sandwich assay was developed using screen-printed carbon electrode strips [97].

Magnetic nanoparticles have frequently been conjugated to aptamers since magnetic properties are highly valuable for purification purposes. An aptamer-based assay for thrombin was constructed by immobilizing the thrombin binding aptamer on a magnetic nanoparticle. A complementary oligonucleotide-to-aptamer sequence was conjugated to gold nanoparticles and then the aptamer was hybridized with the complementary oligonucleotide to form the duplex structure. The Au complementary DNA-aptamer-magnetic nanoparticle complex was then used as a probe for the specific determination of thrombin. In the presence of thrombin, the aptamer forms a G-quaruplex structure, resulting in the dissociation of DNA duplex and the release of the gold-labeled oligonucleotide. This strategy is a good example of sensor systems combining aptamers and multiple nanoparticles [98]. A similar magnetic nanoparticle was employed for direct imaging applications such as carboxymethyl cellulose–modified magnetic nanoparticle grafted with a hepatocellular carcinoma aptamer as a smart imaging probe [69]. The cellulose-derived polymer showed promise for drug delivery by combining cancer cell specificity of EpCAM aptamers and imaging capability of magnetic nanoparticles via MR imaging. Fluorophore labeling of aptamers conjugated to magnetic nanoparticles is another popular design in developing diagnostic systems. The polyA sequence has a strong affinity to gold nanoparticles and the FAM fluorophore molecule, iodide, or thiosulfate ions can act as a displacing agent to separate polyA-tailed aptamer strands from AuNP and simultaneously activate the AuNP-quenched fluorophores by destroying the polyA–AuNP interaction. This FRET signaling can provide magnetic separation-based detection of thrombin. A sandwich-type fluorescence aptasensor for sensitive and specific thrombin detection, protein capture, and separation assay was developed by using aptamer-conjugated magnetic beads and iodide-induced FRET of activatable polyA-based AuNP probes through ligand displacement [47]. An innovative approach to gold conjugation with an aptamer has been recently published, in which an electrochemiluminescence assay was developed [16]. The detection of tumor cells is achieved via aptamer conjugation with bio-bar-code Au-Fe_3O_4 nanoparticles, which function as an amplification station for the strand displacement reaction and rolling circle amplification, and down to 50 CFU detection was demonstrated.

In addition to gold and magnetic NPs, graphene has been another advantageous nanomaterial combined with aptamers for imaging/detection purposes [33]. For label-free detection of cancer cells, a graphene-functionalized aptasensor was developed for the detection of nucleolin surface protein [25]. By modifying graphene with carboxyl moiety, AS1411 aptamers specific to nucleolin protein could be immobilized on graphene nanoparticles. Electrochemical detection via cyclic voltammetry with detection limit of 794 cells/mL was demonstrated. Ma et al. developed a similar electrochemical biosensor for *Salmonella* spp. by using graphene [50]. In this platform, a glassy carbon electrode is modified with GO and aptamers specific to *Salmonella* were immobilized through AuNPs. Cyclic voltammetry measurements have shown that up to 3 CFU/mL can be detected. In another approach, graphene–silver nanoparticle (AgNP) hybrids for the detection of immunoglobulin E were reported [77]. Aptamers specific to IgE were fixed to AuNPs, with affinity to IgE, providing a formation suitable for streptavidin-biotin-mediated binding of AgNP/graphene hybrid. A limit of detection at 3.6 ng/mL based on this platform could be obtained. Wang et al. reported an aptasensor enhanced with AgNPs against adenosine [22]. By using Cy3 dye as a molecular beacon with AgNPs, binding of adenosine to the aptamer provides a close proximity between AgNPs and Cy3 dye, resulting in an increase in the fluorescence signal. Based on this principle, an aptasensor was demonstrated to detect adenosine concentrations from 200 nM to 200 μM linearly. Apart from being a component in the graphene–aptamer conjugate system, AgNPs could also be used by themselves for aptamer conjugation. Dual-aptamer-based sandwich immunosensor for the detection of *S. aureus* was reported on the basis of detection of silver nanoparticles with anodic stripping voltammetry [1]. To detect one of the most important human pathogens, *S. aureus*, the aptamers were conjugated to the magnetic nanoparticle for easy isolation, and another aptamer sequence was conjugated with AgNPs for signal amplification in electrochemical stripping voltammetry read-out. Also, AuNP-based plasmonic assays provide a simpler method to amplify the signal. Therefore, plasmon assays prevail in cellular analysis and are utilized depending on the change in localized surface plasmon resonance (LSPR) [32].

10.3.4 *Bioconjugation Methods*

Aptamers can readily be modified with a variety of chemical groups, which provides the subsequent attachment of an aptamer to various formats of sensors. Bioconjugation is a chemical strategy for covalent modification of biomolecules. Bioconjugation strategies are essential for the discovery of function of biomolecules using chemoselective reactions that are functional under physiological conditions [36, 79]. For example, introduction of traditional functional groups within biomolecules has provided convenient sites for highly chemoselective modification, leading to their immobilization on surfaces and labeling with small molecules. Novel methods have recently been discovered for site-specific covalent conjugations of proteins, aptamers, DNA, RNA, and carbohydrates [36]. Traditional strategies for covalent bioconjugation include the introduction of nonnatural functional groups into biomolecules, followed by site-specific embedding on surfaces via selective chemical reaction. However, poor reactivity has always been the major disadvantage of traditional methods and such chemical modifications have resulted in loss of the biological function of the target biomolecule. Due to high specificity and minimal perturbation on biomolecules, site-specific bioconjugation is favorable to random conjugation methods.

Direct reaction strategies involve nanoparticle surface modification with amine, aldehyde, or active hydrogen groups. These strategies are particularly appropriate for conjugating fluorescence dyes, chelators for nuclear imaging, or drugs. Specific bioconjugation methods have been employed depending on the identity of the biomolecule to be functionalized, the synthetic small molecule to be attached, and the use of the resulting conjugate product.

Several chemical strategies that can be used to attach different organic functional groups (e.g., thioether and orthogonal linkages) to biomolecules have been studied for many years. For example, thiolates act as good nucleophiles in aqueous solutions. Therefore, site-specific labeling to biomolecules via the thiolate group of a cysteine residue is an effective bioconjugation technique [12]. Nucleic acids can readily be conjugated onto gold surfaces through covalent bonds between sulfur and gold atoms (S-Au). Labeling aptamers with sulfhydryl functional groups (SH) forms a stable bond between gold nanoparticles and DNA aptamers [9]. However, the S-Au bond can be broken at high temperatures or the presence of other thiols.

The amide bond between the biomolecules has been extensively used as an attractive bonding system for bioconjugation. For example, a protein can be conjugated with a small molecule or surface having an activated ester. In contrast, the site-specific generation of amides is very challenging. Native chemical ligation and the Staudinger ligation are used as modern ligation methods, thus providing a potential bioconjugation tool for amide bond formation at a specific site in a protein [36, 60]. Similarly, amino-labeled aptamers can be applicable for nanomaterials via chemical conjugation systems.

Click chemistry is a new developing approach that uses only the most practical and reliable chemical transformations in medicinal chemistry. Bioconjugation via click chemistry has been frequently used in proteins and nucleic acids. Click chemistry provides a powerful, attractive alternative to conventional chemistry due to its fast, easy purification, biocompatibility, regiospecificity, reactivity, and high yield [61].

One of the most popular click chemical techniques is known as Staudinger ligation, which is a fast reaction developed by the Bertozzi group in 2000, involving a reaction of azides with triarylphosphines to form a bio-orthogonal product [13]. This particular method has been used as the nontraceless azide–phosphine ligation, which can be performed in both live cells and in vivo. It has some advantages in contrast to cysteine limitation in chemical ligation [42, 60]. Especially, Bertozzi and coworkers pioneered the potential of this reaction as a method for bioconjugation and transformed it into the so-called Staudinger ligation [70]. The reaction between an azide and a phosphine-functionalized group has resulted in stable amide formation [2, 29, 78]. For example, Staudinger ligation has been utilized to label glycans, lipids, DNA, aptamers, and protein surfaces. Thus, this synthetic reaction of azides and phosphines on biomolecules has found widespread applications in the field of chemical biology and biotechnology (e.g., microarrays, glycopeptide constructions, protein engineering) [86].

Both metabolic and mechanistic approaches were also defined by Bertozzi et al., and they have found that some practical limitations affect the Staudinger ligation efficiency. Especially, low chemoselectivity of azide reactants or slow kinetics of the reactions can be considered critical limitations. It should be also noted that basic or very nucleophilic phosphines react faster; however, some aliphatic phosphines undergo air oxidation when they use living systems. The oxidizing extracellular environment is electron poor and hence replete with electrophiles. On the other hand, the reducing environment of the cytosol is electron rich and covered with nucleophilic thiolates that can attack cyclooctyne and its derivatives (including *trans*-cyclooctene) [36].

10.4 Medical Applications

10.4.1 *Pathogen Detection*

Pathogens are contagious microorganisms that are harmful to humans. The detection of pathogenic microorganisms is key to the prevention of infection-related health and safety problems. Current predominant techniques for identification of microbial pathogens rely mostly upon conventional clinical microbiology approaches that suffer from major drawbacks. Standard culturing techniques provide highly reliable pathogen identification but are laborious, time consuming, and expensive. Conventional techniques also are not suitable for managing large numbers of environmental or clinical samples. Thus, these tests do not provide the required information in time, matching the fast infection capabilities of pathogens. For fast determination of the pathogen presence, we need reliable and accurate tools that can permit faster, accurate analytical results. One of the major advantage of aptamers as biorecognition agents is their ability to target and specifically identify bacterial and viral particles without any knowledge of the membrane-associated antigenic determinants or cell surface molecular biomarkers present in that particular microorganism, thanks to the cell-SELEX procedure. The fast adaptation capability of bacteria enables them to survive in a wide variety of environments. Hence, various pathogenic species emerge and pose a threat to human health. Pathogenic bacteria are harmful species that can cause bacterial infections and contagious diseases for serious health complications. The most common pathogens include *Escherichia coli* and *Salmonella* spp. (food poisoning), *Helicobacter pylori* (gastritis and ulcers), *S. aureus* (food poisoning), and *Streptococcus* spp. (pneumonia, meningitis, ear infections, and pharyngitis). The development of accurate sensing systems is essential for maintaining a healthy environment by diagnosis and epidemiological control of pathogenic organism. Other than food-borne pathogens and sanitary water treatment facilities' bacterial populations, bacterial sensing is necessary for the purpose of medical diagnosis. There are three ways of diagnosis: (i) morphological visualization, (ii) specific detection of bacterial cell components, and (iii) whole cell detection [30].

The main challenge in pathogen diagnosis is to distinguish between pathogenic and nonpathogenic strains, which involves specific detection of closely related subspecies. Sensing strategies should be rapid and sensitive for early determination of bacterial strain and proper enumeration of the identified strains. The rapid and accurate detection of pathogenic bacteria is vital for deciding on an appropriate antibiotic treatment to control the spread of the bacteria. Traditional identification of bacteria is based on the morphological features of the cells, which can be observed via simple microscopic observations. Gram staining, culturing, biochemical assays, and PCR amplification–based detection are commonly exploited for highly accurate and sensitive detection.

However, traditional culturing methods or PCR-based techniques require days to hours of detection time for final determination, respectively. In addition, bacterial surface–specific biorecognition probes (e.g., antibodies, aptamers) are used for fast or real-time detection of bacterial types.

Labib et al. selected high-affinity DNA aptamers binding live *Salmonella typhimurium* with a counterselection strategy against heat-killed *Salmonella typhimurium* and a mixture of related pathogens, including *Salmonella enteritidis, E. coli, S. aureus, Pseudomonas aeruginosa*, and *Citrobacter freundii*, to obtain target-specific candidates [46]. A viability impedimetric sensor was fabricated by self-assembly of aptamer molecules on gold-nanoparticle-modified screen-printed carbon electrodes. The sensor has a limit of detection at 600 CFU/mL and a specificity that can distinguish *Salmonella typhymurium* from other *Salmonella* strains. Labib and Brezowski provided recently a detailed overview of developments of electrochemical aptasensors for microbial and viral pathogen detection, viability assessment of microorganisms, bacterial typing, and identification of epitope-specific aptamers [45].

A bloodstream infection, or sepsis, is an extremely serious condition and a fast diagnosis is critical for the life of the patient. Magnetic nanoparticles functionalized with aptamers were designed as biosensors for determination of pathogens in blood samples. Kolesnikov et al. provided a recent and detailed overview of aptamer-based diagnosis of bacterial infections [43]. Aptamer-based detection is nucleic acid–based detection that has many advantages based on the features associated with nucleic acids such as the absence of batch-to-batch variation, reusability due to stability against repetitive denaturation, and ease of modification for conjugation on a variety of surfaces. Biological fluids interfere often with biosensor efficiency, leading to difficulty in direct use of the sensor. However, some properties of specific nanomaterials can be incorporated to overcome this instability of nucleic acids. Association of biorecognition elements with various nanoparticles allows convenient strategies to overcome interferences in biological samples [26]. For example, Ozalp et al. immobilized two different aptamers specifically against *E. coli* and *Salmonella typhymurium* on the surface of magnetic beads and quantified the presence of bacteria using real-time PCR or QCM [64, 65]. In these reports, a fast purification strategy was used to show that even in the presence of 100% milk, detection of approximately 100 CFU/mL for both *E. coli* and *Salmonella typhymurium* could be reached within two hours in spiked milk samples. Similarly, Suh et al. developed the aptamer magnetic pull-down method conjugated to RT-PCR for detection of *Campylobacter jejuni* which is one of the leading food contamination disease [82]. *S. aureus* is a facultative anaerobic, gram-positive bacterium and the most widespread pathogen among hospital-acquired infections. About 20% of the human population is considered to be persistent carriers, 60% are intermittent carriers, and approximately 20% almost never carry *S. aureus*, which is the causative agent for many illnesses, such as minor skin infections or life-threatening diseases, like pneumonia meningitis. Also, elimination of carriage is an attractive preventive approach in surgical patients [41]. The World Health Organization (WHO) reports that microbial infections are the major causes of death worldwide [94]. For example, *S. aureus* infects 500,000 people every year in America, about 20% of which are life-threatening, antibiotic-resistant *S. aureus* infections [40].

Chang et al. developed a rapid, ultrasensitive, and low cost biosensor for *S. aureus* by combining gold nanoparticles [14]. Inflammatory diseases are mediated by the complement factor C5a protein that modulates chemotaxis of CD88-expressing cell lines. A PEGylated spiegelmer aptamer was selected to bind mouse and human C5a with picomolar affinity. This aptamer was demonstrated to inhibit chemotaxis of cells in vitro and efficiently decreased the activation of primary human polymorphonuclear (PMN) leukocytes [34]. The aptamer was suggested as a specific interventional therapy in sepsis, decreasing elevated levels of C5a to prevent multiorgan failure and mortality.

Surface plasmon resonace (SPR) biosensors have been developed as rapid and effective alternative for analyzing biological interactions. Avian influenza (bird flu) is a well-known disease in the poultry industry caused by type A influenza virus. Huge economic losses and health problems have been originated from avian influenza virus (AIV) worldwide. It is an urgent health concern due to the potential outbreaks of highly pathogenic H5N1 influenza in animals and humans. Thus, rapid detection of virus from swab samples is important in the diagnosis of infection. A DNA aptamer as the specific recognition element for developing a portable surface plasmon resonance (SPR) biosensor was selected for rapid detection of AIV H5N1. The biotinylated aptamer sequences immobilized on streptavidin-covered gold surfaces capture the virus to produce surface plasmon signals. The AIV H5N1 in poultry swab samples with concentrations of 0.128 to 12.8 HAU (hemagglutination units) could be detected instantly using this aptasensor, which is comparable to the detection limits of traditional methods [5].

Using a fluorescence assay and electrochemical detection, Kim et al. [37] designed a cocktail of three DNA aptamers binding to the same target. The aptamer sequences were selected for targeting *E. coli* and having different affinities (12.4, 25.4, and 14.2 nM). The detection limit of the cocktail was enhanced to 18 times greater than that of a single aptamer. Wu et al. demonstrated that the salt-induced color change of gold nanoparticles (AuNPs) can be used for the detection of *E. coli* and *Salmonella typhimurium* [95]. Aptamer-AuNPs were mixed with the target bacterial cells, leading to aggregation upon the addition of NaCl. The subsequent color change from red to purple was simply detected either visually or with UV-Vis spectroscopy. This simple colorimetric method had a limit of detection at 10^5 CFU/mL of *E. coli* or *Salmonella typhimurium*, without any labeling or any expensive instrumentation.

10.4.2 *Cancer Diagnosis*

Cancer is a common health problem with major economic consequences. Early and accurate cancer diagnosis provides an effective way to monitor progression and successful treatment of the disease. Numerous attempts around the world resulted in competitive commercialized biosensor devices for the diagnosis and prognosis of cancer. The current biosensors are mostly complicated to operate and relatively expensive outside of hospital and laboratory. For example, enzyme immunoassays capable of quantifying tumor-specific antigens in human serum are available for early diagnosis. Aptamers have been used to develop alternative biosensors that can bind to specific cancer biomarkers for diagnosis of the disease.

Breast cancer is one of the most common cancers among women. The distant metastasis is the major cause of death. Accurate diagnosis of breast cancer cells is important for determining the stage of the disease and deciding on an appropriate therapy with a simple and rapid diagnosis. The lack of probes for accurate identification of tumor cells is the major reason behind the hindrance of early determination of metastasis. Human mucin-1 (MUC1) and folic acid (FA) are the most common markers to monitor the metastatic breast tumors. Elevated level of MUC1 has been considered to be an important indicator in the diagnosis of breast cancer. FA has been demonstrated to preferentially target cancer cells, because the folate receptor is frequently overexpressed on the surface of cancer cells [7]. A sensitive SPR biosensor using magnetic nanoparticle and a two-marker recognition system was developed using the MUC1 aptamer that can capture and signal the presence of breast cancer cells (MCF-7) in a sandwich assay [15]. The additional detection of folic acid receptor (FR) together with MUC1 can provide more accurate diagnostic information, and the treatment can maximize efficacy and minimize toxicity according to the tumor cell type.

Detection of circulating tumor cells (CTCs) plays an important role in cancer diagnosis and prognosis. Conventional methods include tumor biopsy, tissue imaging, and serum marker scanning. CTC detection is especially valuable in predicting patient survival. CTCs originate from primary tumor sites and are released into the bloodstream. The presence of tumor cells in blood has been used as an indication for metastasis, because spreading of CTCs results in tumorigenic occurrence in tissues that are not related to the original cancer tissue. However, CTC detection is highly challenging due to extremely low number of cancer cells in bloodstream compared to very high number of blood cells like red blood cells. The number of CTCs could be as low as a few hundred cells per milliliter of blood. Aptamer-functionalized nanoparticles were frequently reported for detection of tumor cells because they can provide detection of very few numbers of CTCs, which require highly efficient diagnostic tools. An application was reported by Shen et al. who captured non-small-cell lung cancer CTCs from blood and then recovered these for analysis by constructing a capture/recovery platform that consisted of a DNA aptamer conjugated onto silicon nanowire substrates [72].

Although many methods with high sensitivity and selectivity for nondestructive detection of intact cancer cells have been developed, novel approaches are urgently required for rapid, portable, and simple sensors with high sensitivity and specificity detecting intact cancer cells, especially in developing countries and resource-limited and remote regions. Prostate-specific antigen (PSA) is a biomarker to diagnose prostate cancer. Cha et al. developed a rapid aptasensor for diagnosis of prostate cancer [11]. The assay is based on the fact that guanine molecules cannot bind to 3,4,5-trimethoxylphenylglyoxal (TMPG) to emit bright light. Thus, guanine-rich aptamers increase glow chemiluminescence in the presence of the aptamer target. PSA was reported to be determined with a limit of detection at 1 ng/mL, with the reaction of guanine nucleotides of the aptamer sequence with the antigen on the surface of an intact cancer cell. Microfluidic paper-based analytical devices are one of the recent format that can provide low cost portable biosensor. A novel electrochemical lab-on-paper device was designed for in situ monitoring of cell surface glycans. Aptamer-immobilized electrodes were used to capture cancer cells and subsequently quantified by DPV on amicrofluidic paper format [81]. Wang et al. selected DNA aptamer sequences for PC-3 cell lines (a prostate cancer cell type) through the cell-SELEX procedure. The aptamers were fluorescently labeled and used in staining of clinical tissues. When tissue sections were from high risk groups with metastasis, stronger fluorescent staining was observed, which suggested that the selected aptamers bind to proteins related to the prostate cancer [92].

Simple colorimetric assays can be developed using aptamer–nanomaterial conjugates. Gold nanoparticles present unique opportunities for colorimetric systems because of the nature of gold molecules. Medley et al. developed a simple colorimetric assay for cancer cells, using aptamer-conjugated AuNPs, which were prepared by using the thiol-modified aptamer obtained using cell-SELEX for acute leukemia cells [53]. Thiol molecules have spontaneous and strong affinity to gold, which enables self-assembly. Aptamers target AuNPs on the surfaces of cells through target–aptamer interaction. The binding of Apt-AuNPs on the cell surfaces brings gold nanoparticles into close proximity with one another, causing a shift in the extinction spectra, which can be measured for quantifying the number of cells. Similarly, Liu et al. detected Ramos cells with AuNPs conjugated to aptamers [49]. In this study, Ramos cells and Apt-AuNPs were applied to a secondary aptamer immobilized on a strip surface after they bind together. Ramos cells interacted with Apt-AuNPs and migrated over the strip surface to the test region containing the immobilized secondary aptamers, which bind to another region on the cell surface. The accumulation of Apt-AuNPs produces a red color that can be visualized and quantified. The major advantage of colorimetric assays is that the sensor produces a signal that can be observed with the naked eye, but mostly with low sensitivity.

Silica nanoparticles are also commonly employed in developing diagnostic tools with aptamers because they are known for their excellent biocompatibility, easy separation, and broad size range from nanometers to micrometers [85]. Their large surface area and porous structure allowed encapsulation of fluorophores at excessive amounts in order to generate intense fluorescence, which makes them excellent signal enhancers. Aptamer-conjugated, fluorophore-doped silica nanoparticles exhibited high fluorescence intensity with little bleaching, which is essential for a biosensor for cancer cells [90].

Investigation of disease biomarkers is the key to identifying new drugs that can provide better diagnosis and monitor the therapy process. For example, ovarian cancer is a lethal disease with difficulties in diagnosis, and consequently its diagnosis is currently possible only at advanced stages after metastasis has already progressed. This situation reduces the treatment possibilities significantly and early diagnosis is in high demand. Aptamers can be good candidates in these situations because they can readily be selected for any cell type. An aptamer (TOV6) was selected by following cell-SELEX against the ovarian cancer cell line TOV-21G [87]. This study is a representative example for how aptamer selection can lead to identification of a membrane protein marker for specific diagnosis and therapy target of cancers. The selected DNA aptamer was highly selective to differentiate between the ovarian clear cell line (TOV-21G) and the ovarian serous adenocarcinoma (CAOV3) biomarkers at the molecular level. Van Simaeys et al. used the aptamer to pull down the cognate target on the cell surface [88]. The major challenge in similar studies is the hydrophobic nature of membrane proteins and thus the binding interaction between aptamer and its target can be disrupted by detergents ubiquitously used for solubilization of membrane proteins. This challenge could be overcome by incorporating nucleotides in the aptamer sequence for chemical cross-linking and magnetic beads and later mass spectroscopy for identification of the receptor protein on the cell surface of Ramos cells [52]. For example, formaldehyde was used in a technique of cross-linking between aptamer and cell membrane protein target, similar to a well-known technique chromatin immunoprecipitation (ChIP), an experimental method for identification of genomic regions of chromosome binding proteins such as transcription factors [88]. Thus, cross-linking is a method useful to determine protein identity instead of the DNA sequence as in ChIP. Finally, isolated protein samples are fragmented by proteases and mass spectroscopy is used to determine peptide mass fingerprint.

Microfluidic paper-based devices (μ-PAD) are low-cost alternative platforms for portable and disposable biosensors without dependence on sophisticated instrumentation. Paper-based cytodevices have been reported by chemiluminiscence, fluorescent, electrochemical, or Raman scattering. The formats for μ-PAD sensors have been primarily colorimetric or fluorescent signaling. Colorimetric devices are mostly semiquantitative; however, such yes/no biosensors are not adequate for cancer diagnosis. Likewise, fluorescent devices require advanced instrumentation. A recent report demonstrated that electrochemical detection in μ-PAD format could provide an advantageous device for portable and simple cancer detection [80]. Human acute promyelocytic leukemia is a deadly cancer condition that affects the blood and bone marrow. A gold-paper electrode was fabricated by the growth of a gold nanoparticle layer on the paper surface and employed as the working electrode. Leukemia cancer cells (HL-60) cell-targeting aptamers were immobilized on this Au-paper electrode. The electrochemical paper probe was used for cancer cell detection by the horseradish peroxidase–labeled FA integrated as the specific recognition of the folate receptor on the captured HL-60 cell surface by FA and the amplification of electrochemical signal based on the horseradish peroxidase–catalyzed oxidation of *o*-phenylenediamine by peroxide.

Acknowledgments

This chapter has been written as part of the TUBITAK Project 213M315.

References

1. Abbaspour, A., Norouz-Sarvestani, F., Noori, A., and Soltani, N. (2015). Aptamer-conjugated silver nanoparticles for electrochemical dual-aptamer-based sandwich detection of staphylococcus aureus, *Biosens. Bioelectron.*, **68**, pp. 149–155.
2. Agard, N. J., Baskin, J. M., Prescher, J. A., Lo, A., and Bertozzi, C. R. (2006). A comparative study of bioorthogonal reactions with azides, *ACS Chem. Biol.*, **1**(10), pp. 644–648.
3. Ajnai, G., Chiu, A., Kan, T., Cheng, C.-C., Tsai, T.-H., and Chang, J. (2014). Trends of gold nanoparticle-based drug delivery system in cancer therapy, *J. Exp. Clin. Med.*, **6**(6), pp. 172–178.
4. Ali, M. M., Kang, D.-K., Tsang, K., Fu, M., Karp, J. M., and Zhao, W. (2012). Cell-surface sensors: lighting the cellular environment, *Wiley Interdiscip. Rev. Nanomed. Nanobiotechnol.*, **4**(5), pp. 547–561.
5. Bai, H., Wang, R., Hargis, B., Lu, H., and Li, Y. (2012). A spr aptasensor for detection of avian influenza virus H5N1, *Sensors*, **12**(9), pp. 12506–12518.
6. Baumstummler, A., Lehmann, D., Janjic, N., and Ochsner, U. A. (2014). Specific capture and detection of Staphylococcus aureus with high-affinity modified aptamers to cell surface components, *Lett. Appl. Microbiol.*, **59**(4), pp. 422–431.
7. Berois, N., Varangot, M., Aizen, B., Estrugo, R., Zarantonelli, L., Fernández, P., Krygier, G., Simonet, F., Barrios, E., Musé, I., and Osinaga, E. (2000). Molecular detection of cancer cells in bone marrow and peripheral blood of patients with operable breast cancer. Comparison

of CK19, MUC1 and CEA using RT-PCR, *Eur. J. Cancer,* **36**(6), pp. 717–723.

8. Boisgard, R., Kuhnast, B., Vonhoff, S., Younes, C., Hinnen, F., Verbavatz, J. M., Rousseau, B., Furste, J., Wlotzka, B., Dolle, F., Klussmann, S., and Tavitian, B. (2005). In vivo biodistribution and pharmacokinetics of F-18-labelled Spiegelmers: a new class of oligonucleotidic radiopharmaceuticals, *Eur. J. Nucl. Med. Mol. Imaging,* **32**(4), pp. 470–477.
9. Boisselier, E., and Astruc, D. (2009). Gold nanoparticles in nanomedicine: preparations, imaging, diagnostics, therapies and toxicity, *Chem. Soc. Rev.,* **38**(6), pp. 1759–1782.
10. Cao, H.-Y., Yuan, A.-H., Shi, X.-S., Chen, W., and Miao, Y. (2014). Evolution of a gastric carcinoma cell-specific DNA aptamer by live cell-SELEX, *Oncol. Rep.,* **32**(5), pp. 2054–2060.
11. Cha, T., Cho, S., Kim, Y. T., and Lee, J. H. (2014). Rapid aptasensor capable of simply diagnosing prostate cancer, *Biosens. Bioelectron.,* **62**, pp. 31–37.
12. Chalker, J. M., Bernardes, G. J. L., Lin, Y. A., and Davis, B. G. (2009). Chemical modification of proteins at cysteine: opportunities in chemistry and biology, *Chem. Asian J.,* **4**(5), pp. 630–640.
13. Chang, C.-C., Wei, S.-C., Wu, T.-H., Lee, C.-H., and Lin, C.-W. (2013). Aptamer-based colorimetric detection of platelet-derived growth factor using unmodified goldnanoparticles, *Biosens. Bioelectron.,* **42**, pp. 119–123.
14. Chang, Y.-C., Yang, C.-Y., Sun, R.-L., Cheng, Y.-F., Kao, W.-C., and Yang, P.-C. (2013). Rapid single cell detection of Staphylococcus aureus by aptamer-conjugated gold nanoparticles, *Sci. Rep.,* **3**, p. 1863.
15. Chen, H., Hou, Y., Ye, Z., Wang, H., Koh, K., Shen, Z., and Shu, Y. (2014). Label-free surface plasmon resonance cytosensor for breast cancer cell detection based on nano-conjugation of monodisperse magnetic nanoparticle and folic acid, *Sens. Actuators B Chem.,* **201**, pp. 433–438.
16. Chen, M., Bi, S., Jia, X., and He, P. (2014). Aptamer-conjugated bio-bar-code Au–Fe3O4 nanoparticles as amplification station for electrochemiluminescence detection of tumor cells, *Anal. Chim. Acta,* **837**, pp. 44–51.
17. Chen, X., Huang, Y., Duan, N., Wu, S., Xia, Y., Ma, X., Zhu, C., Jiang, Y., and Wang, Z. (2014). Screening and identification of DNA aptamers against T-2 toxin assisted by graphene oxide, *J. Agric. Food Chem.,* **62**(42), pp. 10368–10374.

18. Cho, E. C., Glaus, C., Chen, J., Welch, M. J., and Xia, Y. (2010). Inorganic nanoparticle-based contrast agents for molecular imaging, *Trends Mol. Med.*, **16**(12), pp. 561–573.
19. De Groote, M. A., Nahid, P., Jarlsberg, L., Johnson, J. L., Weiner, M., Muzanyi, G., Janjic, N., Sterling, D. G., and Ochsner, U. A. (2013). Elucidating novel serum biomarkers associated with pulmonary tuberculosis treatment, *PLoS One*, **8**(4), p. e61002.
20. Deng, Q., Watson, C. J., and Kennedy, R. T. (2003). Aptamer affinity chromatography for rapid assay of adenosine in microdialysis samples collected in vivo, *J. Chromatogr. A*, **1005**(1–2), pp. 123–130.
21. Doessing, H., Hansen, L. H., Veedu, R. N., Wengel, J., and Vester, B. (2012). Amplification and re-generation of LNA-modified libraries, *Molecules*, **17**(11), pp. 13087–13097.
22. Du, D., Wang, J., Wang, L., Lu, D., and Lin, Y. (2012). Integrated lateral flow test strip with electrochemical sensor for quantification of phosphorylated cholinesterase: biomarker of exposure to organophosphorus agents, *Anal. Chem.*, **84**(3), pp. 1380–1385.
23. Du, Y., Li, B., and Wang, E. (2013). "Fitting" makes "sensing" simple: label-free detection strategies based on nucleic acid aptamers, *Acc. Chem. Res.*, **46**(2), pp. 203–213.
24. Dubois, C., Campbell, M. A., Edwards, S. L., Wengel, J., and Veedu, R. N. (2012). Stepping towards highly flexible aptamers: enzymatic recognition studies of unlocked nucleic acid nucleotides, *Chem. Comm.*, **48**(44), pp. 5503–5505.
25. Feng, L., Chen, Y., Ren, J., and Qu, X. (2011). A graphene functionalized electrochemical aptasensor for selective label-free detection of cancer cells, *Biomaterials*, **32**(11), pp. 2930–2937.
26. Gedi, V., and Kim, Y.-P. (2014). Detection and characterization of cancer cells and pathogenic bacteria using aptamer-based nano-conjugates, *Sensors*, **14**(10), pp. 18302–18327.
27. Giannetti, A., Tombelli, S., and Baldini, F. (2013). Oligonucleotide optical switches for intracellular sensing, *Anal. Bioanal. Chem.*, **405**(19), pp. 6181–6196.
28. Gold, L., Ayers, D., Bertino, J., Bock, C., Bock, A., Brody, E. N., Carter, J., Dalby, A. B., Eaton, B. E., Fitzwater, T., Flather, D., Forbes, A., Foreman, T., Fowler, C., Gawande, B., Goss, M., Gunn, M., Gupta, S., Halladay, D., Heil, J., Heilig, J., Hicke, B., Husar, G., Janjic, N., Jarvis, T., Jennings, S., Katilius, E., Keeney, T. R., Kim, N., Koch, T. H., Kraemer, S., Kroiss, L., Le, N., Levine, D., Lindsey, W., Lollo, B., Mayfield, W., Mehan, M., Mehler, R., Nelson, S.

K., Nelson, M., Nieuwlandt, D., Nikrad, M., Ochsner, U., Ostroff, R. M., Otis, M., Parker, T., Pietrasiewicz, S., Resnicow, D. I., Rohloff, J., Sanders, G., Sattin, S., Schneider, D., Singer, B., Stanton, M., Sterkel, A., Stewart, A., Stratford, S., Vaught, J. D., Vrkljan, M., Walker, J. J., Watrobka, M., Waugh, S., Weiss, A., Wilcox, S. K., Wolfson, A., Wolk, S. K., Zhang, C., and Zichi, D. (2010). Aptamer-based multiplexed proteomic technology for biomarker discovery, *PLoS One*, **5**(12), p. e15004.

29. Gololobov, Y. G., and Kasukhin, L. F. (1992). Recent advances in the staudinger reaction, *Tetrahedron*, **48**(8), pp. 1353–1406.
30. Gopinath, S. C. B., Tang, T.-H., Chen, Y., Citartan, M., and Lakshmipriya, T. (2014). Bacterial detection: from microscope to smartphone, *Biosens. Bioelectron.*, **60**, pp. 332–342.
31. Gupta, S., Hirota, M., Waugh, S. M., Murakami, I., Suzuki, T., Muraguchi, M., Shibamori, M., Ishikawa, Y., Jarvis, T. C., Carter, J. D., Zhang, C., Gawande, B., Vrkljan, M., Janjic, N., and Schneider, D. J. (2014). Chemically-modified DNA aptamers bind interleukin-6 with high affinity and inhibit signaling by blocking its interaction with interleukin-6 receptor, *J. Biol. Chem.*, **289**(12), pp. 8706–8719.
32. Hernandez, F. J., Dondapati, S. K., Ozalp, V. C., Pinto, A., O'Sullivan, C. K., Klar, T. A., and Katakis, I. (2009). Label free optical sensor for Avidin based on single gold nanoparticles functionalized with aptamers, *J. Biophotonics*, **2**(4), pp. 227–231.
33. Hernandez, F. J., and Ozalp, V. C. (2012). Graphene and other nanomaterial-based electrochemical aptasensors, *Biosensors*, **2**(1), pp. 1–14.
34. Hoehlig, K., Maasch, C., Shushakova, N., Buchner, K., Huber-Lang, M., Purschke, W. G., Vater, A., and Klussmann, S. (2013). A novel C5a-neutralizing mirror-image (l-)aptamer prevents organ failure and improves survival in experimental sepsis, *Mol. Ther.*, **21**(12), pp. 2236–2246.
35. Huenniger, T., Wessels, H., Fischer, C., Paschke-Kratzin, A., and Fischer, M. (2014). just in time-selection: a rapid semiautomated SELEX of DNA aptamers using magnetic separation and BEAMing, *Anal. Chem.*, **86**(21), pp. 10940–10947.
36. Kalia, J., and Raines, R. T. (2010). Advances in bioconjugation, *Curr. Org. Chem.*, **14**(2), pp. 138–147.
37. Kim, Y. S., Chung, J., Song, M. Y., Jurng, J., and Kim, B. C. (2014). Aptamer cocktails: enhancement of sensing signals compared to single use of aptamers for detection of bacteria, *Biosens. Bioelectron.*, **54**, pp. 195–198.

38. Kim, Y. S., Jung, H. S., Matsuura, T., Lee, H. Y., Kawai, T., and Gu, M. B. (2007). Electrochemical detection of 17β-estradiol using DNA aptamer immobilized gold electrode chip, *Biosens. Bioelectron.*, **22**(11), pp. 2525–2531.

39. Kim, Y. S., Lee, S. J., and Gu, M. B. (2008). Electrochemical aptamer-based biosensors, *Biochip J.*, **2**(3), pp. 175–182.

40. Klein, E., Smith, D. L., and Laxminarayan, R. (2007). Hospitalizations and deaths caused by methicillin-resistant Staphylococcus aureus, United States, 1999–2005, *Emerg. Infect. Dis.*, **13**(12), pp. 1840–1846.

41. Kluytmans, J., van Belkum, A., and Verbrugh, H. (1997). Nasal carriage of Staphylococcus aureus: epidemiology, underlying mechanisms, and associated risks, *Clin. Microbiol. Rev.*, **10**(3), pp. 505–520.

42. Kohn, M., and Breinbauer, R. (2004). The Staudinger ligation: a gift to chemical biology, *Angew. Chem., Int. Ed. Engl.*, **43**(24), pp. 3106–3116.

43. Kolesnikov, A. V., Kozyr, A. V., and Shemyakin, I. G. (2012). The prospects for using aptamers in diagnosing bacterial infections, *Mol. Gen. Microbiol. Virol.*, **27**(2), pp. 49–55.

44. Kraemer, S., Vaught, J. D., Bock, C., Gold, L., Katilius, E., Keeney, T. R., Kim, N., Saccomano, N. A., Wilcox, S. K., Zichi, D., and Sanders, G. M. (2011). From SOMAmer-based biomarker discovery to diagnostic and clinical applications: a SOMAmer-based, streamlined multiplex proteomic assay, *PLoS One*, **6**(10), p. e26332.

45. Labib, M., and Berezovski, M. (2014). Electrochemical aptasensors for microbial and viral pathogens, in *Biosensors Based on Aptamers and Enzymes* (Gu, M. B., Kim, H.-S., eds.), **140**, pp. 155–181, Springer Berlin Heidelberg.

46. Labib, M., Zamay, A. S., Kolovskaya, O. S., Reshetneva, I. T., Zamay, G. S., Kibbee, R. J., Sattar, S. A., Zamay, T. N., and Berezovski, M. V. (2012). Aptamer-based viability impedimetric sensor for bacteria, *Anal. Chem.*, **84**(21), pp. 8966–8969.

47. Li, W., Dong, Y., Wang, X., Li, H., and Xu, D. (2015). PolyA-tailed and fluorophore-labeled aptamer-gold nanoparticle conjugate for fluorescence turn-on bioassay using iodide-induced ligand displacement, *Biosens. Bioelectron.*, **66**, pp. 43–49.

48. Liss, M., Petersen, B., Wolf, H., and Prohaska, E. (2002). An aptamer-based quartz crystal protein biosensor, *Anal. Chem.*, **74**(17), pp. 4488–4495.

49. Liu, G., Mao, X., Phillips, J. A., Xu, H., Tan, W., and Zeng, L. (2009). Aptamer–nanoparticle strip biosensor for sensitive detection of cancer cells, *Anal. Chem.*, **81**(24), pp. 10013–10018.

50. Ma, X., Jiang, Y., Jia, F., Yu, Y., Chen, J., and Wang, Z. (2014). An aptamer-based electrochemical biosensor for the detection of Salmonella, *J. Microbiol. Methods*, **98**, pp. 94–98.
51. Mairal, T., Ozalp, V. C., Lozano Sanchez, P., Mir, M., Katakis, I., and O'Sullivan, C. K. (2008). Aptamers: molecular tools for analytical applications, *Anal. Bioanal. Chem.*, **390**(4), pp. 989–1007.
52. Mallikaratchy, P., Tang, Z., Kwame, S., Meng, L., Shangguan, D., and Tan, W. (2007). Aptamer directly evolved from live cells recognizes membrane bound immunoglobin heavy mu chain in Burkitt's lymphoma cells, *Mol. Cell. Proteomics*, **6**(12), pp. 2230–2238.
53. Medley, C. D., Smith, J. E., Tang, Z., Wu, Y., Bamrungsap, S., and Tan, W. (2008). Gold nanoparticle-based colorimetric assay for the direct detection of cancerous cells, *Anal. Chem.*, **80**(4), pp. 1067–1072.
54. Mehan, M., Williams, S., Siegfried, J., Bigbee, W., Weissfeld, J., Wilson, D., Pass, H., Rom, W., Muley, T., Meister, M., Franklin, W., Miller, Y., Brody, E., and Ostroff, R. (2014). Validation of a blood protein signature for non-small cell lung cancer, *Clin. Proteomics*, **11**(1), p. 32.
55. Mehan, M. R., Ayers, D., Thirstrup, D., Xiong, W., Ostroff, R. M., Brody, E. N., Walker, J. J., Gold, L., Jarvis, T. C., Janjic, N., Baird, G. S., and Wilcox, S. K. (2012). Protein signature of lung cancer tissues, *PLoS One*, **7**(4), p. e35157.
56. Meyer, C., Hahn, U., and Rentmeister, A. (2011). Cell-specific aptamers as emerging therapeutics, *J. Nucleic Acids*, **2011**, p. 904750.
57. Murphy, C. J., Gole, A. M., Stone, J. W., Sisco, P. N., Alkilany, A. M., Goldsmith, E. C., and Baxter, S. C. (2008). Gold nanoparticles in biology: beyond toxicity to cellular imaging, *Acc. Chem. Res.*, **41**(12), pp. 1721–1730.
58. Nahid, P., Bliven-Sizemore, E., Jarlsberg, L. G., De Groote, M. A., Johnson, J. L., Muzanyi, G., Engle, M., Weiner, M., Janjic, N., Sterling, D. G., and Ochsner, U. A. (2014). Aptamer-based proteomic signature of intensive phase treatment response in pulmonary tuberculosis, *Tuberculosis*, **94**(3), pp. 187–196.
59. NEB (2014). New England Biolabs® releases next generation "warm-start" reverse transcriptase for nucleic acid detection. https://www.neb.com/about-neb/ news-and-press-releases/2014/11/19/warmstart-rtx (accessed 20.12.2014).
60. Nilsson, B. L., Soellner, M. B., and Raines, R. T. (2005). Chemical synthesis of proteins, *Annu. Rev. Biophys. Biomol. Struct.*, **34**, pp. 91–118.

61. Nwe, K., and Brechbiel, M. W. (2009). Growing applications of "click chemistry" for bioconjugation in contemporary biomedical research, *Cancer Biother. Radiopharm.*, **24**(3), pp. 289–302.

62. Ochsner, U. A., Katilius, E., and Janjic, N. (2013). Detection of Clostridium difficile toxins A, B and binary toxin with slow off-rate modified aptamers, *Diagn. Microbiol. Infect. Dis.*, **76**(3), pp. 278–285.

63. Ostroff, R. M., Mehan, M. R., Stewart, A., Ayers, D., Brody, E. N., Williams, S. A., Levin, S., Black, B., Harbut, M., Carbone, M., Goparaju, C., and Pass, H. I. (2012). Early detection of malignant pleural mesothelioma in asbestos-exposed individuals with a noninvasive proteomics-based surveillance tool, *PLoS One*, **7**(10), p. e46091.

64. Ozalp, V. C., Bayramoglu, G., Erdem, Z., and Arica, M. Y. (2015). Pathogen detection in complex samples by quartz crystal microbalance sensor coupled to aptamer functionalized core–shell type magnetic separation, *Anal. Chim. Acta*, **853**, pp. 533–540.

65. Ozalp, V. C., Bayramoglu, G., Kavruk, M., Keskin, B. B., Oktem, H. A., and Arica, M. Y. (2014). Pathogen detection by core-shell type aptamer-magnetic preconcentration coupled to real-time PCR, *Anal. Biochem.*, **447**, pp. 119–125.

66. Ozalp, V. C., Pedersen, T. R., Nielsen, L. J., and Olsen, L. F. (2010). Time-resolved measurements of intracellular ATP in the yeast Saccharomyces cerevisiae using a new type of nanobiosensor, *J. Biol. Chem.*, **285**(48), pp. 37579–37588.

67. Park, J.-W., Tatavarty, R., Kim, D. W., Jung, H.-T., and Gu, M. B. (2012). Immobilization-free screening of aptamers assisted by graphene oxide, *Chem. Comm.*, **48**(15), pp. 2071–2073.

68. Park, J. H., Jee, M. H., Kwon, O. S., Keum, S. J., and Jang, S. K. (2013). Infectivity of hepatitis C virus correlates with the amount of envelope protein E2: development of a new aptamer-based assay system suitable for measuring the infectious titer of HCV, *Virology*, **439**(1), pp. 13–22.

69. Pilapong, C., Sitthichai, S., Thongtem, S., and Thongtem, T. (2014). Smart magnetic nanoparticle-aptamer probe for targeted imaging and treatment of hepatocellular carcinoma, *Int. J. Pharm.*, **473**(1–2), pp. 469–474.

70. Prescher, J. A., and Bertozzi, C. R. (2005). Chemistry in living systems, *Nat. Chem. Biol.*, **1**(1), pp. 13–21.

71. Rohloff, J. C., Gelinas, A. D., Jarvis, T. C., Ochsner, U. A., Schneider, D. J., Gold, L., and Janjic, N. (2014). Nucleic acid ligands with protein-

like side chains: modified aptamers and their use as diagnostic and therapeutic agents, *Mol. Ther. Nucleic Acids*, **3**, p. e201.

72. Shen, Q., Xu, L., Zhao, L., Wu, D., Fan, Y., Zhou, Y., OuYang, W.-H., Xu, X., Zhang, Z., Song, M., Lee, T., Garcia, M. A., Xiong, B., Hou, S., Tseng, H.-R., and Fang, X. (2013). Specific capture and release of circulating tumor cells using aptamer-modified nanosubstrates, *Adv. Mater.*, **25**(16), pp. 2368–2373.
73. Shi, H., He, X., Cui, W., Wang, K., Deng, K., Li, D., and Xu, F. (2014). Locked nucleic acid/DNA chimeric aptamer probe for tumor diagnosis with improved serum stability and extended imaging window in vivo, *Anal. Chim. Acta*, **812**, pp. 138–144.
74. Shwetha, N., Selvakumar, L. S., and Thakur, M. S. (2013). Aptamer–nanoparticle-based chemiluminescence for p53 protein, *Anal. Biochem.*, **441**(1), pp. 73–79.
75. SomaLogic SOMAmer Technical Note. http://www.somalogic.com/somalogic/media/Assets/PDFs/SSM-017-Rev-2-SOMAmer-Technical-Note-6-24-14.pdf (accessed 20.12.2014).
76. SomaLogic SOMAscan™ Proteomic Assay Technical White Paper, 2013, p. 2.
77. Song, W., Li, H., Liu, H., Wu, Z., Qiang, W., and Xu, D. (2013). Fabrication of streptavidin functionalized silver nanoparticle decorated graphene and its application in disposable electrochemical sensor for immunoglobulin E, *Electrochem. Comm.*, **31**, pp. 16–19.
78. Staudinger, H., and Meyer, J. (1919). New organic compounds of phosphorus. III. Phosphinemethylene derivatives and phosphinimines, *Helv. Chim. Acta*, **2**, pp. 635–646.
79. Stephanopoulos, N., and Francis, M. B. (2011). Choosing an effective protein bioconjugation strategy, *Nat. Chem. Biol.*, **7**(12), pp. 876–884.
80. Su, M., Ge, L., Ge, S., Li, N., Yu, J., Yan, M., and Huang, J. (2014). Paper-based electrochemical cyto-device for sensitive detection of cancer cells and in situ anticancer drug screening, *Anal. Chim. Acta*, **847**, pp. 1–9.
81. Su, M., Ge, L., Kong, Q., Zheng, X., Ge, S., Li, N., Yu, J., and Yan, M. (2015). Cyto-sensing in electrochemical lab-on-paper cyto-device for in-situ evaluation of multi-glycan expressions on cancer cells, *Biosens. Bioelectron.*, **63**, pp. 232–239.
82. Suh, S. H., Dwivedi, H. P., and Jaykus, L.-A. (2014). Development and evaluation of aptamer magnetic capture assay in conjunction with real-time PCR for detection of Campylobacter jejuni, *LWT: Food Sci. Technol.*, **56**(2), pp. 256–260.

83. Szeitner, Z., Lautner, G., Nagy, S. K., Gyurcsanyi, R. E., and Meszaros, T. (2014). A rational approach for generating cardiac troponin I selective Spiegelmers, *Chem. Comm.*, **50**(51), pp. 6801–6804.

84. Tan, Y., Guo, Q., Xie, Q., Wang, K., Yuan, B., Zhou, Y., Liu, J., Huang, J., He, X., Yang, X., He, C., and Zhao, X. (2014). Single-walled carbon nanotubes (SWCNTs)-assisted cell-systematic evolution of ligands by exponential enrichment (cell-SELEX) for improving screening efficiency, *Anal. Chem.*, **86**(19), pp. 9466–9472.

85. Tao, Z. (2014). Mesoporous silica-based nanodevices for biological applications, *RSC Adv.*, **4**(36), pp. 18961–18980.

86. van Berkel, S. S., van Eldijk, M. B., and van Hest, J. C. M. (2011). Staudinger ligation as a method for bioconjugation, *Angew. Chem., Int. Ed. Engl.*, **50**(38), pp. 8806–8827.

87. Van Simaeys, D., López-Colón, D., Sefah, K., Sutphen, R., Jimenez, E., and Tan, W. (2010). Study of the molecular recognition of aptamers selected through ovarian cancer cell-SELEX, *PLoS One*, **5**(11), p. e13770.

88. Van Simaeys, D., Turek, D., Champanhac, C., Vaizer, J., Sefah, K., Zhen, J., Sutphen, R., and Tan, W. (2014). Identification of cell membrane protein stress-induced phosphoprotein 1 as a potential ovarian cancer biomarker using aptamers selected by cell systematic evolution of ligands by exponential enrichment, *Anal. Chem.*, **86**(9), pp. 4521–4527.

89. Wang, K., Huang, J., Yang, X., He, X., and Liu, J. (2013). Recent advances in fluorescent nucleic acid probes for living cell studies, *Analyst*, **138**(1), pp. 62–71.

90. Wang, L., and Tan, W. (2005). Multicolor FRET silica nanoparticles by single wavelength excitation, *Nano Lett.*, **6**(1), pp. 84–88.

91. Wang, R. E., Zhang, Y., Cai, J., Cai, W., and Gao, T. (2011). Aptamer-based fluorescent biosensors, *Curr. Med. Chem.*, 18(27), pp. 4175–4184.

92. Wang, Y., Luo, Y., Bing, T., Chen, Z., Lu, M., Zhang, N., Shangguan, D., and Gao, X. (2014). DNA aptamer evolved by cell-SELEX for recognition of prostate cancer, *PLoS One*, **9**(6), pp. 1–10.

93. Webber, J. P., Stone, T. C., Katilius, E., Smith, B. C., Gordon, B., Mason, M. D., Tabi, Z., Brewis, I. A., and Clayton, A. (2014). Proteomics analysis of cancer exosomes using a novel modified aptamer-based array (SOMAscan^{TM}) platform, *Mol. Cell. Proteomics*, **13**(4), pp. 1050–1064.

94. WHO Antimicrobial resistance: global report on surveillance, 2014, http://www.who.int/drugresistance/en/.

95. Wu, W. H., Li, M., Wang, Y., Ouyang, H. X., Wang, L., Li, C. X., Cao, Y. C., Meng, Q. H., and Lu, J. X. (2012). Aptasensors for rapid detection of Escherichia coli O157:H7 and Salmonella typhimurium, *Nanoscale Res. Lett.*, **7**(1), p. 658.
96. Yang, X., Wang, Y., Wang, K., Wang, Q., Wang, P., Lin, M., Chen, N., and Tan, Y. (2014). DNA aptamer-based surface plasmon resonance sensing of human C-reactive protein, *RSC Adv.*, **4**(58), pp. 30934–30937.
97. Yeh, F.-Y., Liu, T.-Y., Tseng, I. H., Yang, C.-W., Lu, L.-C., and Lin, C.-S. (2014). Gold nanoparticles conjugates-amplified aptamer immunosensing screen-printed carbon electrode strips for thrombin detection, *Biosens. Bioelectron.*, **61**, pp. 336–343.
98. Zheng, J., Cheng, G.-F., He, P.-G., and Fang, Y.-Z. (2010). An aptamer-based assay for thrombin via structure switch based on gold nanoparticles and magnetic nanoparticles, *Talanta*, **80**(5), pp. 1868–1872.
99. Zhou, J., and Rossi, J. J. (2011). Cell-specific aptamer-mediated targeted drug delivery, *Oligonucleotides*, **21**(1), pp. 1–10.
100. Zichi, D., Eaton, B., Singer, B., and Gold, L. (2008). Proteomics and diagnostics: let's get specific, again, *Curr. Opin. Chem. Biol.*, **12**(1), pp. 78–85.

Chapter 11

Molecular Imaging Utilizing Aptamer-Targeted Probes

Nicholas Fletcher[a] and Kristofer Thurecht[a,b]

[a] *Centre for Advanced Imaging, the University of Queensland, Brisbane 4072, Australia*

[b] *ARC Centre of Excellence in Convergent Bio-Nano Science and Technology, Monash University, 399 Royal Parade, Parkville 3052, Victoria, Australia*

n.fletcher1@uq.edu.au

11.1 Introduction

Molecular imaging, the visualization, characterization, and measurement of biological processes at the molecular and cellular levels in humans and other living systems [39], is becoming an increasingly common tool used in a variety of fields ranging from basic biology to translational medicine. This is largely due to the ability of molecular imaging to visualize and quantify molecular events or disease targets, both spatially and temporally within living organisms. The noninvasive nature of this approach allows for the acquisition of information in vivo, providing more relevant data on intact systems than alternate in vitro/ex vivo diagnostic assays [8].

While conventional imaging uses nonspecific contrast agents to visualize anatomical and pathological features, molecular imaging

Aptamers: Tools for Nanotherapy and Molecular Imaging
Edited by Rakesh N. Veedu

ISBN 978-981-4669-83-2 (Hardcover), 978-981-4669-84-9 (eBook)
www.panstanford.com

makes use of targeted imaging probes. These probes have a high affinity for a molecular target or antigen and incorporate an imaging modality to provide localization information about the probe. Molecular imaging probes can be targeted toward a variety of biomolecules such as disease-specific markers, making this approach of great interest for disease diagnosis and therapy, particularly in oncology [62]. There are also a variety of imaging modalities that may be incorporated into such probes, the most common of which are optical imaging (luminescence/fluorescence), targeted ultrasound, X-ray computed tomography (CT), molecular magnetic resonance imaging (MRI), and nuclear imaging (single-photon emission computed tomography [SPECT] and positron emission tomography [PET]).

Molecular imaging probes may be produced using several approaches. These involve either direct conjugation of a targeting agent to an imaging moiety or attachment of targeting agents to larger constructs such as nanocrystals or polymers [4, 8, 14]. In many cases, particularly for polymers with numerous functional groups, these larger constructs are desirable due to the potential for incorporation of multiple imaging moieties for enhanced diagnostics or therapeutics for targeted delivery. In recent years, the field of molecular imaging has rapidly expanded, and examples of probes comprising a range of materials, targeted toward a multitude of targets and incorporating a variety of imaging modalities have been reported [4, 8, 14].

Targeting agents for molecular imaging probes must possess a high affinity for a specific molecular target. Targeting biomolecules such as antibodies, peptides, and aptamers are commonly used for a variety of applications. Of these biomolecules, antibodies and aptamers have the highest affinity and specificity for their targets; however, compared to antibodies, aptamers have several advantages [19]. These include ease of synthesis, higher production reproducibility, better thermal and chemical stability, more versatile chemistry, and lower immunogenicity [27]. Thus, peptide and oligonucleotide aptamers make ideal high affinity targeting agents for molecular imaging probes. The use of aptamers to target imaging probes is a field of growing interest, and aptamers that are either coupled covalently or physically to an imaging moiety have been produced, which recognize a variety of targets including proteins, phospholipids, sugars, nucleic acids, and whole cells [27].

This chapter discusses the field of molecular imaging utilizing aptamer-targeted probes. It begins with the general approaches for aptamer-targeted imaging probe design, with the remainder of the chapter divided into sections describing the use of aptamers as targeting agents for the various imaging modalities. For each modality, a background of the technique involved and the approaches used to construct molecular imaging agents are presented. Examples of aptamer-targeted probes for each modality are also discussed.

11.2 Design of Aptamer-Targeted Imaging Probes

As discussed in other chapters of this book, aptamers show strong and specific binding to molecular targets, making them useful biomolecules for producing molecular imaging probes. However, aptamers alone do not possess all the properties required for molecular imaging and require functionalization with reporter molecules. The structure of these imaging moieties is dependent on the desired imaging modality, as is discussed in each section below; however, there are several commonalities in imaging probe design.

Aptamers utilized for targeting molecular imaging probes have thus far been almost exclusively DNA and RNA oligonucleotides. While peptide aptamers have been extensively developed, they are as yet underrepresented as targeting agents for molecular imaging probes [38]. The general design principles for aptamer-targeted molecular imaging probes therefore center on oligonucleotide aptamers.

Oligonucleotides are synthetically derived, allowing chemical modifications that provide many advantages in imaging probe design. One of the most common approaches to probe design is incorporation of 5′ or 3′ functional groups such as amines or thiols via phosphoramidite chemistry during synthesis. These functional groups enable simple ligation chemistries for conjugation with imaging moieties or macromolecules that incorporate carboxylates or maleimides, respectively [24].

As well as covalent attachment of aptamers to imaging constructs, physical association has also been well explored. These approaches utilize either electrostatic association of the negatively charged oligonucleotides to a positively charged construct [36], or binding of the oligonucleotides to complementary oligonucleotide sequences [64]. Regions of the aptamer sequence not involved with target binding may be varied, enabling extension of oligonucleotides to incorporate sequences for binding to complementary strands. These complementary strands are often short "capture" strands, which are covalently immobilized on the imaging construct [30]. Alternatively, the aptamer may be designed to possess regions complementary to either regions of the same strand, or other free oligonucleotide sequences [43], to create more complex constructs that are often responsive to target binding (as described later in this chapter).

Physical association may also be utilized to deliver therapeutics by incorporating domains for therapeutic binding. Repetitive cytosine-guanine sequences, for example, are known to bind doxorubicin [2]. Appending such sequences to aptamers enables aptamer-targeted therapeutic delivery. While this chapter does not aim to review all aptamer-mediated drug delivery examples, in many instances, the constructs are designed such that molecular imaging and therapeutic delivery are combined.

One factor that can also be varied during probe design is the number of targeting aptamers per imaging construct. An approach used to increase binding avidity and targeting efficacy is to create multivalent constructs with multiple aptamers, resulting in enhanced cellular binding [20]. This increased avidity is a functional benefit of using larger multivalent constructs compared to monomeric aptamer-targeted probes.

While oligonucleotide aptamers have many advantages as targeting agents, their structure also possesses disadvantages such as susceptibility to endogenous nucleases in vivo. As molecular imaging applications necessitate probe stability in vivo, increasing aptamer stability has been a research target of interest. Common strategies utilized to increase nuclease stability include the incorporation of nonnatural nucleic acids such as locked nucleic acids [52], 5-(*N*-benzylcarboxyamide)-2'-deoxyuridines [34, 37],

2′-fluoropyrimidines [60, 59], or inverted 3′ thymidine-capping nucleotides [5, 59]. Such nonnatural nucleic acids can be incorporated during the systematic evolution of ligands by exponential enrichment (SELEX) selection process [59] or substituted into existing sequences to increase stability [52].

While there are some generalities in aptamer-targeted molecular imaging probe design, the specifics are dependent on the imaging technique to be used. Techniques that have been explored for aptamer-targeted probes include optical, ultrasound, X-ray CT, MRI, and nuclear tomographic imaging.

11.3 Optical Imaging

Optical imaging involves detection of photons emitted by a molecular imaging probe in the ultraviolet to infrared wavelength range. This can be further divided into the two broad areas of luminescence and fluorescence. Luminescence involves the emission of photons generated by a chemical or biological process. In preclinical imaging, this is most commonly an enzymatic process. This technique generally involves genetically modifying a target cell line of interest to express a bioluminescent reporter, either endogenously or by co-expression with a gene of interest. This type of approach is extensively used as a route to monitor gene expression and xenograft tumor growth in response to potential therapies. While this technique is commonly used in the preclinical setting, it does not form a part of aptamer-targeted molecular imaging and will thus not be discussed in detail here. For further information, the reader is directed to several recent reviews [11, 45].

In contrast to luminescence, fluorescence requires the absorption of light from an external source to produce photons. In fluorescence, incident light causes excitation of an electron within a fluorophore to a higher quantum state. Some energy is then lost due to nonradiative decay; and when the electron returns to its ground state, light is emitted at a longer (lower-energy) wavelength than the incident photon [33]. The wavelength of light that will excite electrons of a particular fluorophore is determined by the structure of the molecule. This means that individual fluorophores have specific excitation and emission wavelengths, allowing multiplexing if several fluorophores are used.

Fluorescence imaging is one of the most commonly used techniques in in vitro and preclinical imaging due to its simplicity in equipment requirements, lack of ionizing radiation, and ease of translation from in vitro to in vivo studies. While luminescence is generally engineered into a system as a "reporter" for cellular activity, fluorescent probes may be used to detect target molecules within an intact system. Fluorescence imaging achieves micrometer resolution in vitro and millimeter resolution in vivo [23]. However, quantification of in vivo signals is confounded by tissue absorption, which also limits imaging depths to only centimeters (discussed below) [32]. Fluorescence microscopes are most commonly used for in vitro probe imaging applications, while for in vivo applications detection generally incorporates small animal planar scanners. In each case, the fluorophore is excited by a laser of appropriate wavelength and the emitted light is then detected, generally by a charge-coupled device (CCD) camera [33]. The fluorescence image itself provides little anatomical or positional information and is therefore commonly overlaid with a bright-field image for microscopy, or X-ray/reflectance image for in vivo imaging.

One factor that influences the selection of fluorophores for imaging probes is the desired application. While a wide variety of fluorophores are applicable for in vitro studies, shorter wavelengths of light are highly attenuated by tissue absorption and scattering in in vivo studies. This can be partially overcome by moving to fluorophores that absorb and emit light at longer wavelengths, where tissue attenuation is lower. However, the choice of wavelength is bounded at the upper end by the absorption of infrared light by water. This has led to a recent shift toward using near-infrared (NIR; 700–900 nm) fluorophores, such as cyanine and Alexa Fluor dyes, for in vivo optical imaging, where attenuation by tissue and water is the lowest [32].

The simplest method for producing a fluorescent molecular imaging probe is to directly conjugate a fluorophore to a targeting molecule such as an antibody, peptide, or aptamer. Commonly used fluorophores include organic dyes and the more photostable quantum dots (QDs; nanocrystals of semiconducting material). As discussed above, the synthetic nature of aptamer production makes them well suited to chemical modifications and many aptamer-targeted fluorescent probes have been reported. To date, the majority of the reported constructs have been directed toward detection and imaging of cancerous cells.

Probes produced by direct conjugation of small-molecule dyes have predominantly incorporated DNA and RNA aptamers such as those that recognize integrin $\alpha v\beta 3$ [41] and angiogenin [35] as markers for angiogenesis, or nucleolin [34] and epithelial cell adhesion molecules [55] that are overexpressed in many tumor types. In these examples, the aptamers were directly conjugated with cyanine or fluorescein fluorophores, usually through terminal amines incorporated during synthesis. These constructs showed highly specific binding of cellular targets in vitro and enabled visualization of probe internalization on binding (Fig. 11.1). Such results make these aptamer-targeted optical imaging probes promising candidates for future preclinical in vivo studies of carcinoma pathogenesis and treatment.

More recent studies of organic fluorophore-labeled aptamers have incorporated the DNA aptamer TD05, which targets B-cell

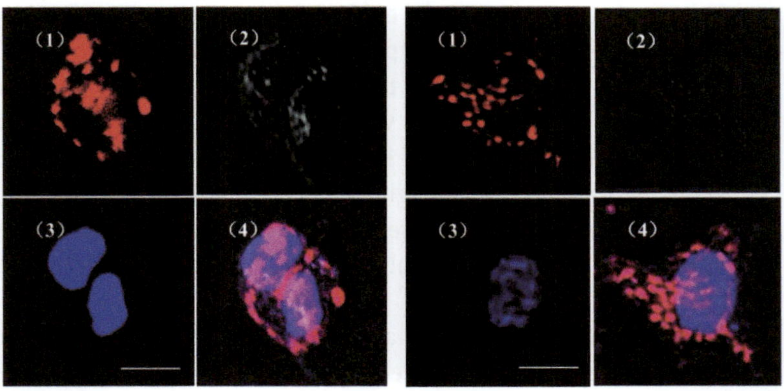

Figure 11.1 Intracellular distribution of internalized angionenin–aptamer conjugate in HUVE (left) and MCF-7 (right) cells visualized with confocal microscopy. In each series (1) is the fluorescence image of Cy5-labelled aptamer conjugate, (2) is the bright-field image, (3) is the fluorescence image of nuclear staining, and (4) is the overlaid image. Reproduced with permission from Ref. [35]. Copyright (2008), American Chemical Society.

lymphoma [54, 52]. These cyanine5 (Cy5)-labeled aptamers showed rapid and highly specific binding of lymphoma cells both in vitro and in vivo, demonstrating the potential of these constructs for in vivo studies requiring real-time molecular imaging. Furthermore, by modifying the aptamer sequence to incorporate locked nucleic acids and 3′-3′-thymidine capping, the authors enhanced the serum stability and imaging window of the aptamer-Cy5 probe [52]. Furthermore, this probe also incorporated an imaging "switch" on target binding, as discussed below. Such sequential modifications of existing aptamer probes allows for ongoing improvement and production of enhanced molecular imaging probes.

As well as small-molecule fluorophores, aptamers have been directly conjugated to QDs to produce optical molecular imaging probes. Aptamers utilized for these probes have included the RNA aptamer A10 [2], which recognizes prostate-specific membrane antigen (PSMA), which is associated with prostate cancer; the DNA aptamer AS1411 [31], which binds to nucleolin transmembrane protein; and a DNA aptamer that targets mucin 1 [49], a cell surface protein overexpressed in many cancers. These aptamer-targeted constructs showed specific targeting of cancerous cells in vitro, demonstrating their applicability as molecular imaging probes. Furthermore, in several cases, targeted therapeutic delivery vehicles were produced by incorporation of therapeutics into the QD construct [2, 49]. For example, this was achieved by attaching the chemotherapeutic doxorubicin via an acid-degradable hydrazone linkage to the QD of the mucin 1 aptamer–QD construct [49]. This construct allowed both the imaging and treatment of cancer in both in vitro and in vivo models. These dual imaging–therapeutic delivery constructs have great potential in future studies of therapeutic delivery and efficacy.

Rather than direct covalent attachment, an alternate approach for constructing aptamer-targeted QDs involving biorecognition has been explored. In this approach, aptamers are conjugated with biotin at either the 5′ or 3′ terminus. The QD is then functionalized with streptavidin and the highly specific and strong binding between biotin and streptavidin results in physical association of the aptamer and QD. Optical imaging constructs assembled using this approach have incorporated the A9 RNA aptamer, which targets PSMA [9], or the TLS9a DNA aptamer, which targets liver hepatoma cells [68]. Both constructs demonstrated suitable stability for in vitro imaging and showed specific targeting of tumor cells. These biotin–streptavidin-associated constructs possess the added advantage of simplicity of assembly. For example, a single streptavidin-coated QD may easily be coated with any one of a variety of biotin functionalized aptamers, or vice versa, depending on the desired targets and imaging parameters. This may be useful in the rapid production of future probes for various targets.

One evolving area of aptamer-targeted agents involves the conjugation of aptamers to polymeric constructs [1, 57]. However, to date, this has predominantly been targeted toward therapeutic delivery by drug-loaded polymeric nanoparticles [1]. Aptamer-targeted polymeric nanoparticles for molecular imaging are a current research target of interest. One example where this has been achieved utilized a branched polymer construct composed predominantly of polyethylene glycol as the core [12]. The branched architecture of this construct provided multiple functionalities for incorporation of targeting and imaging moieties. Both a peptide aptamer specific for heat shock protein 70, which is overexpressed in many cancers, and Rhodamine-B were conjugated to the same polymeric construct [12]. The aptamer–polymer construct showed enhanced targeting to melanoma cells both in vitro and in vivo compared to a small-molecule-targeted construct. Furthermore, the polymers also incorporated MRI-active ^{19}F-containing monomers, providing potential for future multimodal studies of tumor targeting.

In addition to direct fluorescence imaging of probes, responsive constructs may be designed to monitor binding (or release) of a molecular target. Such a system is comprised of a fluorophore and an acceptor (quencher or second fluorophore) that absorbs light emitted from the initial fluorophore by Förster resonance energy transfer (FRET). Rather than the relaxation of fluorophore electrons producing a detectable fluorescence signal, the energy is instead transferred to the FRET acceptor, resulting in the fluorescent probe being in an off state [10]. Importantly, the transfer to acceptor is distance dependent (<10 nm). By causing an increase in the distance between fluorophore and acceptor by conformation change

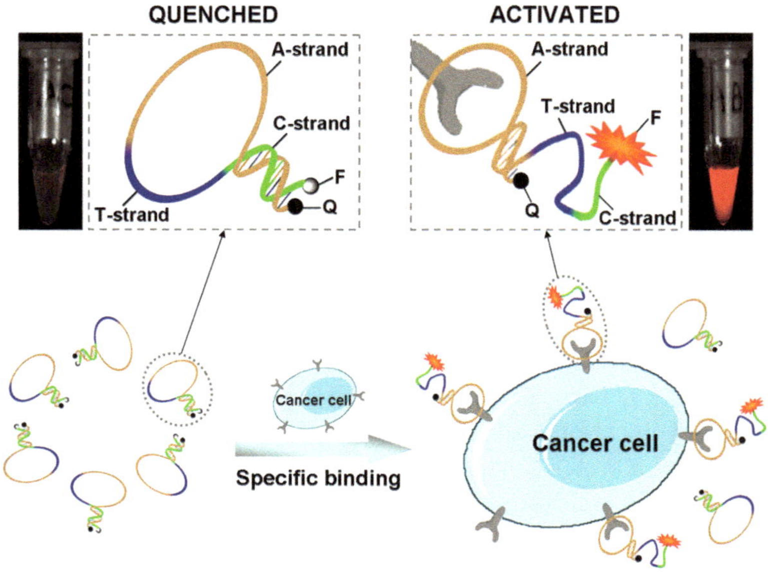

Figure 11.2 Schematic illustrating aptamer-targeted responsive fluorescent probe binding to target cells. The probe is comprised of a cancer-targeted aptamer sequence (A), a poly-T linker (T), and a short DNA sequence (C) complementary to a part of the A-strand, with a fluorophore (F) and a quencher (Q) attached at either terminus. Reproduced from Ref. [53] with permission of PNAS.

or molecular dissociation (release), an on state is generated where no transfer occurs and fluorescence is detected.

Production of FRET-based switchable oligonucleotide constructs generally involves the extension of the aptamer at one termini by a sequence complementary to the opposite aptamer termini (Fig. 11.2). A fluorophore and appropriate quencher are then conjugated to opposing termini of the oligonucleotide. On binding of these complementary regions, a hairpin structure is formed, which brings the quencher and fluorophore in close proximity, producing an *off* state. In these constructs, fluorescence is then switched *on* upon binding of a probe to the target molecule, providing a lower background in acquired images and a direct indication of the presence of target molecules. Such constructs are commonly called *aptamer beacons* and have been produced on the basis of aptamers for several targets, including thrombin [16], MRC kinase [58], and lymphoblastic leukemia (sgc8) [53]. These aptamer beacons show enhanced contrast and lower background compared to always-on aptamer-targeted imaging probes with demonstrated efficacy both in vitro and in vivo. The simplicity in design and several demonstrated examples illustrates the potential for development of aptamer beacons for a variety of targets.

In another approach, a FRET imaging pair may also be used to demonstrate co-localization of two cell surface proteins targeted by two different probes. For example, by labeling an RGD peptide with a QD and the AS1411 DNA aptamer with Cy3, Kang et al. [25] showed integrin $\alpha_v\beta_3$ and nucleolin transmembrane protein to be within approximately 10 nanometers of each other on the cell surface. This ability to determine protein co-localization within intact systems may be key in future studies of protein–protein interactions and cell surface receptors.

The direct and noninvasive detection of therapeutic release is key in the future development of targeted therapies [6]. In the case where the therapeutic molecule is also fluorescent, FRET between therapeutic and imaging probe may be utilized to directly image drug delivery. The fluorescence profile of doxorubicin makes it an appropriate therapeutic for such an approach in combination with QDs. Aptamer-targeted QDs have been utilized as doxorubicin delivery vehicles. In these constructs, FRET between the QD and doxorubicin results in an off state, while the therapeutic is bound [2, 49]. Upon release of the therapeutic at the target site, the distance between QD and doxorubicin increases, causing a restoration of fluorescence to result in an on state. Such an approach has demonstrated specific doxorubicin delivery to cancer cells both in vitro and in vivo. These studies provide valuable information about therapeutic delivery by targeted constructs and may potentially lead to future clinical applications.

While fluorescence-based probes are predominantly targeted toward preclinical imaging due to limitations in tissue penetration and quantification, molecular imaging probes using clinically applicable modalities such as ultrasound, X-ray CT, MR, and nuclear imaging have also been developed.

11.4 Ultrasound Imaging

Ultrasound imaging is one of the most commonly used clinical imaging modalities owing to its ease of use, safety, and low cost [26]. In ultrasound imaging, high-frequency sound waves are transmitted into the tissues of interest. Sound waves are scattered from interfaces between tissue components of dissimilar density and compressibility. Reflected sound waves are detected by a transducer, and the measured intensity and time taken to travel through tissues is used to reconstruct an image. Ultrasound imaging produces images at micrometer resolution in real time and has demonstrated utility in the visualization of anatomical features. It has, however, only a useful penetration depth of a few centimeters and also suffers from poor contrast between tissues.

One common approach to increase the utility of this technique is to employ ultrasound contrast agents, which have a density and compressibility substantially different from that of blood and tissue [15]. These contrast agents are typically small bubbles of either gas or perfluorocarbons on the nanometer to micrometer scale. Due to the large size of these contrast agent bubbles they are not able to pass from the circulatory system to other tissues and are therefore limited to imaging targets within the vasculature. The bubbles are generally stabilized by phospholipids, polysaccharides, proteins, or polymers. Tethering of these bubbles to a targeting moiety such as a peptide, antibody, or aptamer then produces targeted imaging probes for ultrasound imaging.

Aptamer-conjugated ultrasound contrast agents are a relatively nascent area of study. The first example of an aptamer-targeted ultrasound contrast agent utilized the sgc8c DNA aptamer targeted to human acute lymphoblastic leukemia (CCRF-CEM) cells [61]. Conjugation of the 5′ thiol–modified aptamer to the surface of perfluoropropane-filled nanobubbles used thiol-maleimide coupling chemistry and provided good binding of the construct to target cells in vitro, with enhanced ultrasound contrast observed.

Switchable ultrasound contrast agents have also been constructed on the basis of aptamer recognition of target proteins. Reversibly cross-linking the surface of a microbubble with a DNA aptamer that recognizes thrombin produced a contrast agent that gave enhanced ultrasound contrast in the presence of the target protein [43]. Recognition of thrombin resulted in dissociation of the aptamer from capture oligonucleotides on the bubble surface, and a change in bubble compressibility. These switchable aptamer-cross-linked microbubbles have further been evaluated for in vivo ultrasound detection of thrombosis with promising results (Fig. 11.3) [42]. While aptamer-targeted contrast agents for ultrasound are relatively

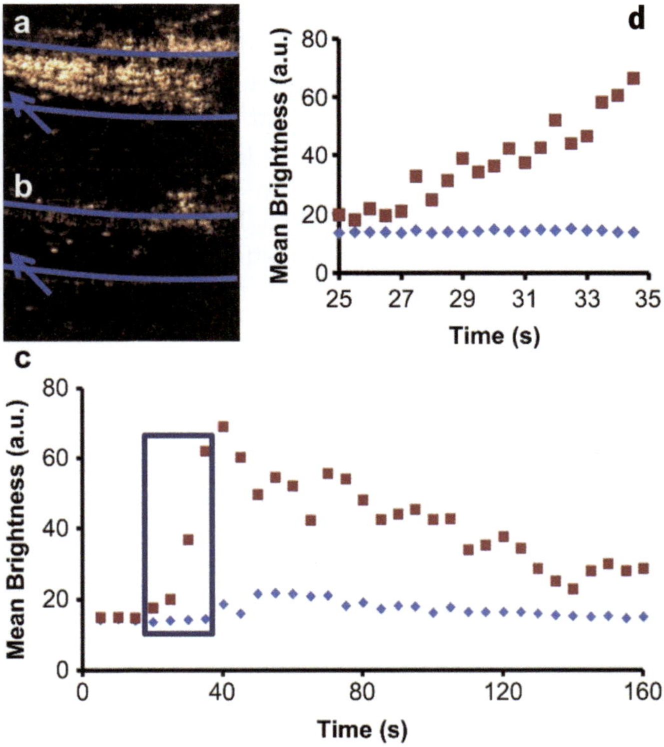

Figure 11.3 In vivo contrast-enhanced ultrasound imaging of deep venous thrombosis. Activation of (a) thrombin-aptamer and (b) scrambled-aptamer-targeted nanobubbles at the clot site in cadence mode. Arrows indicate the probable clot site. Outlines indicate vena cava walls. (c) The acoustic signal generated within the clot site over time for thrombin-aptamer-targeted (red) and scrambled-aptamer-targeted (blue) nanobubbles and (d) enlargement of the area of interest highlighted in blue in (c). Reproduced from Ref. [42], Copyright (2013), with permission from Elsevier.

underdeveloped, the potential for clinical application makes this an appealing research space. Aptamer-targeted contrast agents for X-ray CT imaging have also been investigated as potential clinically relevant imaging agents.

11.5 X-Ray CT Imaging

X-ray CT is a commonly used imaging technique in clinical practice. This technique rapidly generates anatomical images of patients with micrometer resolution and no practical limitation on tissue imaging depth. In X-ray CT, X-rays are directed through the patient, with the image produced at the detector by the differing levels of X-ray absorption by tissues within the body, analogous to conventional 2D X-ray imaging [51]. In CT imaging, however, the X-ray source and detector are rotated around the patient and images at multiple angles are then reconstructed using computed tomography to produce a 3D image.

While X-ray CT produces high-resolution images of dense tissues such as bone, the technique suffers from poor contrast between soft tissues due to low attenuation of X-rays. One approach to overcome this poor contrast is to use X-ray CT contrast agents. These agents are generally composed of high-atomic-weight elements, such as iodine, barium, or gold, which show high attenuation of X-rays [46, 67]. CT contrast agents are generally nonspecific and are often used to image vasculature or the digestive tract, following injection or ingestion respectively. Consequently, these contrast agents also require large quantities of material to produce sufficient image contrast.

By targeting these materials to specific tissues, high local concentrations of the contrast agent may be achieved, producing sufficient X-ray CT contrast for imaging. This approach can be utilized as a molecular imaging technique. To date there have been few reported examples of targeted X-ray CT contrast agents [13, 23]. Notably, it is only in the past few years that an aptamer-targeted X-ray CT contrast agent has been developed [30]. In this example, the PSMA-specific RNA aptamer A9 was bound to the surface of a gold nanoparticle via recognition of a covalently attached capture oligonucleotide (Fig. 11.4). This construct showed PSMA-specific

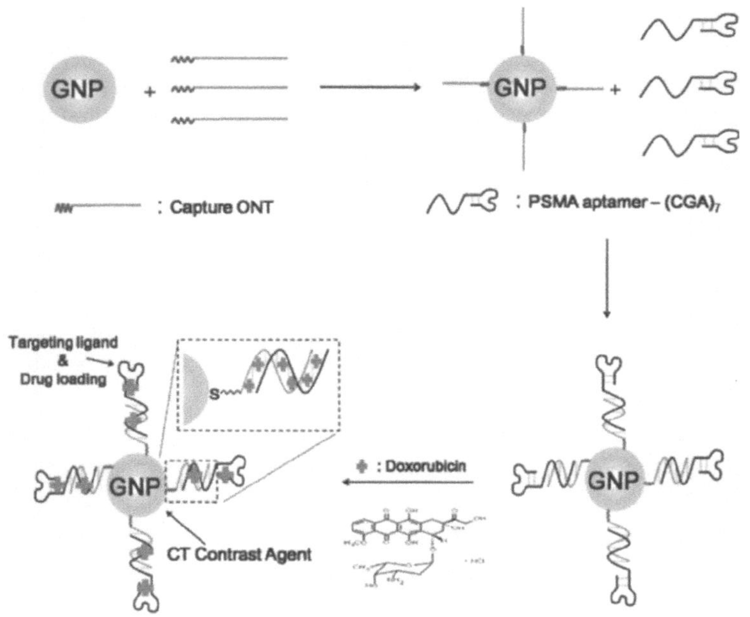

Figure 11.4 Schematic illustration of the assembly of doxorubicin-loaded aptamer-conjugated gold nanoparticles targeted to cells expressing PSMA. Adapted with permission from Ref. [30]. Copyright (2010), American Chemical Society.

targeting of prostate cancer cells in vitro and produced enhanced X-ray CT contrast in target cells. Furthermore, by extension of the aptamer sequences to include a repetitive cytosine-guanine-adenine sequence at the 3′ end of the oligonucleotide, the aptamer-targeted molecular imaging agent also selectively delivered doxorubicin to PSMA-expressing cells.

While aptamer-targeted CT contrast agents have to date not been extensively investigated, molecular imaging probes for MRI, which is also clinically relevant, have been much better developed.

11.6 Magnetic Resonance Imaging

MRI is a clinical imaging technique commonly used to produce anatomical and physiological information. MRI provides micrometer resolution with isotropic tissue penetration through a region of interest. In an MRI acquisition, a large external magnetic field is applied to the subject, which causes MR-active nuclei (most commonly ^{1}H in water) to become aligned [40]. During the acquisition, a radio-frequency pulse is applied to the subject, which perturbs the aligned nuclei. Receivers in the instrument then detect the MR-active nuclei as they relax to their aligned state. Nuclei undergo relaxation in two ways, spin–lattice (T_1) and spin–spin (T_2) relaxation. The rates of T_1 and T_2 relation are highly dependent on the environment of the nuclei, and the contrast in the image is generated by differing rates of magnetic resonance nuclei relaxation between tissues.

While tissues possess some endogenous magnetic resonance contrast, injectable MRI contrast agents are commonly used to highlight regions of interest or biological abnormalities on the basis of differing tissue uptake. These contrast agents are paramagnetic species that usually increase the relaxivity of nearby MR-active nuclei, resulting in decreased T_1 or T_2 relaxation times. Commonly used contrast agents include gadolinium chelates, superparamagnetic iron oxide nanoparticles (SPIONs), and manganese oxide nanoparticles [23]. As with contrast agents used in other imaging modalities, by coupling these contrast agents with a targeting moiety such as an aptamer, a molecular imaging probe may be produced.

Gadolinium (Gd(III)) contrast agents are most effective at decreasing local proton T_1 relaxation times and give positive contrast in T_1-weighted images. Multiple aptamer-targeted gadolinium-based molecular imaging probes have been reported. These probes have utilized either a gadolinium chelator conjugated to the 5′ amine of an aptamer-complementary strand [63] or gadolinium-doped strontium hydroxyapatite nanoparticles to which the aptamer was electrostatically associated [36]. In both cases, the aptamer was used to target the probe as well as create a responsive construct for imaging or therapeutic delivery. For example, contrast enhancement by an adenosine-recognizing DNA aptamer probe was switched off upon binding of the target molecule [63]. This was due to changes in probe conformation and release of the complementary strand to which the gadolinium chelate was attached. This dissociation caused a change in the rotational correlation time of the chelated gadolinium and a consequent increase in the local proton T_1 relaxation time.

While gadolinium-based MRI contrast agents are currently the most commonly used in clinical settings, many recent preclinical studies have focused on developing targeted SPION imaging probes. SPION contrast agents are most effective at decreasing T_2 relaxation times of nearby MR-active nuclei and consequently give negative contrast in T_2-weighted MRI images. Work by Lu and coworkers has focused on producing dextran-coated SPION probes for small-molecule and protein sensing [64, 65]. In these examples, attachment of the aptamer to the surface of the SPIONs resulted in association of the probe with the desired target. Interestingly, Yigit et al. [64] showed that by using an adenosine aptamer to cross-link SPIONS via base pairing with surface-bound capture strands, a responsive MRI contrast agent was generated. On binding of adenosine, conformational changes resulted in dissociation of SPION clusters that gave increased T_2 values and enhanced signal in T_2-weighted images.

As well as molecular sensing, aptamer-conjugated SPIONs have also been utilized to both isolate and track mesenchymal stem cells (MSCs) postinjection [50]. Harvested MSCs are a promising source of material for cellular cardiomyoplasty, and tracking their localization is of interest for developing future therapies. For this application, a biotinylated porcine-MSC-specific DNA aptamer (G-8) was bound to the surface of streptavidin-coated SPIONs. This construct showed specific cellular recognition and enabled immunomagnetic isolation of MSCs from bone marrow. Furthermore, probe-labeled MSCs were imaged posttransplantation using MRI, enabling noninvasive tracking of cellular migration.

In addition to tracking of injected cells, aptamer-targeted SPION probes have been developed by Jon and coworkers for prostate cancer imaging and therapeutic delivery [60, 66]. These probes were composed of PSMA targeting A9 or A10 RNA aptamers attached to the surface of polyethylene glycol coated SPIONs. The probes targeted PSMA-expressing cells with high specificity and generated

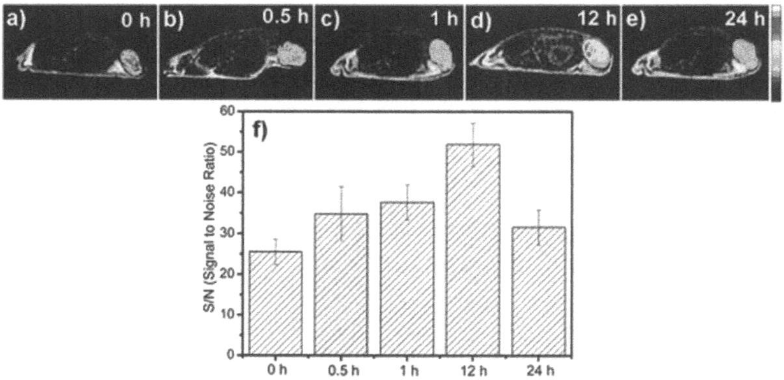

Figure 11.5 T_1-weighted magnetic resonance images of a tumor-bearing mouse model injected with aptamer-targeted manganese oxide nanocrystals at time points (a) 0 h to (e) 24 h, and (f) the corresponding signal-to-noise ratio of the tumor. Reproduced from Ref. [18] with permission from The Royal Society of Chemistry.

changes in contrast sufficient for MRI, both in vitro and in vivo. Furthermore, these probes were demonstrated to bind and deliver doxorubicin to PSMA-expressing cells with high specificity. These PSMA-targeting molecular imaging probes are highly promising for future clinical studies.

Aptamer-targeted iron oxide and manganese oxide nanocrystal MRI probes have also been developed by other groups for imaging of cancers. The targets of these probes have included integrin $\alpha_v\beta_3$ as a marker for angiogenesis [37]; nucleolin (aptamer AS1411), which is up-regulated in many cancer types [18]; and vascular endothelial growth factor receptor 2 (VEGFR2), which is overexpressed in glioblastoma [29]. In each case, covalent attachment of the aptamer via conjugation of 5′ terminal amines to the nanocrystal surface gave highly specific targeting of cancer cell lines in vitro and enabled in vivo imaging of model tumors (Fig. 11.5).

The range of successfully demonstrated aptamer-targeted MRI probes with high specificity and confirmed in vivo efficacy makes these promising molecular imaging tools for future clinical applications. Another clinically relevant molecular imaging modality, which has been coupled with aptamer targeting, is nuclear imaging.

11.7 Nuclear Tomographic Imaging

Nuclear tomographic imaging (SPECT and PET) is the most readily clinically translatable molecular imaging modality due to isotropic tissue penetration, high sensitivity, and millimeter-scale resolution. This is further complemented in PET by being highly quantitative, allowing accurate measurement of biodistribution and pharmacokinetics of labeled materials [3]. Nuclear tomographic imaging is based on the detection of radionuclide-labeled materials, as is discussed in further detail below. This approach in isolation provides limited anatomical information and in the clinic is therefore commonly coupled with X-ray (2D or CT) or MRI for signal co-registration. This requirement for instrumentation to acquire anatomical information is a limiting factor in the utility of radiotracer imaging as this necessitates the use of highly specialized equipment.

Molecular imaging probe design for nuclear tomographic imaging generally involves the covalent attachment of a chelator to a targeting molecule of interest. In the case of aptamers this is regularly accomplished by conjugation to the $5'$ terminal amine of an oligonucleotide aptamer. These chelators then strongly bind radionuclides and allow nuclear imaging of the molecular target. While both SPECT and PET are based on detecting radionuclide decay, the process by which the signal is generated and detected in each approach differs, leading to dissimilarities in imaging capabilities.

11.7.1 *SPECT*

SPECT imaging utilizes radionuclides that undergo gamma decay to produce γ-rays at an energy level that is characteristic of the isotope [28]. These γ-rays are photons of sufficiently high energy to pass relatively unimpeded through biological tissue, thereby producing images that are independent of tissue depth. SPECT isotopes include ^{123}I, ^{111}In, and ^{99m}Tc, which have decay half-lives in the range of hours to days, allowing imaging over longer times than PET isotopes (discussed below). A γ-scintillation camera is used to detect γ-ray emissions and can collect planar 2D images or be rotated around a subject to collect images at multiple angles. In the latter case, CT is then used to reconstruct a 3D image. Notably, the characteristic energy levels of emitted γ-rays allows for multiplexing of SPECT imaging probes by using several radioisotopes. However, to date this has not been explored for aptamer-targeted probes.

The first reported aptamer-targeted molecular imaging probe was a neutrophil elastase DNA aptamer conjugated with a peptide "cage" for ^{99m}Tc chelation [7]. Using this probe, the authors obtained 2D in vivo images of inflammation in a rat model. Notably, a higher target-to-background ratio was obtained for the aptamer-targeted probe compared to an antibody-based construct for the same target.

Following on from this initial report, there have been several examples of aptamer-targeted probes involving chelation of ^{99m}Tc. The first of these utilized the TTA1 RNA aptamer, which binds to tenascin-C, a protein overexpressed in many solid tumors [17]. This probe incorporated mercaptoacetyl glycine (MAG_2), which is a commonly used chelator for ^{99m}Tc and maintained good radionuclide binding after conjugation. In vivo analysis of tumor targeting in xenograft models showed rapid aptamer probe clearance and good tumor uptake. This enabled 2D imaging of tumors using a planar γ-scintillation camera (Fig. 11.6).

Subsequent work has developed several ^{99m}Tc-labeled probes using DNA aptamers that target MUC1, a cell surface protein overexpressed in many cancers, especially breast cancer [56]. The first of these probes incorporated a cyclen-based ligand for radionuclide binding that was conjugated to multiple aptamers [5]. Notably, in vitro and in vivo studies showed multivalent aptamer probes to have enhanced tumor targeting and pharmacokinetic properties compared to monovalent equivalents. However, the ligand was ultimately deemed to be a relatively poor chelator for metal ions and unsuitable for molecular imaging probes. In a following report, aptamers for MUC1 were conjugated to MAG_2; however, low tumor uptake and high clearance were seen in a breast cancer xenograft model, making this a nonideal imaging probe [47].

11.7.2 *PET*

Radionucleotides used in PET imaging include ^{11}C, ^{18}F, ^{68}Ga, ^{64}Cu, ^{124}I, and ^{89}Zr, which undergo radioactive decay resulting

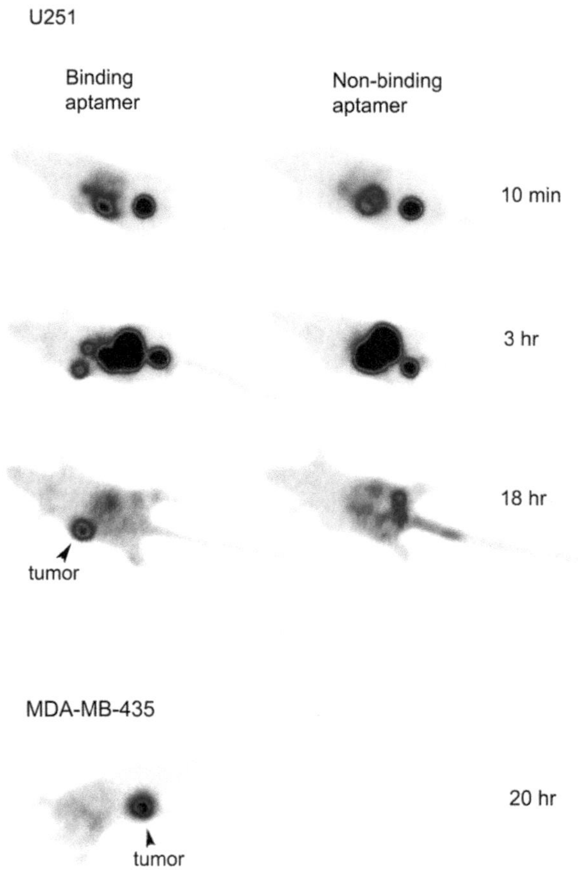

Figure 11.6 In vivo imaging of ^{99m}Tc-labeled binding TTA1 aptamer and nonbinding aptamer in xenograft tumor models by a γ-scintillation camera. This research was originally published in *JNM*. Ref. [17]. Figure 6. © by the Society of Nuclear Medicine and Molecular Imaging, Inc.

in positron emission. These isotopes have decay half-lives in the range of minutes to hours, providing shorter imaging times than SPECT isotopes. Emitted positrons travel a short distance through surrounding tissue, losing kinetic energy until interacting with an electron, annihilating both particles. The annihilation event results in the emission of two 511 keV γ-rays travelling in almost completely opposite directions. Similar to γ-rays of

SPECT isotopes, these photons travel relatively easily through biological tissues to reach the instrument detector. PET imaging uses rings of scintillation detectors arranged around the subject. These detectors record coincident pairs of photons on opposite sides of the ring, allowing tomographic reconstruction of a 3D PET image. Importantly, by correcting for several factors such as attenuation of signal by tissue and detector efficiency [3], data collected in a PET image is usually quantitative, allowing local concentrations of radiolabeled materials to be accurately determined.

Compared to aptamer-targeted SPECT probes, the field of aptamer-targeted PET probes is less developed. The first reported aptamer-targeted PET probe utilized the RNA A10-3.2 aptamer for PSMA [48]. This study investigated a variety of chelators for ^{64}Cu as well as varied radiolabeling solution conditions. The authors showed good radionuclide binding by the construct with promising in vitro results through selective binding to PSMA-expressing cells.

A recent study utilized a tenascin-C-binding DNA aptamer either conjugated directly to an ^{18}F-containing small molecules or labeled with ^{64}Cu via the chelator 1,4,7-triazacyclononane-1,4,7-triacetic acid (NOTA) [22]. Both radiolabeled constructs showed selective targeting to tenascin-C-expressing tumors coupled with rapid clearance from the blood through the kidneys. This combination of tumor targeting and fast clearance gave a high signal-to-noise ratio, beneficial in molecular imaging, making these constructs of interest for future clinical applications.

Given the recent promising results for aptamer-targeted nuclear imaging, it seems likely that similar molecular imaging probes will soon progress to the clinical setting, especially in the field of oncology. Such progress will be contingent on factors such as strong radioisotope chelation, efficient targeting, and rapid blood clearance.

11.8 Multimodal Imaging

Each molecular imaging modality possesses its own advantages and disadvantages. No single modality is perfect and able to obtain all relevant information for disease diagnosis or investigation of biological processes [8]. For example, fluorescence offers multiplexing and rapid acquisitions but suffers from high attenuation by tissue, limiting quantitation; ultrasound, MRI, and CT offer high-resolution anatomical information but suffer from low sensitivity; and nuclear imaging offers extremely sensitive detection but limited anatomical information. The combination of multiple imaging modalities, termed "multimodal imaging," offers synergistic advantages over each technique in isolation.

As described in the preceding sections, the combination of X-ray and fluorescence imaging, especially in preclinical studies, has now become commonplace. In these examples, the X-ray image is used to provide anatomical registration of the molecular-targeted fluorescence image. Multimodal imaging has also become an aim for ongoing medical research, with combinations such as PET-CT and PET-MRI becoming increasingly common in the clinic. In the same way that fluorescence and X-ray modalities may be combined, so, too, can MRI/CT with PET to provide co-registration of PET tracer accumulation. Such combinations are especially useful in clinical diagnosis as the PET signal provides information on the presence of molecular targets, while the MRI/CT image allows high-resolution delineation of anatomical features such as tumors. X-ray fluorescence multimodal imaging is commonly used in fluorescent molecular imaging probes, but to date there have been limited examples in other modalities such as nuclear imaging, where it may offer the greatest benefit.

In these examples the second modality (X-ray/MRI/CT) is used purely for co-registration of the molecular imaging probe signal. Another approach involves combining moieties for multiple imaging modalities within a single construct. This allows for a truly multimodal approach, whereby the benefits of each modality individually may be combined, providing a more complete view of intact systems. Aptamer-targeted probes for multimodal imaging are currently being developed. In the only example of an aptamer-targeted multimodal imaging construct to date, Hwang et al. reported a multimodal imaging probe for nucleolin [21] (Fig. 11.7). In this example, the DNA aptamer AS1411 was conjugated to an MR-active cobalt-ferrite nanoparticle that was also fluorescently labeled with rhodamine and functionalized with a NOTA chelator.

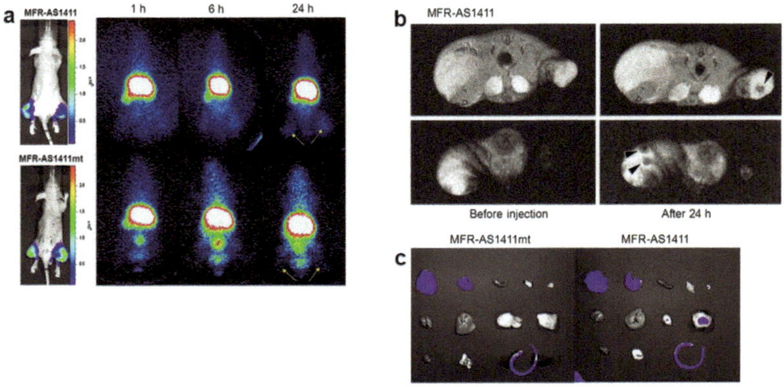

Figure 11.7 Aptamer-targeted multimodal imaging probe in a tumor-bearing mouse model. In vivo tumor imaging using (a) γ-scintillation camera and (b) MRI, as well as (c) ex vivo imaging using fluorescence. AS1411 is the targeted aptamer, while AS1411mt is the mutated nontargeting control. Arrows highlight tumor locations in (a) and sites of enhanced MRI contrast in (b). Ex vivo organs in (c) in order from upper left to lower right were intestine, liver, spleen, muscle, fat, kidney, stomach, right tumor, left tumor, heart, lung, and tail. This research was originally published in *JNM*. Ref. [21]. Figure 4. © by the Society of Nuclear Medicine and Molecular Imaging, Inc.

Radiolabeling of this probe with ^{67}Ga then allowed detection by fluorescent imaging, MRI, and a γ-scintillation camera. This construct showed good tumor uptake and rapid blood clearance in a mouse tumor model. Probes such as this with good tumor targeting and versatile imaging hold great promise as future clinical diagnostic tools.

11.9 Conclusion

The field of aptamer-targeted molecular imaging is a rapidly emerging research field. Aptamer-targeted imaging probes have already been demonstrated for a variety of targets with the utilization of a range of modalities and high levels of success in preclinical models. In many cases, direct comparisons between probes targeted by aptamers and alternate targeting agents showed enhanced targeting by the aptamer conjugate. As discussed, aptamers possess many properties such as high specificity and better stability than alternate biomolecule targeting species that makes them appealing for incorporation into imaging probes. Furthermore, their synthetic production and ability to be easily generated for a wide variety of targets makes aptamers extremely attractive targeting agents for molecular imaging probes.

To date, a high proportion of aptamer-targeted imaging probes have involved detection by optical modalities, which predominantly find application in preclinical settings. However, the recent shift toward clinically relevant modalities such as ultrasound, X-ray, CT, MRI, and nuclear imaging is promising for their future development. Each modality possesses strengths and weaknesses, and the design of probes for each requires specific design considerations as outlined. The combination of multiple imaging modalities within a single construct may further enhance the utility of aptamer-targeted imaging probes in disease diagnosis, monitoring, and therapeutic delivery. Given the potential of aptamer-targeted imaging agents and the recent adoption of their use in targeted therapeutics, it seems likely that aptamer-targeted molecular imaging agents will soon find utility in clinical applications [44].

References

1. Aravind, A., Yoshida, Y., Maekawa, T., and Kumar, D. S. (2012). Aptamer-conjugated polymeric nanoparticles for targeted cancer therapy, *Drug Delivery Transl. Res.*, **2**, pp. 418–436.
2. Bagalkot, V., Zhang, L., Levy-Nissenbaum, E., Jon, S., Kantoff, P. W., Langer, R., and Farokhzad, O. C. (2007). Quantum dot: aptamer conjugates for synchronous cancer imaging, therapy, and sensing of drug delivery based on Bi-fluorescence resonance energy transfer, *Nano Lett.*, **7**, pp. 3065–3070.
3. Basu, S., Zaidi, H., Holm, S., and Alavi, A. (2011). Quantitative techniques in PET-CT imaging, *Curr. Med. Imaging Rev.*, **7**, pp. 216–233.
4. Boase, N. R. B., Blakey, I., and Thurecht, K. J. (2012). Molecular imaging with polymers, *Polym. Chem.*, **3**, pp. 1384–1389.
5. Borbas, K. E., Ferreira, C. S. M., Perkins, A., Bruce, J. I., and Missailidis, S. (2007). Design and synthesis of mono- and multimeric targeted

radiopharmaceuticals based on novel cyclen ligands coupled to Anti-MUC1 aptamers for the diagnostic imaging and targeted radiotherapy of cancer, *Bioconjug. Chem.*, **18**, pp. 1205–1212.

6. Brannon-Peppas, L., and Blanchette, J. O. (2012). Nanoparticle and targeted systems for cancer therapy, *Adv. Drug Delivery. Rev.*, **64**, Suppl, pp. 206–212.
7. Charlton, J., Sennello, J., and Smith, D. (1997). In vivo imaging of inflammation using an aptamer inhibitor of human neutrophil elastase, *Chem. Biol.*, **4**, pp. 809–816.
8. Chen, Z.-Y., Wang, Y.-X., Lin, Y., Zhang, J.-S., Yang, F., Zhou, Q.-L., and Liao, Y.-Y. (2014). Advance of Molecular Imaging Technology and Targeted Imaging Agent in Imaging and Therapy, *BioMed Res. Int.*, **2014**, p. 12.
9. Chu, T. C., F. Shieh, L. A. Lavery, M. Levy, R. Richards-Kortum, B. A. Korgel and A. D. Ellington (2006). Labeling tumor cells with fluorescent nanocrystal-aptamer bioconjugates, *Biosens. Bioelectron.*, **21**, pp. 1859–1866.
10. Clegg, R. M. (1995). Fluorescence resonance energy transfer, *Curr. Opin. Biotechnol.*, **6**, pp. 103–110.
11. Close, D. M., Xu, T. T., Sayler, G. S., and Ripp, S. (2011). In vivo bioluminescent imaging (BLI): noninvasive visualization and interrogation of biological processes in living animals, *Sensors*, **11**, pp. 180–206.
12. Coles, D. J., B. E. Rolfe, N. R. B. Boase, R. N. Veedu and K. J. Thurecht (2013). Aptamer-targeted hyperbranched polymers: towards greater specificity for tumours in vivo, *Chem. Commun.*, **49**, pp. 3836–3838.
13. Cormode, D. P., Naha, P. C., and Fayad, Z. A. (2014). Nanoparticle contrast agents for computed tomography: a focus on micelles, *Contrast Media Mol. Imaging*, **9**, pp. 37–52.
14. Fuchs, A. V., Gemmell, A. C., and Thurecht, K. J. (2015). Utilising polymers to understand diseases: advanced molecular imaging agents, *Polym. Chem.*, **6**, pp. 868–880.
15. Gessner, R., and Dayton, P. A. (2010). Advances in molecular imaging with ultrasound, *Mol. Imaging*, **9**, pp. 117–127.
16. Hamaguchi, N., Ellington, A., and Stanton, M. (2001). Aptamer beacons for the direct detection of proteins, *Anal. Biochem.*, **294**, pp. 126–131.
17. Hicke, B. J., Stephens, A. W., Gould, T., Chang, Y. F., Lynott, C. K., Heil, J., Borkowski, S., Hilger, C. S., Cook, G., Warren, S., and Schmidt, P. G. (2006). Tumor targeting by an aptamer, *J. Nucl. Med.*, **47**, pp. 668–678.
18. Hu, H., Dai, A., Sun, J., Li, X., Gao, F., Wu, L., Fang, Y., Yang, H., An, L., Wu, H., and Yang, S. (2013). Aptamer-conjugated $Mn_3O_4@SiO_2$ core-shell

nanoprobes for targeted magnetic resonance imaging, *Nanoscale*, **5**, pp. 10447–10454.

19. Hu, M. D., and Zhang, K. H. (2013). The application of aptamers in cancer research: an up-to-date review, *Future Oncol.*, **9**, pp. 369–376.
20. Huang, Y. F., Chang, H. T., and Tan, W. H. (2008). Cancer cell targeting using multiple aptamers conjugated on nanorods, *Anal. Chem.*, **80**, pp. 567–572.
21. Hwang, D. W., Ko, H. Y., Lee, J. H., Kang, H., Ryu, S. H., Song, I. C., Lee, D. S., and Kim, S. (2010). A nucleolin-targeted multimodal nanoparticle imaging probe for tracking cancer cells using an aptamer, *J. Nucl. Med.*, **51**, pp. 98–105.
22. Jacobson, O., Yan, X., Niu, G., Weiss, I. D., Ma, Y., Szajek, L. P., Shen, B., Kiesewetter, D. O., and Chen, X. (2015). PET imaging of tenascin-C with a radiolabeled single-strand DNA aptamer, *J. Nucl. Med.,* **56**, pp. 616–621.
23. James, M. L., and Gambhir, S. S. (2012). A molecular imaging primer: modalities, imaging agents, and applications, *Physiol. Rev.*, **92**, pp. 897–965.
24. Kalia, J., and Raines, R. T. (2010). Advances in bioconjugation, *Curr. Org. Chem.*, **14**, pp. 138–147.
25. Kang, W. J., Ko, M. H., Lee, D. S., and Kim, S. (2009). Bioimaging of geographically adjacent proteins in a single cell by quantum dot-based fluorescent resonance energy transfer, *Proteomics Clin. Appl.*, **3**, pp. 1383–1388.
26. Kaufmann, B. A., and Lindner, J. R. (2007). Molecular imaging with targeted contrast ultrasound, *Curr. Opin. Biotechnol.*, **18**, pp. 11–16.
27. Keefe, A. D., Pai, S., and Ellington, A. (2010). Aptamers as therapeutics, *Nat. Rev. Drug Discovery*, **9**, pp. 537–550.
28. Khalil, M. M., Tremoleda, J. L., Bayomy, T. B., and Gsell, W. (2011). Molecular spect imaging: an overview, *Int. J. Mol. Imaging*, **2011**, p. 796025.
29. Kim, B., Yang, J., Hwang, M., Choi, J., Kim, H, O., Jang, E., Lee, J. H., Ryu, S. H., Suh, J. S., Huh, Y. M., and Haam, S. (2013). Aptamer-modified magnetic nanoprobe for molecular MR imaging of VEGFR2 on angiogenic vasculature, *Nanoscale Res. Lett.*, **8**, pp. 1–10.
30. Kim, D., Jeong, Y. Y., and Jon, S. (2010). A drug-loaded aptamer-gold nanoparticle bioconjugate for combined CT imaging and therapy of prostate cancer, *ACS Nano*, **4**, pp. 3689–3696.

31. Ko, M. H., Kim, S., Kang, W. J., Lee, J. H., Kang, H., Moon, S. H., Hwang do, W., Ko, H. Y., and Lee, D. S. (2009). In vitro derby imaging of cancer biomarkers using quantum dots, *Small*, **5**, pp. 1207–1212.

32. Kovar, J. L., Simpson, M. A., Schutz-Geschwender, A., and Olive, D. M. (2007). A systematic approach to the development agents for optical imaging of mouse of fluorescent contrast cancer models, *Anal. Biochem.*, **367**, pp. 1–12.

33. Leblond, F., Davis, S. C., Valdes, P. A., and Pogue, B. W. (2010). Pre-clinical whole-body fluorescence imaging: review of instruments, methods and applications, *J. Photochem. Photobiol. B-Biol.*, **98**, pp. 77–94.

34. Lee, K. Y., Kang, H., Ryu, S. H., Lee, D. S., Lee, J. H., and Kim, S. (2010). Bioimaging of nucleolin aptamer-containing 5-(N-benzyl-carboxyamide)-2′-deoxyuridine more capable of specific binding to targets in cancer cells, *J. Biomed. Biotechnol.*, p. 9.

35. Li, W., Yang, X. H., Wang, K. M., Tan, W. H., He, Y., Guo, Q. P., Tang, H. X., and Liu, J. B. (2008). Real-time imaging of protein internalization using aptamer conjugates, *Anal. Chem.*, **80**, pp. 5002–5008.

36. Li, Z., Liu, Z., Yin, M., Yang, X., Yuan, Q., Ren, J., and Qu, X. (2012). Aptamer-capped multifunctional mesoporous strontium hydroxyapatite nanovehicle for cancer-cell-responsive drug delivery and imaging, *Biomacromolecules*, **13**, pp. 4257–4263.

37. Lim, E.-K., Kim, B., Choi, Y., Ro, Y., Cho, E.-J., Lee, J. H., Ryu, S.-H., Suh, J.-S., Haam, S., and Huh, Y.-M. (2014). Aptamer-conjugated magnetic nanoparticles enable efficient targeted detection of integrin $\alpha v\beta 3$ via magnetic resonance imaging, *J. Biomed. Mater. Res. A*, **102**, pp. 49–59.

38. Liu, K. C., Lin, B. S., and Lan, X. P. (2013). Aptamers: a promising tool for cancer imaging, diagnosis, and therapy, *J. Cell. Biochem.*, **114**, pp. 250–255.

39. Mankoff, D. A. (2007). A definition of molecular imaging, *J. Nucl. Med.*, **48**, pp. 18N–21N.

40. McRobbie, D. (2007) *MRI from Picture to Proton*, 2nd Ed. (Cambridge University Press, Cambridge, UK).

41. Mi, J., Zhang, X. W., Giangrande, P. H., McNamara, J. O., Nimjee, S. M., Sarraf-Yazdi, S., Sullenger, B. A., and Clary, B. M. (2005). Targeted inhibition of alpha v beta 3 integrin with an RNA aptamer impairs endothelial cell growth and survival, *Biochem. Biophys. Res. Commun.*, **338**, pp. 956–963.

42. Nakatsuka, M. A., Barback, C. V., Fitch, K. R., Farwell, A. R., Esener, S. C., Mattrey, R. F., Cha, J. N., and Goodwin, A. P. (2013). In vivo

ultrasound visualization of non-occlusive blood clots with thrombin-sensitive contrast agents, *Biomaterials*, **34**, pp. 9559–9565.

43. Nakatsuka, M. A., Mattrey, R. F., Esener, S. C., Cha, J. N., and Goodwin, A. P. (2012). Aptamer-crosslinked microbubbles: smart contrast agents for thrombin-activated ultrasound imaging, *Adv. Mater.*, **24**, pp. 6010–6016.
44. Ni, X., Castanares, M., Mukherjee, A., and Lupold, S. E. (2011). Nucleic acid aptamers: clinical applications and promising new horizons, *Curr. Med. Chem.*, **18**, pp. 4206–4214.
45. Paley, M. A., and Prescher, J. A. (2014). Bioluminescence: a versatile technique for imaging cellular and molecular features, *MedChemComm*, **5**, pp. 255–267.
46. Pasternak, J. J., and Williamson, E. E. (2012). Clinical pharmacology, uses, and adverse reactions of iodinated contrast agents: a primer for the non-radiologist, *Mayo Clin. Proc.*, **87**, pp. 390–402.
47. Pieve, C. D., Perkins, A. C., and Missailidis, S. (2009). Anti-MUC1 aptamers: radiolabelling with 99mTc and biodistribution in MCF-7 tumour-bearing mice, *Nucl. Med. Biol.*, **36**, pp. 703–710.
48. Rockey, W. M., Huang, L., Kloepping, K. C., Baumhover, N. J., Giangrande, P. H., and Schultz, M. K. (2011). Synthesis and radiolabeling of chelator–RNA aptamer bioconjugates with copper-64 for targeted molecular imaging, *Bioorg. Med. Chem.*, **19**, pp. 4080–4090.
49. Savla, R., Taratula, O., Garbuzenko, O., and Minko, T. (2011). Tumor targeted quantum dot-mucin 1 aptamer-doxorubicin conjugate for imaging and treatment of cancer, *J. Controlled Release*, **153**, pp. 16–22.
50. Schafer, R., Wiskirchen, J., Guo, K., Neumann, B., Kehlbach, R., Pintaske, J., Voth, V., Walker, T., Scheule, A. M., Greiner, T. O., Hermanutz-Klein, U., Claussen, C. D., Northoff, H., Ziemer, G., and Wendel, H. P. (2007). Aptamer-based isolation and subsequent imaging of mesenchymal stem cells in ischemic myocard by magnetic resonance imaging, *Rofo. Fortschr. Geb. Rontgenstr. Neuen Bildgeb. Verfahr.*, **179**, pp. 1009–1015.
51. Schambach, S. J., Bag, S., Schilling, L., Groden, C., and Brockmann, M. A. (2010). Application of micro-CT in small animal imaging, *Methods*, **50**, pp. 2–13.
52. Shi, H., He, X., Cui, W., Wang, K., Deng, K., Li, D., and Xu, F. (2014). Locked nucleic acid/DNA chimeric aptamer probe for tumor diagnosis with improved serum stability and extended imaging window in vivo, *Anal. Chim. Acta*, **812**, pp. 138–144.
53. Shi, H., He, X., Wang, K., Wu, X., Ye, X., Guo, Q, Tan, W., Qing, Z., Yang, X., and Zhou, B. (2011). Activatable aptamer probe for contrast-enhanced

in vivo cancer imaging based on cell membrane protein-triggered conformation alteration, *Proc. Natl. Acad. Sci. U S A*, **108**, pp. 3900–3905.

54. Shi, H., Tang, Z. W., Kim, Y., Nie, H. L., Huang, Y. F., He, X. X., Deng, K., Wang, K. M., and Tan, W. H. (2010). In vivo fluorescence imaging of tumors using molecular aptamers generated by cell-SELEX, *Chem.: Asian J.*, **5**, pp. 2209–2213.
55. Shigdar, S., Lin, J., Yu, Y., Pastuovic, M., Wei, M., and Duan, W. (2011). RNA aptamer against a cancer stem cell marker epithelial cell adhesion molecule, *Cancer Sci.*, **102**, pp. 991–998.
56. Singh, R., and Bandyopadhyay, D. (2007). MUC1: a target molecule for cancer therapy, *Cancer Biol. Ther.*, **6**, pp. 481–486.
57. Thomas, S. W., Joly, G. D., and Swager, T. M. (2007). Chemical sensors based on amplifying fluorescent conjugated polymers, *Chem. Rev.*, **107**, pp. 1339–1386.
58. Tok, J., Lai, J., Leung, T., and Li, S. F. Y. (2010). Molecular aptamer beacon for myotonic dystrophy kinase-related Cdc42-binding kinase alpha, *Talanta*, **81**, pp. 732–736.
59. Tong, R., Coyle, V. J., Tang, L., Barger, A. M., Fan, T. M., and Cheng, J. J. (2010). Polylactide nanoparticles containing stably incorporated cyanine dyes for in vitro and in vivo imaging applications, *Microsc. Res. Tech.*, **73**, pp. 901–909.
60. Wang, A. Z., Bagalkot, V., Vasilliou, C. C., Gu, F., Alexis, F., Zhang, L., Shaikh, M., Yuet, K., Cima, M. J., Langer, R., Kantoff, P. W., Bander, N. H., Jon, S. Y., and Farokhzad, O. C. (2008). Superparamagnetic iron oxide nanoparticle-aptamer bioconjugates for combined prostate cancer imaging and therapy, *ChemMedChem*, **3**, pp. 1311–1315.
61. Wang, C. H., Huang, Y. F., and Yeh, C. K. (2011). Aptamer-conjugated nanobubbles for targeted ultrasound molecular imaging, *Langmuir*, **27**, pp. 6971–6976.
62. Weissleder, R., and Pittet, M. J. (2008). Imaging in the era of molecular oncology, *Nature*, **452**, pp. 580–589.
63. Xu, W. C., and Lu, Y. (2011). A smart magnetic resonance imaging contrast agent responsive to adenosine based on a DNA aptamer-conjugated gadolinium complex, *Chem. Commun.*, **47**, pp. 4998–5000.
64. Yigit, M. V., Mazumdar, D., Kim, H. K., Lee, J. H., Dintsov, B., and Lu, Y. (2007). Smart "turn-on" magnetic resonance contrast agents based on aptamer-functionalized superparamagnetic iron oxide nanoparticles, *ChemBioChem*, **8**, pp. 1675–1678.

65. Yigit, M. V., Mazumdar, D., and Lu, Y. (2008). MRI detection of thrombin with aptamer functionalized superparamagnetic iron oxide nanoparticles, *Bioconjug. Chem.*, **19**, pp. 412–417.
66. Yu, M. K., Kim, D., Lee, I. H., So, J. S., Jeong, Y. Y., and Jon, S. (2011). Image-guided prostate cancer therapy using aptamer-functionalized thermally cross-linked superparamagnetic iron oxide nanoparticles, *Small*, **7**, pp. 2241–2249.
67. Yu, S. B., and Watson, A. D. (1999). Metal-based X-ray contrast media, *Chem. Rev.*, **99**, pp. 2353–2377.
68. Zhang, J., Jia, X., Lv, X. J., Deng, Y. L., and Xie, H. Y. (2010). Fluorescent quantum dot-labeled aptamer bioprobes specifically targeting mouse liver cancer cells, *Talanta*, **81**, pp. 505–509.

Chapter 12

First Therapeutic Aptamer: VEGF-Targeting Macugen

Marissa Leonard[a,b] **and Xiaoting Zhang**[a,b,c]

[a] *Department of Cancer Biology, Vontz Center for Molecular Studies, University of Cincinnati College of Medicine, Cincinnati, OH 45267, USA*

[b] *Graduate Program in Cancer and Cell Biology, Vontz Center for Molecular Studies, University of Cincinnati College of Medicine, Cincinnati, OH 45267, USA*

[c] *Cincinnati Cancer Center, University of Cincinnati College of Medicine, Cincinnati, OH 45267, USA*

zhangxt@ucmail.uc.edu

Age-related macular degeneration (AMD) is the leading cause of vision loss in Western patients over the age of 65. Decades of research have discovered vascular endothelial growth factor (VEGF) as the key agent in the development of AMD. Starting from the mid-1990s, researchers began to isolate RNA aptamers to target VEGF with high specificity for therapeutic applications. Following further chemical and structural modifications, an RNA aptamer with VEGF-targeting abilities was eventually developed and later succeeded in clinical trials. This RNA aptamer, named "Macugen," ultimately became the first FDA-approved therapeutic aptamer and has since benefited countless AMD patients. The following chapter will describe the identification and development of the Macugen aptamer, its cellular effects, pharmacokinetics, and

Aptamers: Tools for Nanotherapy and Molecular Imaging
Edited by Rakesh N. Veedu

ISBN 978-981-4669-83-2 (Hardcover), 978-981-4669-84-9 (eBook)
www.panstanford.com

ultimate successes in clinical trials. In addition, future perspectives for RNA aptamer-based therapies like Macugen for broadened disease treatment will be discussed.

12.1 Introduction

12.1.1 *Neovascular Age-Related Macular Degeneration*

Age-related macular degeneration (AMD) affects nearly 14 million people globally, remaining the leading cause of irreversible vision loss in Western patients over 65 years of age [10]. AMD is categorized into two main subtypes, the early stage dry form and the late, more severe wet form. Dry AMD is characterized by accumulation of reactive oxygen-free radicals and lipid peroxide, leading to chronic inflammation and thinning of retinal pigment epithelium cells [26]. Almost 20% of dry AMD converts to the more serious wet form of AMD and the onset of wet AMD, also known as choroidal neovascularization (CNV), accounts for nearly 90% of cases of severe vision loss due to AMD [10]. Research over the past decades has indicated that the development of wet AMD or CNV is mainly due to the highly elevated level of vascular endothelial growth factor (VEGF) triggered by chronic inflammation [26]. High levels of retinal VEGF can stimulate vascular permeability, resulting in an aberrant leak of blood vessel fluids or blood into the macula at the center of the retina, and consequently vision loss in the center field of view [25]. Thus, the increased level of VEGF is considered to be responsible for the development of wet AMD, making VEGF a main focus of the research for targeted therapy of this disease.

12.1.2 *Vascular Endothelial Growth Factor and Its Receptors*

VEGF (also known as VEGF-A) is a member of a growth factor gene family that also includes VEGF-B, VEFG-C, VEGF-D, and placental growth factor (PGF). VEGFs are the key signaling proteins involved in both de novo vasculogenesis and angiogenesis from preexisting vasculature. The major activity of VEGF in promoting angiogenesis is through stimulating the proliferation of vascular endothelial cells (ECs) and preventing ECs from apoptosis [16]. In addition, VEGF plays important roles in vascular permeability and induces tumor- and disease-driving vascular leakage [13, 32]. As mentioned earlier, this VEGF-induced vascular leakage is a significant contributor to the progression of the dry AMD to the more severe wet form of the disease.

The human VEGF gene contains 8 exons and 7 introns and can generate 4 major variants ($VEGF_{121}$, $VEGF_{165}$, $VEGF_{189}$, and $VEGF_{206}$) through alternative splicing [11]. With a 45 kDa size, VEGF is a heparin-binding homodimeric glycoprotein that binds 3 related class V receptor tyrosine kinases, VEGFR-1, VEGFR-2, and VEGFR-3 [27]. This receptor class is characterized by 7 immunoglobulin (Ig)-like domains in the extracellular domain (ECD), a single transmembrane domain, and a tyrosine kinase domain [34]. VEGFR-1 binds to family members VEGF-A, VEGF-B, and PIGF. VEGFR-2 also binds to VEGF-A, along with VEGF-C and VEGF-D. VEGF-A, however, does not bind to the third receptor, VEGFR-3, which only binds to VEGF-C and VEGF-D [9]. In addition to these receptors, VEGF can also interact with various co-receptors like heparin sulfate proteoglycans and neuropilins [9]. Importantly, the high-affinity binding of heparin to $VEGF_{165}$ regulates its mitogenic activity in addition to its VEGFR receptor phosphorylation and activation [6]. Upon ligand binding, VEGFRs form homodimers or heterodimers, which can then activate kinase-mediated autophosphorylation and multiple downstream signaling cascades [29]. These signaling pathways are predominantly involved in the regulation of processes including vascular permeability, vasculogenesis, and lymphangiogenesis [20]. As an essential regulator of the angiogenic and vasculogenic processes, the upregulated or aberrant expression of VEGF plays well-established roles in tumorigenesis, inflammatory disorders, and, a focus of this chapter, ocular neovascular diseases such as AMD [15].

12.1.3 *Targeting VEGF in Age-Related Macular Degeneration*

Nearly three decades ago, the importance of high levels of VEGF in ocular neovascular disorders became apparent. Early studies measuring VEGF concentration in ocular fluid from patients undergoing intraocular surgery found that 69 of 136 samples from patients with diabetic retinopathy tested positive for VEGF. In addition, 29 of 38 samples from patients with iris neovascularization expressed VEGF, in contrast to only 2 of 31 samples from patients with no neovascular syndromes expressing VEGF [15]. Furthermore, this group found that both VEGF and vitreous fluid containing VEGF stimulated growth of retinal ECs in vitro [3]. A similar study measured the vitreous levels of VEGF in eyes of patients with and without proliferative diabetic retinopathy and showed that vitreous VEGF was at significantly higher concentrations in eyes of patients with the disease. Significantly, these diseased samples have a median VEGF concentration much higher than the known concentration required for maximum in vitro proliferation of vascular ECs [1].

Once the correlation between ocular disease and VEGF expression is established through studies like these, researchers began to examine both the necessity and the sufficiency of VEGF expression in causing ocular disease. A research team from Massachusetts Eye and Ear Infirmary injected $VEGF_{165}$ into the vitreous of six healthy primate eyes and inactivated control recombinant $VEGF_{165}$ into 40 healthy control eyes. They found that while none of the control-receiving subjects exhibited any retinal vascular changes, all six, or 100%, of the eyes injected with bioactive $VEGF_{165}$ developed severe forms of retinopathy [35]. Following up with this study, a lab from the National Eye Institute used Long–Evans rat models and injected an adenovirus vector encoding $VEGF_{165}$ into the subretinal space of the rat eye. They found there was a clear induction of CNV shortly after injection that was sustained up to four weeks [7].

To complement the experimental approaches utilized in these examples, studies also began to investigate how the inhibition of VEGF activity could affect ocular neovascularization [28]. After observing a significant correlation between intraocular expression of VEGF and the occurrence of ischemia-induced retinal neovascularization in mice, a Beetham Eye Institute research group used VEGF-neutralizing chimeric proteins generated via the conjugation of the ECD of human or mouse VEGF receptors with IgG [4]. Compared to controls, they observed a dramatic inhibition of in vitro retinal EC growth stimulation by VEGF and an impressive 100% reduction of retinal neovascularization in vivo by the chimeric proteins [4].

Soon after, the Massachusetts Eye and Ear Infirmary research group intravitreously injected a neutralizing anti-VEGF monoclonal antibody into the eyes of nonhuman primates (cynomolgus monkeys) that had previously undergone laser retinal vein occlusion to induce retinal ischemia. They found that none of the eight primates receiving the anti-VEGF antibody developed iris neovascularization, while five of the eight control-receiving primates developed the disorder [2]. Following this study, researchers optimized the anti-VEGF monoclonal antibody by keeping only its antigen-binding fragment. They found, unlike its full-length version, the shortened antibody could successfully cross the internal limiting membrane and access subretinal space in the same cynomolgus monkey experimental model [24]. This optimized the anti-VEGF antibody, termed "rhuFab VEGF," and enhanced both the retinal/choroidal distribution and overall antineovascularization efficacy compared to the full-length antibodies [24].

These studies not only demonstrated the prominent roles of VEGF in ocular neovascularization but also demanded for clinically applicable therapeutic agents capable of specifically targeting and inhibiting VEGF. One challenge researchers were facing at the time was that these agents needed to be able to specifically target the $VEGF_{165}$ isoform, because research indicated that $VEGF_{165}$ and $VEGF_{121}$ isoforms are both upregulated during normal or physiological vascularization [21]. Unfortunately, VEGFR-1/IfG fusion protein and anti-VEGF monoclonal antibodies inhibited all VEGF isoforms, including $VEGF_{121}$, and resulted in repression of both normal and disease causing neovascularization, strongly hindering the usefulness of these therapeutic agents [21, 28]. This phenomenon, often labeled an "off-target effect," generated the need for an aptamer-based, highly specific, targeted therapy to inactivate solely $VEGF_{165}$ for the treatment of ocular neovascular disease AMD.

12.2 Development of Macugen (Pegaptanib Sodium)

12.2.1 *Isolation of RNA Aptamers Targeting $VEGF_{165}$*

With the abundance of evidence identifying $VEGF_{165}$ as a promising therapeutic target to treat AMD, in vitro studies began to isolate aptamers able to specifically inhibit $VEGF_{165}$. As early as 1994, the Jellinek et al. isolated 6 aptamer candidates via the Tuerek and Gold SELEX procedure [22]. From a starting random pool of 10^{14} RNA oligonucleotides comprising 30 random nucleotides flanked with two constant regions on the either end, they isolated 6 subgroups of RNA aptamer candidates on the basis of structural similarities. Further truncation experiments were performed to determine the minimal sequence required to maintain high-level binding affinities for $VEGF_{165}$ [22]. The resulting truncated RNA aptamers were between 29 and 36 nucleotides in length and were able to not only competitively bind to the heparin-binding domain of VEGF but also inhibit VEGF from binding to its cell surface receptors [22].

The next step taken to further develop these anti-VEGF aptamers was to increase their biological functions and therapeutic potentials by increasing their in vivo stability. This was carried out by Green et al. through the use of chemical modifications, including the substitution of ribopurines with 2′-O-methylpurine (2′-O-Me) nucleotides [19]. These replacements led to an approximate 17-fold enhancement in VEGF binding affinity and a dramatic increase in stability compared to its unmodified form [19]. To further increase VEGF-binding affinities, the Ruckmen et al. group substituted the previously employed 2′-NH_2-pyrimidine-substituted oligo libraries with 2′-F-pyrimidine-replaced versions [31]. These modified oligos exhibited greater duplex stability than the 2′-OMe versions previously used, along with a significantly improved $VEGF_{165}$ binding affinity [31]. Ruckman et al. also further truncated these modified aptamer candidates to the minimal length and obtained 3 specific aptamers, t22-OMe, t2-OMe, and t44-OMe, with strikingly enhanced biological efficacy, stability, affinity, and specificity for $VEGF_{165}$ [28, 31]. Ultimately, by linking 40 kDa polyethylene glycol molecules to the t44-OMe aptamer, its ability to inhibit $VEGF_{165}$ activity was greatly improved, much more so than the other two aptamers. This aptamer then became the prototype aptamer for continued modification and development into the eventually Food and Drug Administration (FDA)-approved Macugen.

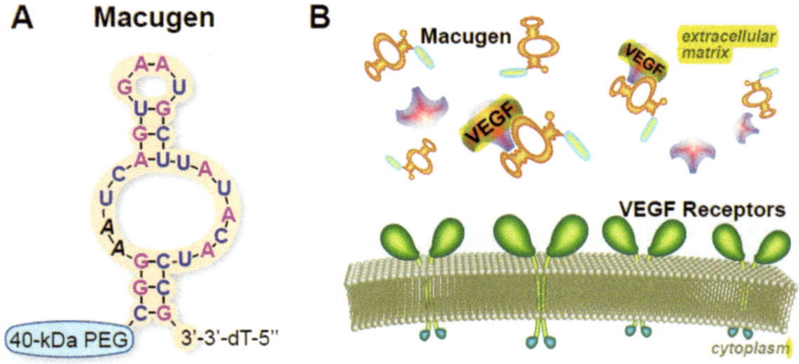

Figure 12.1 Macugen structure and target binding. (A) Nucleotide sequence and basic folding structure of Macugen. Magenta indicates 2′-O-methylpurines, while blue indicates 2′-F-pyrimidines. Unmodified nucleotides are shown in black. A 40 kDa PEG attachment linked at the 5′ end of the aptamer is also represented. (B) Mechanism of Macugen-mediated inhibition of VEGF receptor binding. Macugen aptamer binds VEGF at its U14 position and blocks interaction with its corresponding VEGF receptors.

12.2.2 *Cellular Effects of Macugen*

The final clinical version Macugen following the above-outlined biochemical modifications consists of a 3D RNA molecule of 28 nucleotides in length. The oligonucleotide sequence is flanked by a 20 kd monomethoxy polyethylene glycol unit covalently linked to an amino group lysine residue at the 5′ end [23], as shown in Fig. 12.1. Macugen's molecular weight is about 50 kd, with an osmolality ranging from 280 to 360 mOsm/kg and a pH between 6 and 7 [8, 23]. The dissociation constant of Macugen for $VEGF_{165}$ is ~50 pmol/L, providing a high binding affinity for optimal in vitro and in vivo functions [37].

In this well-characterized form, the effect of Macugen on VEGF-induced cellular responses was then tested. Results from cell culture studies employing human umbilical vein ECs clearly demonstrated the ability of the aptamer to inhibit VEGF-mediated signal transduction, calcium mobilization, VEGFR-2 receptor phosphorylation, Phospholipase Cγ phosphorylation, and cellular proliferation [8, 23]. In addition, it was found that pretreatment of these ECs with Macugen only inhibited proliferative responses to the $VEGF_{165}$ isoform but not $VEGF_{121}$, which is also an active mitogen in human umbilical vein ECs [8]. This is further supported by the structural evidence that binding between VEGF and Macugen requires close association between $VEGF_{165}$ cystine-137 in the heparin-binding domain, a residue that is absent in the $VEGF_{121}$ isoform [15].

12.2.3 *Pharmacokinetics of Macugen*

Macugen demonstrated remarkable stability in human plasma at ambient temperature for longer than 18 hours. Its stability and pharmacokinetics was further tested in animal models, including rabbits and rhesus monkeys [36, 37]. It was found that 24 hours after the point of injection of Macugen into the eyes of rabbits, the aptamer resided predominantly in vitreous fluid, aqueous fluid, and the retina [37]. Intravenous or subcutaneous administration of Macugen (1 mg/kg) in rhesus monkeys showed the half-lives of the aptamer were 9.3 hours postintravenous injection and 12 hours postsubcutaneous injection, with no apparent toxicities observed [5]. Interestingly, one study found that bioactive Macugen was still present in the vitreous humor of the eyes of monkeys, up to 28 days after receiving a 0.5 mg intravitreous dose [12]. Results from these studies allowed for the progression from Macugen evaluation in animals to highly anticipated clinical trials, as discussed in the following section.

12.3 Clinical Trials

12.3.1 *Clinical Trials in Patients with Neovascular AMD*

Following extensive successes of Macugen in in vitro and animal studies, phase I and II clinical trials were carried out to evaluate the safety and efficacy of the aptamer therapy for the treatment of CNV. In 1998, the pharmaceutical company Eyetech clinically tested the drug and found that of 15 wet AMD patients, 80% showed visual stabilization or improvement after 3 months [14]. This same company then further tested Macugen's therapeutic potential by monitoring the visual improvement and stabilization in 21 patients with subfoveal CNV secondary to AMD. This study was able to demonstrate that Macugen can enhance visual stability and improvement in greater than 85% of patients. In addition, compared to photodynamic therapy (PDT) alone, the combination of Macugen with PDT improved vision in participants by 60% [38]. Like in CNV, Macugen also underwent phase II clinical trials, treating patients with diabetic macular edema. Results demonstrated an overall improvement in vision with Macugen treatment, with a decrease in edema and an approximate 50% reduction in the requirement of laser therapy, indicating a high potential for Macugen for the treatment of diabetic macular edema patients, in addition to wet AMD cases. These exciting findings enabled a fast-track designation by the FDA for them to continue assessing Macugen treatment in the highly anticipated phase III trials.

12.3.2 *Phase III Clinical Trials of Macugen (VISION Trials)*

Known as the VEGF Inhibition Study in Ocular Neovascularization (VISION) trial, a phase III, two-part dose-ranging study spread across 117 test sites around the world was conducted in 2001, which included 1208 patients [18, 33]. Patients aged 50 years and older with various subtypes of CNV were recruited for the study. Each study participant was randomly assigned either a placebo, or sham, injection or an intravitreal injection of 0.3 mg, 1 mg, or 3 mg of Macugen every 6 weeks for 48 weeks total [18, 33]. All arms of the study were allowed treatment with PDT and Disudyne to ensure placebo patients were treated equivalently to usual neovascular AMD care [33]. The first portion of the trial enrolled 586 patients from August 2001 through July 2002, while the second portion enrolled 622 patients from October 2001 through August 2002 [18]. The efficacy determinant was measured by the maintenance of visual acuity, as shown by loss of fewer than 15 letters on the study eye chart. Follow-up exams took place in weeks 6, 12, 18, 24, 30, 42, 48, and 54 [18].

At as early as 6-weeks' postinjection, significant differences in effectiveness between the sham injection and Macugen administration were apparent. Compared to patients administered 0.3 mg or 1 mg of Macugen, sham-injected patients were twice as likely to suffer from severe vision loss, as indicated by a loss of $\geq$30 letters, or 6 lines, on the study eye chart. In weeks 30 and 54, a clear regression was observed in the growth of total lesion areas, size of CNV, and vascular leakage severity in Macugen-receiving patients. Interestingly, there were no observed differences in efficiency between 0.3 mg, 1 mg, and 3 mg of Macugen. In addition, there was no evidence to indicate that any subtype of CNV lesions or lesion size may affect treatment responses [33]. Moreover, two-year treatment data shows that patients who maintained the Macugen treatment regimen for two entire years experienced lower levels of vision loss than those who discontinued the treatment [38].

12.3.3 *Benefits and Safety of Macugen*

Macugen has since become one of the preferred treatments for wet AMD, alongside with other angiogenesis inhibitors like Eylea and Lucentis. Patients treated with Macugen experience numerous benefits that include greater improvement of visual acuity and reduction in the risk level of vision loss in patients with both wet AMD and macular edema. Though there are some adverse effects observed following Macugen treatment, they are mainly caused by the injection procedure used to administer the drug. These include some cases of infections, retinal detachment, eye irritation, and eye pain. Reported minor adverse reactions associated with Macugen administration included increased levels of vitreous floaters, inflammation level, and vitreous opacities [18]. More serious Macugen-associated injection-related adverse effects reported were endophthalmitis, retinal detachment, and traumatic injury to the lens; but the occurrence of these events was pretty rare and not statistically relevant [18]. Overall, Macugen is pretty safe for the treatment of neovascular AMD and diabetic macular edema. Importantly, the treatment response occurs quickly and is maintained for as long as two years, although a regression in the response level could endure during the second year of treatment [30]. In addition, long-term tolerability of this drug appears to be beneficial for the treatment of ocular neovascularization disorders. With these, the benefits of the treatment significantly outweigh the potential adverse side effects and make Macugen, the first therapeutic aptamer, a success in the treatment of AMD and macular edema patients.

12.4 Future Perspectives and Conclusion

This chapter described the historical invention and pioneering success of the first therapeutic aptamer Macugen, which, without a doubt, will open the door for many more applications for aptamers in the future. What makes Macugen such a success for AMD patient treatments is its ability to target a specific protein directly involved with molecular pathogenesis of the disease, its specific targeting ability that rivals monoclonal antibody, its long-lasting therapeutic effects, and high level of safety. The fast-track FDA approval of Macugen and its subsequent success further indicate the high potential that aptamers hold in the clinical setting. Currently, there are hundreds of aptamers being developed and characterized preclinically, not only in disease therapeutics, but also in disease detection, biomarker discovery, and even imaging technologies [17]. The flexibility and easy conjugation nature of the RNA aptamer to a plethora of small moieties like small interfering RNAs (siRNAs), nanoparticle carriers, and even photothermal molecules truly open up unlimited possibilities in the ways they can be employed. The continued advancement of RNA nanotechnology will likely not only produce more aptamers and uncover more applications but also lead to a reduction in the cost and time for production, making RNA aptamers more available and accessible for both basic science and clinical use. We can fully anticipate an exciting and promising future with more and more RNA aptamer-based tools and treatments to be created and improved for patient benefit for many generations to come.

Acknowledgments

We thank members of X. Zhang Laboratories for valuable comments and suggestions. We thank Glenn Doerman for illustration artwork. This study was supported by the Department of Defense Idea Award,

Susan G. Komen for the Cure Foundation Career Catalyst Grant, the American Cancer Society Research Scholar Award, and the Cincinnati Cancer Center Focused Pilot grant to X. Z.

References

1. Adamis, A. P., Miller, J. W., Bernal, M. T., D'Amico, D. J., Folkman, J., Yeo, T. K., and Yeo, K. T. (1994). Increased vascular endothelial growth factor levels in the vitreous of eyes with proliferative diabetic retinopathy, *Am. J. Ophthalmol.*, **118**(4), pp. 445–450.
2. Adamis, A. P., Shima, D. T., Tolentino, M. J., Gragoudas, E. S., Ferrara, N., Folkman, J., D'Amore, P. A., Miller, J. W. (1996). Inhibition of vascular endothelial growth factor prevents retinal ischemia-associated iris neovascularization in a nonhuman primate, *Arch. Ophthalmol.*, **114**(1), pp. 66–71.
3. Aiello, L. P., Avery, R. L., Arrigg, P. G., Keyt, B. A., Jampel, H. D., Shah, S. T., Pasquale, L. R., Thieme, H., Iwamoto, M. A., Park, J. E., et al. (1994). Vascular endothelial growth factor in ocular fluid of patients with diabetic retinopathy and other retinal disorders, *N. Engl. J. Med.*, **331**(22), pp. 1480–1487.
4. Aiello, L. P., Pierce, E. A., Foley, E. D., Takagi, H., Chen, H., Riddle, L., Ferrara, N., King, G. L., and Smith, L. E. (1995). Suppression of retinal neovascularization in vivo by inhibition of vascular endothelial growth factor (VEGF) using soluble VEGF-receptor chimeric proteins, *Proc. Natl. Acad. Sci. U S A*, **92**(23), pp. 10457–10461.
5. Asahara, T., Takahashi, T., Masuda, H., Kalka, C., Chen, D., Iwaguro, H., Inai, Y., Silver, M., and Isner, J. M. (1999). VEGF contributes to postnatal neovascularization by mobilizing bone marrow-derived endothelial progenitor cells, *EMBO J.*, **18**(14), pp. 3964–3972.
6. Ashikari-Hada, S., Habuchi, H., Kariya, Y., and Kimata, K. (2005). Heparin regulates vascular endothelial growth factor165-dependent mitogenic activity, tube formation, and its receptor phosphorylation of human endothelial cells. Comparison of the effects of heparin and modified heparins, *J. Biol. Chem.*, **280**(36), pp. 31508–31515.
7. Baffi, J., Byrnes, G., Chan, C. C., and Csaky, K. G. (2000). Choroidal neovascularization in the rat induced by adenovirus mediated expression of vascular endothelial growth factor, *Invest. Ophthalmol. Vis. Sci.*, **41**(11), pp. 3582–3589.

8. Bell, C., Lynam, E., Landfair, D. J., Janjic, N., and Wiles, M. E. (1999). Oligonucleotide NX1838 inhibits VEGF165-mediated cellular responses in vitro, *In Vitro Cell. Dev. Biol. Anim.*, **35**(9), pp. 533–542.
9. Cebe-Suarez, S., Zehnder-Fjallman, A., and Ballmer-Hofer, K. (2006). The role of VEGF receptors in angiogenesis; complex partnerships, *Cell. Mol. Life Sci.*, **63**(5), pp. 601–615.
10. Chappelow, A. V., and Kaiser, P. K. (2008). Neovascular age-related macular degeneration: potential therapies, *Drugs*, **68**(8), pp. 1029–1036.
11. Cross, M. J., Dixelius, J., Matsumoto, T., and Claesson-Welsh, L. (2003). VEGF-receptor signal transduction, *Trends Biochem. Sci.*, **28**(9), pp. 488–494.
12. Drolet, D. W., Nelson, J., Tucker, C. E., Zack, P. M., Nixon, K., Bolin, R., Judkins, M. B., Farmer, J. A., Wolf, J. L., Gill, S. C., and Bendele, R. A. (2000). Pharmacokinetics and safety of an anti-vascular endothelial growth factor aptamer (NX1838) following injection into the vitreous humor of rhesus monkeys, *Pharm. Res.*, **17**(12), pp. 1503–1510.
13. Dvorak, H. F., Detmar, M., Claffey, K. P., Nagy, J. A., van de Water, L., and Senger, D. R. (1995). Vascular permeability factor/vascular endothelial growth factor: an important mediator of angiogenesis in malignancy and inflammation, *Int. Arch. Allergy Immunol.*, **107**(1–3), pp. 233–235.
14. Eyetech Study, G. (2003). Anti-vascular endothelial growth factor therapy for subfoveal choroidal neovascularization secondary to age-related macular degeneration: phase II study results, *Ophthalmology*, **110**(5), pp. 979–986.
15. Ferrara, N. (2004). Vascular endothelial growth factor: basic science and clinical progress, *Endocr. Rev.*, **25**(4), pp. 581–611.
16. Ferrara, N., Gerber, H. P., and LeCouter, J. (2003). The biology of VEGF and its receptors, *Nat. Med.*, **9**(6), pp. 669–676.
17. Germer, K., Leonard, M., and Zhang, X. (2013). RNA aptamers and their therapeutic and diagnostic applications, *Int. J. Biochem. Mol. Biol.*, **4**(1), pp. 27–40.
18. Gragoudas, E. S., Adamis, A. P., Cunningham, E. T., Jr., Feinsod, M., Guyer, D. R., and VISiONCT Group (2004). Pegaptanib for neovascular age-related macular degeneration, *N. Engl. J. Med.*, **351**(27), pp. 2805–2816.
19. Green, L. S., Jellinek, D., Bell, C., Beebe, L. A., Feistner, B. D., Gill, S. C., Jucker, F. M., and Janjic, N. (1995). Nuclease-resistant nucleic acid ligands to vascular permeability factor/vascular endothelial growth factor, *Chem. Biol.*, **2**(10), pp. 683–695.

20. Hicklin, D. J., and Ellis, L. M. (2005). Role of the vascular endothelial growth factor pathway in tumor growth and angiogenesis, *J. Clin. Oncol.*, **23**(5), pp. 1011–1027.
21. Ishida, S., Usui, T., Yamashiro, K., Kaji, Y., Amano, S., Ogura, Y., Hida, T., Oguchi, Y., Ambati, J., Miller, J. W., Gragoudas, E. S., Ng, Y. S., D'Amore, P. A., Shima, D. T., and Adamis, A. P. (2003). VEGF164-mediated inflammation is required for pathological, but not physiological, ischemia-induced retinal neovascularization, *J. Exp. Med.*, **198**(3), pp. 483–489.
22. Jellinek, D., Green, L. S., Bell, C., and Janjic, N. (1994). Inhibition of receptor binding by high-affinity RNA ligands to vascular endothelial growth factor, *Biochemistry,* **33**(34), pp. 10450–10456.
23. Kourlas, H., and Schiller, D. S. (2006). Pegaptanib sodium for the treatment of neovascular age-related macular degeneration: a review, *Clin. Ther.*, **28**(1), pp. 36–44.
24. Krzystolik, M. G., Afshari, M. A., Adamis, A. P., Gaudreault, J., Gragoudas, E. S., Michaud, N. A., Li, W., Connolly, E., O'Neill, C. A., and Miller, J. W. (2002). Prevention of experimental choroidal neovascularization with intravitreal anti-vascular endothelial growth factor antibody fragment, *Arch. Ophthalmol.*, **120**(3), pp. 338–346.
25. Lopez, P. F., Sippy, B. D., Lambert, H. M., Thach, A. B., and Hinton, D. R. (1996). Transdifferentiated retinal pigment epithelial cells are immunoreactive for vascular endothelial growth factor in surgically excised age-related macular degeneration-related choroidal neovascular membranes, *Invest. Ophthalmol. Vis. Sci.*, **37**(5), pp. 855–868.
26. Mu, Y., Zhao, M., and Su, G. (2014). Stem cell-based therapies for age-related macular degeneration: current status and prospects, *Int. J. Clin. Exp. Med.*, **7**(11), pp. 3843–3852.
27. Neufeld, G., Cohen, T., Gengrinovitch, S., and Poltorak, Z. (1999). Vascular endothelial growth factor (VEGF) and its receptors, *FASEB J.*, **13**(1), pp. 9–22.
28. Ng, E. W., Shima, D. T., Calias, P., Cunningham, E. T., Jr., Guyer, D. R., and Adamis, A. P. (2006). Pegaptanib, a targeted anti-VEGF aptamer for ocular vascular disease, *Nat. Rev. Drug Discovery,* **5**(2), pp. 123–132.
29. Olsson, A. K., Dimberg, A., Kreuger, J., and Claesson-Welsh, L. (2006). VEGF receptor signalling: in control of vascular function, *Nat. Rev. Mol. Cell. Biol.*, **7**(5), pp. 359–371.
30. Pegaptanib Sodium Multi-center Study Group (2011). [Long-term efficacy and safety profile of pegaptanib sodium for age-related macular degeneration with choroidal neovascularization: evaluation of extended

phase II clinical trial], *Nippon Ganka Gakkai Zasshi.*, **115**(2), pp. 122–133.

31. Ruckman, J., Green, L. S., Beeson, J., Waugh, S., Gillette, W. L., Henninger, D. D., Claesson-Welsh, L., and Janjić. N. (1998). 2'-Fluoropyrimidine RNA-based aptamers to the 165-amino acid form of vascular endothelial growth factor (VEGF165). Inhibition of receptor binding and VEGF-induced vascular permeability through interactions requiring the exon 7-encoded domain, *J. Biol. Chem.*, **273**(32), pp. 20556–20567.
32. Senger, D.R., Galli, S. J., Dvorak, A. M., Perruzzi, C. A., Harvey, V. S., and Dvorak, H. F. (1983). Tumor cells secrete a vascular permeability factor that promotes accumulation of ascites fluid, *Science,* **219**(4587), pp. 983–985.
33. Sivaprasad, S. (2008). Role of pegaptanib sodium in the treatment of neovascular age-related macular degeneration, *Clin. Ophthalmol.*, **2**(2), pp. 339–346.
34. Stuttfeld, E., and Ballmer-Hofer, K. (2009). Structure and function of VEGF receptors, *IUBMB Life*, **61**(9), pp. 915–922.
35. Tolentino, M. J., Miller, J. W., Gragoudas, E. S., Jakobiec, F. A., Flynn, E., Chatzistefanou, K., Ferrara, N., and Adamis, A. P. (1996). Intravitreous injections of vascular endothelial growth factor produce retinal ischemia and microangiopathy in an adult primate, *Ophthalmology*, **103**(11), pp. 1820–1828.
36. Tucker, C. E., Chen, L. S., Judkins, M. B., Farmer, J. A., Gill, S. C., and Drolet, D. W. (1999). Detection and plasma pharmacokinetics of an anti-vascular endothelial growth factor oligonucleotide-aptamer (NX1838) in rhesus monkeys, *J. Chromatogr. B Biomed. Sci. Appl.*, **732**(1), pp. 203–212.
37. Vinores, S. A. (2003). Technology evaluation: pegaptanib, Eyetech/Pfizer, *Curr. Opin. Mol. Ther.*, **5**(6), pp. 673–679.
38. Vinores, S. A. (2006). Pegaptanib in the treatment of wet, age-related macular degeneration, *Int. J. Nanomedicine,* **1**(3), pp. 263–268.

Chapter 13

Aptamers: Scope, Limitations, and Future Prospects

John G. Bruno
OTC Biotech, LLC,4100 NW Loop 410, Suite 230, San Antonio, TX 78229, USA
john.bruno@otcorp.com

This chapter summarizes the range of recent DNA and RNA aptamer applications for in vitro and in vivo diagnostics especially imaging techniques and associated therapeutics or combined theranostics while acknowledging current limitations of aptamers. In addition, proposed methods for modifying aptamers to overcome these limitations are discussed.

13.1 Introduction

Aptamers are relatively short (60–200 bases) oligonucleotides composed of DNA or RNA [66] or, more recently, xeno nucleic acids (XNAs) [76, 95, 111] having modified backbones and unnatural monomers for greater stability in vivo. Aptamer development is analogous to an artificial in vitro immune system. Much like B lymphocyte clones are selected for high affinity and specificity to target antigens in vivo and stimulated to proliferate, so, too,

Aptamers: Tools for Nanotherapy and Molecular Imaging
Edited by Rakesh N. Veedu

ISBN 978-981-4669-83-2 (Hardcover), 978-981-4669-84-9 (eBook)
www.panstanford.com

are aptamers selected by affinity techniques and proliferated by polymerase chain reaction (PCR) through iterative rounds of selection and amplification. In essence, aptamers, which obviate host animals, were invented to compete with antibodies in a variety of research and industrial applications [66]. While aptamers have not truly lived up to their potential for various technical and business reasons [30, 31], aptamers have found special and promising niches for in vivo imaging and therapeutic research and development, which serve as the focus of this chapter. Combining advances in nanotechnology, such as novel dye and drug-loaded nanoparticles (NPs) or dendrimers [94, 127] of various types with advances in aptamer stability, pharmacokinetic half-lives is providing hope for effective new coupled diagnostic and therapeutic technologies (theranostics) to treat cancers [65, 72], cardiovascular diseases [4, 91], deep tissue infections, and a variety of other internal maladies. This chapter begins by examining the current scope of aptamer applications in diagnostics and medicine and builds toward future theranostic prospects involving aptamers, while acknowledging aptamer weaknesses and suggesting means to overcome these deficiencies.

13.2 Scope of Aptamer Technology

13.2.1 *Range of in vitro Diagnostic Applications*

To date aptamers have been used in virtually every conceivable way that antibodies have been used for in vitro diagnostic assays and biosensors. The breadth of aptamer usage has included lateral flow test strips [14, 24], blotting [29], enzyme-linked immunosorbent assay (ELISA)-like (enzyme-linked aptasorbent assay [ELASA] or enzyme-linked oligonucleotide assay [ELONA]) microplate assays [15, 17–20, 26, 28, 71], electrochemiluminescence [28], surface plasmon resonance (SPR) [113], surface-enhanced Raman spectroscopy (SERS) [8, 55], and magnetic bead-based capture coupled with fluorescence [26], mass spectral [90], or PCR-based [124] detection. One assay motif in which aptamers may be better suited and therefore superior to antibodies is fluorescence resonance energy transfer (FRET) including aptamer beacons (involving a fluorophore and a quencher dye) [16, 21–23, 27]. Aptamers are more flexible than internally disulfide-linked antibodies and are easily modified at particular loci during chemical synthesis [66]. Conversely, aptamers can now be internally disulfide linked as well to stabilize desirable loop structures (i.e., potential binding sites) when flexibility may be detrimental. Regardless of whether aptamers ever become effective competitors versus antibodies in commercial in vitro diagnostic markets or not, the body of in vitro diagnostic research has laid the scientific foundation on which new imaging techniques have been built.

13.2.2 *Aptamer Applications for in vivo Diagnostic Imaging*

Aptamers coupled to a variety of NPs, including highly fluorescent quantum dots (QDs) or other reporter molecules by themselves (naked) or doped into NPs, are the subject of intense research and development in medical imaging. The strong interest in aptamers from the imaging community is due to limitations of current computed tomography (CT)-iodinated contrast agents such as lack of targeting ability, short imaging time windows, and renal toxicity [72]. Aptamers also possess traits, which are desirable for in vivo imaging such as high affinities, specificity, lack of toxicity or immunogenicity, small size, and ease of production by chemical synthesis [12, 36, 58, 66].

Aptamers have been chemically conjugated to gold NPs and fluorescent gold NPs [2, 5, 9, 12, 72, 83, 121] for CT imaging, iron or manganese oxide magnetic nanoparticles (MNPs) for magnetic resonance imaging (MRI) [59, 61, 63], radioisotope surface-labeled or filled single-walled carbon nanotubes (SWCNTs) [9, 34, 54, 60, 87], and numerous QD-aptamer in vivo fluorescence imaging reports exist [34, 82, 99]. Aptamer-"flower" or aptamer-"popcorn" NPs [9, 83] and aptamer–dendrimer conjugates have also been used for effective imaging. Gold nanorods (NRs) and carbon nanotubes (CNTs) and their aptamer conjugates have also been used for photoacoustic or optoacoustic imaging [121]. Several types of NPs (especially gold NRs and CNTs) exhibit thermal expansion and contraction upon pulsed excitation with lasers, leading to acoustic vibrations, which are detectable even deep in tissues [2, 9, 62, 83, 121]. Results generally appear quite promising in terms of sensitive and specific detection of tumors and other targets in tissues, which is, in turn, fueling increased research and innovation.

Perhaps the greatest innovations are emerging from combinations of two or more different aptamers on NPs for enhanced target specificity [67] as well as combinations of aptamer-NPs with several different detection modalities (multimodal; for example CT, MRI, and radionuclide detection together) [63] for enhanced certainty in imaging. In particular, Hwang et al. [63] have developed a multimodal ^{67}Ga and Rhodamine-labeled silica-cobalt-ferrite MNP coupled to the AS1411 nucleolin aptamer and demonstrated superimposable fluorescent, radioisotopic, and magnetic resonance tumor imaging both in vitro and in vivo. Finally, the theranostic combinations of aptamer-NP imaging with specific "magic bullet"-like drug delivery are particularly exciting [34, 65, 72, 94], because physicians can now see and validate in real-time targeted drug delivery to tumors or other diseased regions in the body.

13.2.3 *Therapeutic Applications*

Like antibodies, simple aptamers and aptamer-toxin, -Fc (tail of IgG), -C1q (complement), and other conjugates have been developed and used to effectively bind and neutralize or inhibit lethal viruses [10, 24, 37, 52, 77, 104, 108], biotoxins [13, 27, 46, 79, 114, 116], and venoms, including snake and spider venoms [98, 122]; kill cancer cells [106]; and kill antibiotic-resistant bacteria [19, 69] and other target cells in vitro. The use of NPs coupled to aptamers enables delivery of larger and more concentrated toxic payloads to target cells. NPs can protect the drug payload from its environment while in transit to target cells and offer slow timed release [35, 54, 60, 87, 109, 125].

Dendrimers [94, 127] and porous silica NPs [57, 67, 81, 110] coupled to aptamers have been popular vehicles for experimental drug delivery with promising results as well. Generation 4 to 7 (G4–G7) polyamidoamine (PAMAM) dendrimers can hold small-molecule drugs like sponges in their "branches" or have aptamers covalently coupled at the ends of their branches [94, 127]. Silica NPs have even recently been designed with aptamer caps, plugs, or gates to block and then release drugs from within the NPs [92].

SWCNTs have also served as drug delivery vessels coupled to aptamers [109, 125]. In most cases drugs have been attached to the surface of SWCNTs, which limits the available space for aptamer surface attachment, but SWCNTs may also be filled with drugs internally [34, 54, 60, 87, 125] and capped. The newer filling techniques could have the added benefit of solubilizing hydrophobic drugs (in the internal graphene environment of the CNT), which might otherwise be poorly soluble and poorly absorbed in vivo. Internal drug or dye loading protects the payload from the environment while freeing space on the CNT surface for attachment of targeting agents such as aptamers.

Perhaps the most fascinating aspect of CNT drug release is that it can be controlled remotely by microwave or radio-frequency (RF) fields, which heat the CNTs to release drugs [96]. CNTs alone (without drug loading) as well as QDs, gold NPs, and MNPs with or without aptamers can be used in RF fields to kill target cells by hyperthermia [48–51, 62, 91, 96]. It may seem counterintuitive or surprising that SWCNTs can absorb and emit heat in an RF field since they are composed of rolled graphene and electrons should be shared equally in nonpolar carbon–carbon bonds (i.e., have low dielectric values). However, SWCNTs are known to be either metallic or semiconducting depending on their (n, m) carbon indices or geometries. The metallic and semiconducting nature of CNTs probably explains their ability to absorb microwaves and engage in RF-induced hyperthermic killing of target cells [48, 60, 91].

MNPs have the added advantage of being potentially concentrated to a tumor or other target by an external (ex vivo) magnetic field prior to hyperthermia induction by RF waves [34, 59, 74, 105, 115]. Superparamagnetic iron oxide nanoparticles (SPIONs) consist of α-Fe_2O_3 (hematite), γ-Fe_2O_3 (maghemite), and Fe_3O_4 (magnetite) of 1–200 nm or greater diameters [115]. SPIONs of 10–100 nm have the greatest lifetimes in vivo because they are large enough to avoid renal clearance, but small enough to evade phagocytosis [115]. SPIONs currently have a large diversity of surface coatings [115] for chemical functionalization and aptamer attachment, if desired. MNPs can be used for magnetic hyperthermia (MHT) in conjunction with external alternating magnetic fields (AMFs) of 88.9 kHz to 13.56 MHz [59, 115]. By comparison, most microwave ovens utilize RF radiation of 915 MHz or 2450 MHz. The concept is to heat targeted tumors to between 41°C and 43°C because some types of tumor cells are more sensitive to moderately elevated temperatures than surrounding tissues, which do not suffer much collateral damage [33, 38]. These temperatures bring about a number of changes detrimental to the cell cycle and can induce apoptosis [101].

QDs themselves are composed of toxic metals such as cadmium and selenium and are therefore cytotoxic to a wide variety of cell types [7, 97, 103, 112, 119], even without an electromagnetic field. QDs are thought to be strong free-radical (reactive oxygen species, or ROS; for example, singlet oxygen) generators, especially when coupled with photosensitizer ligand dyes such as Rose Bengal or chlorin e6 [30, 47, 103, 120]. Regardless of the exact mechanisms, MNPs, CNTs, and QDs are clearly potential cytotoxic components to pair with aptamers in highly lethal conjugates, and each type of NP has the added advantage of enabling deep tissue imaging to determine if the conjugate has reached its target in vivo.

13.2.4 *Current Clinical Trials*

The enigmatic demise of Archemix, Inc. (a major, but defunct, aptamer therapeutics company) several years ago despite closing a large deal with Baxter for rights to its hemophilia aptamers was a clear blow against aptamer clinical trials. And the recent termination of Regado Bioscience's phase III trial enrollment for its Revolixys kit composed of pegnivacogin, an anticoagulant aptamer against factor IX, and its complementary oligonucleotide was another major hit against aptamer clinical trials. However, if one searches the National Institutes of Health (NIH) clinical trials site (https://clinicaltrials.gov/) with the search term "aptamer," the site returns 5 active clinical studies that are recruiting human subjects among a total of 28 active or completed studies. These clinical studies have ranged from phase I to III and span treatments for macular degeneration, idiopathic vasculopathy, Hippel–Lindau disease, diabetic macular edema, acute myeloid leukemia, type 2 diabetes, and lupus and are used in stem cell transplantation, thus offering some hope that aptamers are still being seriously considered as pharmaceuticals.

In particular, among the macular degeneration drugs, two new aptamers made by Opthotech, Inc. called Fovista® (an anti–platelet-derived growth factor [PDGF] aptamer) and Zimura® (an anticomplement [C5] aptamer) appear to be progressing well in phase II and III trials and may someday compete with or augment Macugen® (to date the only Food and Drug Administration [FDA]-approved aptamer for wet age-related macular degeneration). For details about the handful of aptamers currently in clinical trials, the reader is referred to several recent articles [6, 40, 68, 78].

13.3 Physical Limitations of Aptamers

13.3.1 *Stability and Pharmacokinetics*

Aptamers and any oligonucleotides (e.g., antisense, small interfering RNA [siRNA]) used in vivo face two major challenges, (1) endo- and exonuclease degradation and (2) rapid clearance by the kidneys, liver, etc., unless weight is added. The first challenge has been counteracted in a number of ways by essentially modifying the aptamer's backbone, sugar moieties, or bases to present substrates that endo- and exonucleases cannot fit into their active sites. In particular, aptamer phosphate backbones have been modified with sulfur atoms at nonbridging oxygens in the phosphate groups (i.e., phosphorothioates or *thioaptamers*) [73, 85]. The deoxyribose and ribose sugars have been modified at the 2′ positions with fluoro and O-methyl groups while the 3′ and 5′ ends have been blocked with inverted bases [66], biotin [43], proteins [25, 43] or polyethylene glycol (PEG) [11, 39, 41] to slow or halt enzymatic degradation and add weight.

The second challenge can be overcome by adding weight and size to the aptamer conjugate. Of course, adding NPs will help in this regard, but Dougan et al. [43] have shown that a simple 3′ biotin–streptavidin coupling greatly increased the pharmacokinetic (PK) half-life of aptamers in blood by slowing kidney clearance and blocking Exonuclease I activity. Replacing the nitrogen bases or nucleotides with unnatural bases is another strategy to enhance aptamer PK, while also perhaps enhancing aptamer affinity and specificity [31, 100, 107]. The downside to the use of artificial bases is that aptamers are no longer recognized as "self" by the immune system and may easily become immunogenic [58, 66]. A newer class of xeno-nucleic acids (XNAs) is also being investigated for development of aptamers [76, 95, 111]. XNAs include locked nucleic acids (LNAs), peptide nucleic acids (PNAs), and even more exotic artificial nucleic acids, which would likely be immunogenic, but may exhibit excellent stability against endo- and exonucleases because they bear little resemblance to the native DNA and RNA substrates [76, 95]. Spiegelmers represent a similar "indigestible" class of aptamers made with levorotatory, or L forms, of the nucleotides, which exhibit excellent stability versus serum nucleases.

Nobel laureate Kary Mullis has patented aptamer-alpha-gal sugar conjugates for his company Altermune, LLC, which noncovalently bind to the 1% of anti-alpha-gal (galactose-alpha-1,3-galactose) circulating antibodies in human serum [71, 88] to increase weight and retard organ clearance [39, 43, 58] of the aptamer-alpha-gal complex. Binding to circulating antibodies also imparts the functionality of the antibody's Fc tail to activate complement and allow phagocytosis or antibody-dependent cellular cytotoxicity (ADCC) [18, 19, 30, 71, 88]. The only potential deficiency of Altermune's approach is that the aptamer-alpha-gal conjugate must find and bind its cognate antibody before being eliminated by the kidneys or liver [39, 43, 58, 70]. While some aptamer-alpha-gal-antibody complexes will undoubtedly form, some aptamer-alpha-gal conjugates will probably also be lost to the kidneys before encountering protective antibodies. It, therefore, makes greater sense to the author to inoculate with premade heavy aptamer-3′-protein covalent conjugates as he has published [25] and patented (US Patent Nos. 7910297, 8318920, and 8389710).

Regardless of potential inefficiency in vivo, the author and his colleagues are also currently testing an approach similar to that of Dr. Mullis, which was pioneered by Manoharan et al. in 2002 [86], involving attachment of ibuprofen or other nonsteroidal anti-inflammatory drugs (NSAIDs, including aspirin, naproxen) to effectively "hitch a ride" on serum albumin and other serum proteins [64, 75, 117, 128]. Serum albumin is more abundant than the 1% circulating anti-alpha-gal antibodies [71, 88], but only exhibits a K_a of $\sim 10^6$ for ibuprofen [64, 75, 117, 128]. This may be good in terms of offloading aptamer-3′-ibuprofen conjugates at sites of inflammation. After developing a 98% efficient ibuprofen-*N*-hydroxysuccinimide (NHS) conjugation approach for adding ibuprofen to the 3′ amino end of a DNA oligonucleotide (oligo or aptamer) [31], the author was able to show enhanced survival of that aptamer-3′-ibuprofen conjugate in rat blood for 30 min versus an unconjugated oligo, as illustrated in Fig. 13.1.

The ELISA-like (ELASA) plate assay results shown in Fig. 13.1 simply involved overnight adherence of untreated rat serum and serum treated with biotin-5′-DNA oligo-(60-mer)-3′-ibuprofen to polystyrene 96-well plates in 0.1 M $NaHCO_3$ (pH 8.5) overnight at 4°C followed by decanting, washing in phosphate buffered saline (PBS), binding streptavidin-peroxidase, and development in 2,2′-azino-bis(3-ethylbenzothiazoline-6-sulfonic acid) (ABTS). In particular, two male Sprague–Dawley rats per group were injected i.v. (0.18 mg/kg) with 5′-biotin-DNA oligo-3′-ibuprofen conjugate, unconjugated oligo-5′-biotin (no ibuprofen on the oligo), or no oligo. Blood samples were collected 30 min postinjection, they were centrifuged, and serum samples were immediately frozen at −80°C. The average A_{405nm} values for each group along with standard deviations ($N = 4$ or 8) were determined (Fig. 13.1), and each of the groups was compared to the others using a one-tailed Student's *t*-test at $a = 0.05$ and Excel statistical software. As Fig. 13.1 demonstrates, there was a statistically significant greater amount of ibuprofen conjugated oligo 30 min after injection versus the oligo without attached ibuprofen.

13.3.2 *Size and Charge*

Aptamers are much smaller than antibodies. IgG is approximately 122 × 139 Å versus the theoretical thrombin aptamer size of 21 × 25 Å [80]. This size differential may be a limitation when aptamers are immobilized on surfaces for target capture or biosensing, but it may be overcome by use of lengthy linkers or tethers. Smaller size may also be an advantage for penetrating crevices in complex cell topologies where larger antibodies may not fit.

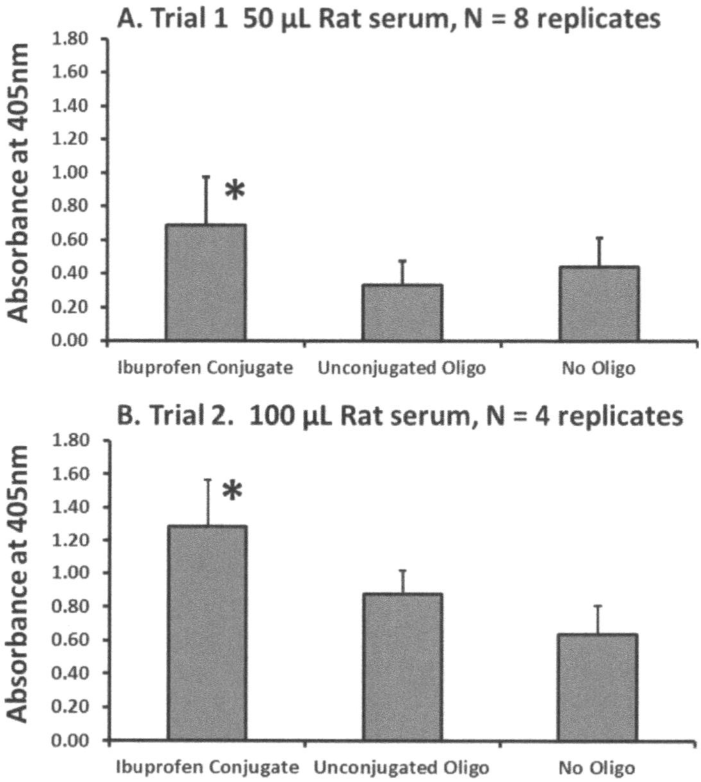

Figure 13.1 Pooled ELISA-like (ELASA) microplate assay results for two separate trials of the aptamer-3′-ibuprofen conjugate rat serum albumin PK comparison to the unconjugated aptamer (unconjugated oligo) and uninjected controls (no oligo). Panel A shows results for 50 μL of serum coating wells ($N = 8$), and panel B shows results of 100 μL coating of wells ($N = 4$). The asterisks indicate statistically different results vs. the other two groups ($p < 0.01$) by t-testing for the ibuprofen conjugates.

Clearly, the highly anionic phosphate backbone of aptamers may make them repel anionic phospholipid bilayer cell membranes. However, aptamers can flip in solution and reorient their single-stranded (ss) regions to expose more cationic nitrogenous bases, and divalent cationic (Ca^{2+} or Mg^{2+}) buffers or higher salt concentrations in general may overcome negative charge repulsions to facilitate binding.

13.3.3 *Morphologic Temperature Dependence*

The flexibility of aptamers in solution allows them great conformational freedom for induced fit to their targets, but it also leaves them susceptible to large changes in shape due to temperature. Temperatures exceeding the melting temperature could completely linearize an aptamer and would make it completely incapable of binding its target. Some aptamers are more temperature stable than others, as revealed by secondary stem-loop computer modeling with the free Internet UNAFold software using the temperatures shown in Fig. 13.2 and physiologic salt concentration of 145 mM Na^+. The author laboratory selects all aptamers intended for in vivo use at 37°C because of the effect shown in Fig. 13.2. But if dramatic changes in the secondary or tertiary structure occur as a result of temperature and the user wishes to lock a loop structure in place, that can now be accomplished by means of disulfide bonds

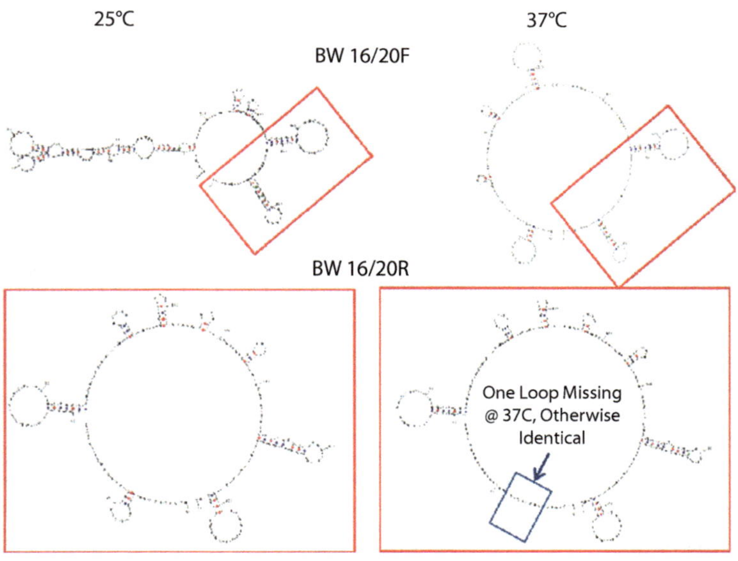

Figure 13.2 Dramatic temperature dependence of secondary stem-loop structure of some aptamers between room and body temperatures is shown in the top row of images. Relative temperature independence of other aptamer secondary structures is shown in the bottom row. Red boxes enclose relatively unchanged regions of these 200 base aptamers.

in the proximal stem regions using thiol groups introduced during aptamer DNA synthesis.

13.3.4 *Monomer Diversity*

Aptamers have always suffered a shortage of monomer diversity by being limited to four different nucleotides (A, C, G, T, or U). Aptamers have compensated by having greater overall library diversity (generally $\sim 10^{15}$ different oligonucleotide sequences in the starting pool) [66] than the antibody repertoire (diversity of 10^9 maximally). But, new technologies such as Artificial Expanded Genetic Information Systems (AEGIS) [100, 107] and X-Aptamer technology from AM Biotechnologies, Inc. [56] are expanding the monomer (artificial base) menu and appear to lead to even greater affinity and specificity [31].

13.3.5 *Length and Specificity for Complex Antigens*

In the past, commercially available DNA oligonucleotide aptamer length was limited to $\sim$70 bases due to the low yields of lengthier aptamer templates by chemical synthesis [30]. However, advances in chemical synthesis have enabled routine aptamer template lengths up to 200 bases (e.g., Ultramers® by Integrated DNA Technologies, Inc. Coralville, IA). Longer templates mean that the degenerate (randomized) midregion of aptamer templates can be lengthened to better mimic the 110 amino acid complementarity determining regions (CDR) of antibodies, which harbor three hypervariable antigen-binding regions per CDR per heavy or light chain (12 hypervariable binding sites per antibody molecule) [30]. By analogy, the author's group has lengthened the degenerate region of their aptamer DNA template to 164 randomized bases with flanking 18 base constant primer regions. So, logically, more binding regions should emerge with longer aptamers. The author has seen numerous complex loop structures in these longer multivalent aptamers (e.g., Fig. 13.2) that probably confer greater specificity for complex multi-epitope targets such as cell surfaces [1, 30, 84, 126].

13.3.6 *Mass Production by Chemical Synthesis*

Aptamers are inexpensive on the milligram scale by chemical synthesis (~50 cents per base), but their price jumps to ~$2000 per gram by parallel chemical synthesis [58]. The author has direct experience with this aspect of aptamers, having paid $20,000 for 10 grams of an aptamer for animal testing once. This cost may be prohibitive for aptamers to be considered competitive as pharmaceuticals. Thus, inexpensive mass production of aptamers is needed and may be provided by enzymatic methods such as PCR or asymmetric PCR (predominately single-stranded PCR enabled by adding a 100:1 ratio of one primer over the other). Companies such as Vandalia Research (Huntington, WV, USA) are laying the foundation for less expensive mass production of aptamers and other medicinal oligonucleotides via large-scale PCR [25, 89]. PCR with Taq polymerase also conveniently adds a 3′ adenine overhang, which the author's group has used to add proteins to the 3′ end, using a variety of linkers [25], thereby enabling mass production of stable aptamer 3′ protein (Fc, C1q, albumin, etc.) conjugates (US Patent Nos. 7910297, 8318920, and 8389710).

13.4 Future Prospects for Aptamers

13.4.1 *Analgesic and Antishock Aptamers*

The author is currently working through an NIH SBIR grant (No. 2 R43 GM101712-01) on what may be a future commercial application of aptamers in the area of long-term pain relief for burn and cancer patients. Pain is often mediated by small molecules such as prostaglandins. NSAIDs inhibit the cyclooxygenase-2 (COX-2) enzyme so as to reduce prostaglandin production after tissue damage and thereby temporarily ameliorate pain. Similarly, the author is working with Professor Kenneth Hargreaves (Chair of Endodontics at the Univ. of Texas Health Science Center in San Antonio) who discovered that oxidized linoleic acid metabolites (OLAMs) known as hydroxyoctadecadienoic acids (HODES) mediate heat or burn pain and some types of cancer pain and shock [53, 93]. Hargreaves and Bruno recently patented (US Patent Application

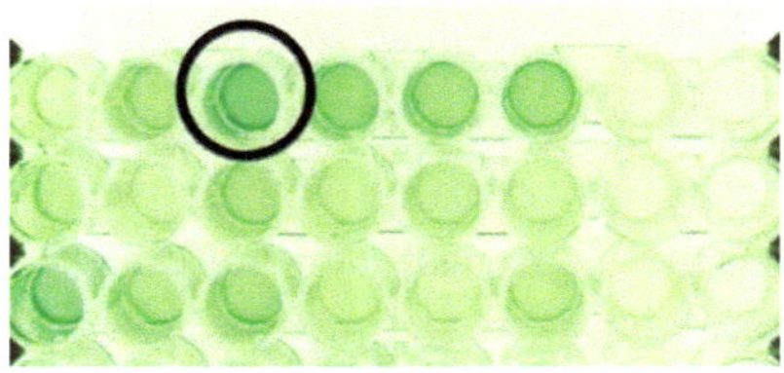

Figure 13.3 Strong ELASA reaction of the circled 13-HODE-9R aptamer versus immobilized 13-HODE.

No. 14/029,440; Notice of Allowance granted Jan. 6, 2015) several aptamer DNA sequences having high affinity to 9-HODE and 13-HODE, respectively, as illustrated by the relative affinity ranking (darkest-green color) in a simple ELASA microplate assay from the author's laboratory for one of the HODEs (Fig. 13.3). The anti-HODE aptamers inhibit electrical activity from excited isolated nerves in patch clamp experiments (Fig. 13.4) and when rats are injected in their hind limbs and held against a 43°C metal block, they do not recoil, suggesting strongly that they cannot feel the heat pain generated by the block.

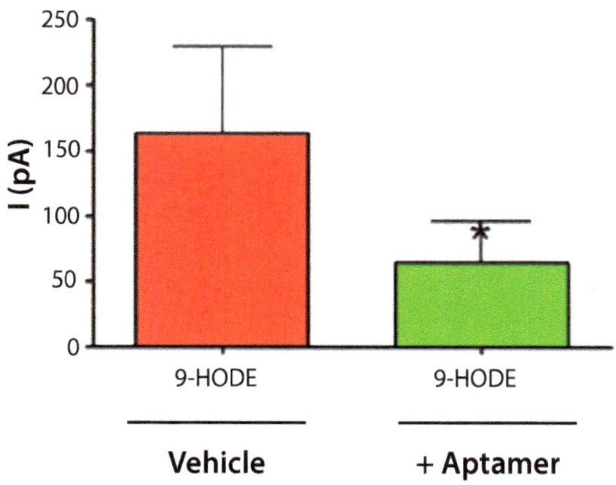

Figure 13.4 Results of preliminary rat nerve patch clamp studies demonstrating that a 9-HODE aptamer significantly diminishes the electrical activity (current in picoAmps on the *y* axis) stimulated by 9-HODE versus 9-HODE alone (i.e., the aptamer binds and blocks the activity of 9-HODE).

13.4.2 *Futuristic Aptamer Medical Applications*

The author has recently published a review article [31] that highlights certain other potential futuristic medical applications of aptamers such as the potential to induce stem cell differentiation or transdifferentiation, much like Richard Lerner's laboratory at Scripps has demonstrated with antibody induction of stem cell differentiation [42, 118, 123]. In this way, aptamers might replace more expensive recombinant growth factors in regenerative medicine or be used to discover previously unknown differentiation inducing surface receptors and pathways through whole cell SELEX. The author's review article [31] also covers potential use of aptamers to rapidly clot blood for diffuse life-threatening hemorrhages. Prof. Tan's group at the University of Florida has recently demonstrated the feasibility of such rapid blood clotting [3].

Finally, aptamers may find a place as structural or targeting components in the very futuristic concept of nanorobotics or the use of "nanobots" for internal medical applications such as nanosurgery to remove tumors, parasites, or cardiovascular plaques [32]. Recently George Church's group at Harvard has taken a first step in this area by using aptamers as targeting molecules and locks on DNA origami-fashioned barrels, which can be opened to deliver a therapeutic drug payload to target cells [44]. In Church's proof-of-concept experiments, antibody fragments were effectively targeted and delivered to cancer cells by aptamers resulting in target cell death by apoptosis [44].

13.5 Summary

This chapter has attempted to summarize the salient aspects of aptamers, their uses, and current limitations. Like all pharmaceuticals or affinity reagents (antibodies, peptides, receptors, ligands, etc.), aptamers have their strengths and weaknesses. Before passing judgment on aptamers, it would be wise to remember that aptamers are still relatively new ($\sim$25 years since invention) versus antibodies and still being continually improved by a small community of aptamer proponents. While aptamers may never supplant antibodies, their ability to obviate animal hosts and be easily selected and reproduced in vitro with very high fidelity, affinity, specificity, and when modified, good in vivo stability, and PK probably means that they will find niche applications in diagnostic imaging and therapeutics. In short, despite some setbacks, the future still seems quite promising for aptamers and aptamer-NP conjugates as diagnostic imaging reagents and effective therapeutics or theranostic agents.

Acknowledgments

Funding was provided by NIH SBIR grant no. 1 R43 GM101712-01 and DARPA SBIR contract no. W911NF-14-P-0037. The author thanks Professor Kenneth Hargreaves and his laboratory at the University of Texas Health Sciences Center in San Antonio (Dental School) for assistance in acquisition of some of the data presented herein.

References

1. Ahmad, K. M., Xiao, Y., and Soh, H. T. (2012). Selection is more intelligent than design: improving the affinity of a bivalent ligand through directed evolution, *Nucleic Acids Res.*, **40**, pp. 11777–11783.
2. Ai, J., Xu, Y., Lou, B., Li, D., and Wang, E. (2014). Multifunctional AS1411-functionalized fluorescent gold nanoparticles for targeted cancer cell imaging and efficient photodynamic therapy, *Talanta*, **118**, pp. 54–60.
3. Altman, M. O., Chang, Y. M., Xiong, X., and Tan, W. (2013). Modifying cellular properties using artificial aptamer-lipid receptors, *Sci. Rep.*, **3**, p. 3343.
4. Annapragada, A. (2015). Advances in nanoparticle imaging technology for vascular pathologies, *Ann. Rev. Med.*, **66**, pp. 177–193.
5. Aravind, A., Mathew, A., and Kumar, D. S. (2013). Nano-based aptamers mediated imaging and drug delivery in cancer, in *Biomedical Applications of Aptamers* (Bruno, J. G., ed.), pp. 143–179, Nova Science Publishers, New York.
6. Ashrauzzaman, M. (2014). Aptamers as both drugs and drug-carriers, *BioMed. Res. Int.*, **2014**, Article ID 697923, 21 pages.

7. Bakalova, R., Ohba, H., Zhelev, Z., Ishikawa, M., and Baba, Y. (2004). Quantum dots as photosensitizers? *Nat. Biotechnol.*, **22**, pp. 1360–1361.

8. Barhon, F., Bardliving, C., Phifer, A., Bruno, J. G., and Batt, C. (2013). New aptasensor based on polymer-gold nanoparticles composite microspheres for the detection of malathion using surface-enhanced Raman spectroscopy, *Ind. Biotechnol.*, **9**, pp. 42–49.

9. Beqa, L., Fan, Z., Singh, A. K., Senapati, D., and Ray, P. C. (2011). Gold nano-popcorn attached SWCNT hybrid nanomaterial for targeted diagnosis and photothermal therapy of human breast cancer cells, *ACS Appl. Mater. Interfaces*, **3**, pp. 3316–3324.

10. Binning, J. M., Wang, T., Luthra, P., Shabman, R. S., Borek, D. M., Liu, G., Xu, W., Leung, D. W., Basler, C. F., and Amarasinghe, G. K. (2013). Development of RNA aptamers targeting Ebola virus VP35, *Biochemistry*, **52**, pp. 8406–8419.

11. Boomer, R. M., Lewis, S. D., Healy, J. M., Kurz, M., Wilson, C., and McCauley, T. G. (2005). Conjugation to polyethylene glycol polymer promotes aptamer biodistribution to healthy and inflamed tissues. *Oligonucleotides*, **15**, pp. 183–195.

12. Borkowski, S., and Dinkelborg, L. M. (2006). Aptamers for in vivo imaging, in *The Aptamer Handbook* (Klussmann S., ed.), pp. 343–362, WILEY-VCH Verlag GmbH & Co., Weinheim, Germany.

13. Bruno J. G. (2014). Anti-soman derivative DNA aptamers protect acetylcholinesterase in neuro-2a cells in vitro, *Am. J. Neuroprot. Neuroregen.*, **6**, pp. 43–46.

14. Bruno, J. G. (2014). Application of DNA aptamers and quantum dots to lateral flow test strips for detection of foodborne pathogens with improved sensitivity versus colloidal gold, *Pathogens*, **3**, pp. 341–355.

15. Bruno, J. G., Carrillo, M. P., Phillips, T., and King, B. (2008). Development of DNA aptamers for cytochemical detection of acetylcholine, *In Vitro Cell. Dev. Biol. Anim.*, **44**, pp. 63–72.

16. Bruno, J. G., Carrillo, M. P. (2012). Development of aptamer beacons for rapid presumptive detection of *Bacillus* spores, *J. Fluoresc.*, **22**, pp. 915–924.

17. Bruno, J. G., Carrillo, M. P., Cadieux, C. L., Lenz, D. L., Cerasoli, D. M., and Phillips, T. (2009). DNA aptamers developed against a soman derivative cross-react with methylphosphonic acid but not with the flanking hydrophobic groups, *J. Mol. Recognit.*, **22**, pp. 197–204.

18. Bruno, J. G., Carrillo, M. P., and Crowell, R. (2009). Preliminary development of DNA aptamer-Fc conjugate opsonins, *J. Biomed. Mater. Res. A.*, **90**, pp. 1152–1161.
19. Bruno, J. G., Carrillo, M. P., and Phillips, T. (2008). In vitro antibacterial effects of anti-lipopolysaccharide DNA aptamer-C1qrs complexes, *Folia Microbiol. (Praha)*, **53**, pp. 295–302.
20. Bruno, J. G., Carrillo, M. P., Phillips, T., and Edge, A. (2011). Discrimination of recombinant from natural human growth hormone using DNA aptamers, *J. Biomol. Tech.*, **22**, pp. 27–36.
21. Bruno, J. G., Carrillo, M. P., Phillips, T., and Edge, A. (2012). Serum inverts and improves the fluorescence response of an aptamer beacon to various vitamin D analytes, *Luminescence*, **27**, pp. 51–58.
22. Bruno, J. G., Carrillo, M. P., Phillips, T., Hanson, D., and Bohmann, J. A. (2011). DNA aptamer beacon assay for C-telopeptide and handheld fluorometer to monitor bone resorption, *J. Fluoresc.*, **21**, pp. 2021–2033.
23. Bruno, J. G., Carrillo, M. P., Phillips, T., Vail, N. K., and Hanson, D. (2008). Competitive FRET-aptamer-based detection of methylphosphonic acid: a common nerve agent metabolite, *J. Fluoresc.*, **18**, pp. 867–876.
24. Bruno, J. G., Carrillo, M. P., Richarte, A. M., Phillips, T., Andrews, C., and Lee, J. S. (2012). Development, screening, and analysis of a small DNA aptamer library potentially useful for diagnosis and passive immunity of arboviruses, *BMC Res. Notes*, **5**, p. 633.
25. Bruno, J. G., and Crowell, R. (2008). Selective glutaraldehyde-mediated coupling of proteins to the 3′ adenine terminus of polymerase chain reaction products, *J. Biomol. Tech.*, **19**, pp. 177–183.
26. Bruno, J. G., Phillips, T., Montez, T., Garcia, A., Sivils, J. C., Mayo, M. W., and Greis, A. (2015). Development of a fluorescent enzyme-linked DNA aptamer-magnetic bead sandwich assay and portable fluorometer for ultrasensitive and rapid *Listeria* detection, *J. Fluoresc.*, **25**, pp. 173–183.
27. Bruno, J. G., Richarte, A. M., Carrillo, M. P., Edge, A. (2012). An aptamer beacon responsive to botulinum toxins, *Biosens. Bioelectron.*, **31**, pp. 240–243.
28. Bruno, J. G., Richarte, A. M., and Phillips, T. (2014). Preliminary development of a DNA aptamer-magnetic bead capture electrochemiluminescence sandwich assay for brain natriuretic peptide, *Microchem. J.*, **115**, pp. 32–38.

29. Bruno, J. G., Richarte, A. M., Savage, A. A., and Sivils, J. C. (2014). Development and characterization of DNA aptamers which bind kinesins from *Leishmania* promastigotes, *J. Bionanosci.,* **8**, pp. 195–206.
30. Bruno, J. G. (2013). A review of therapeutic aptamer conjugates with emphasis on new approaches, *Pharmaceuticals*, **6**, pp. 340–357.
31. Bruno, J. G. (2015). Predicting the uncertain future of aptamer-based diagnostics and therapeutics, *Molecules,* **20**, pp. 6866–6887.
32. Burgess, R. (2012). Future concepts in nanomedicine, in *Understanding Nanomedicine: An Introductory Textbook* (Burgess, R., ed.), pp. 399–424, Pan Stanford Publishing, Singapore.
33. Cavaliere, R., Ciocatto, E. C., Giovanella, B. C., Heidelberger, C., Johnson, R. O., Margottini, M., Mondovi, B., Moricca, G., and Rossi-Fanelli, A. (1967). Selective heat sensitivity of cancer cells. Biochemical and clinical studies, *Cancer*, **20**, pp. 1351–1381.
34. Chen, M. L., He, Y. J., Chen, X. W., and Wang, J. H. (2012). Quantum dots conjugated with Fe_3O_4-filled carbon nanotubes for cancer-targeted imaging and magnetically guided drug delivery, *Langmuir*, **28**, pp. 16469–16476.
35. Chen, X., Chen, H., Tripisciano, C., Jedrzejewska, A., Rümmeli, M. H., Klingeler, R., Kalenczuk, R. J., Chu, P. K., and Borowiak-Palen, E. (2011). Carbon-nanotube-based stimuli-responsive controlled-release system, *Chemistry*, **17**, pp. 4454–4459.
36. Chen, Z. Y., Wang Y. X., Lin Y., Zhang J. S., Yang, F., Zhou, Q. L., and Liao, Y. Y. (2014). Advance of molecular imaging technology and targeted imaging agent in imaging and therapy, *BioMed. Res. Int.*, **2014**, Article ID 819324.
37. Cheng, C., Dong, J., Yao, L., Chen, A., Jia, R., Huan, L., Guo, J., Shu, Y., and Zhang, Z. (2008). Potent inhibition of human influenza H5N1 virus by oligonucleotides derived by SELEX, *Biochem. Biophys. Res. Commun.*, **366**, pp. 670–674.
38. Christophi, C., Winkworth, A., Muralihdaran, V., and Evans, P. (1998). The treatment of malignancy by hyperthermia, *Surg. Oncol.*, **7**, pp. 83–90.
39. Cload, S. T., McCauley, T. G., Keefe, A. D., Healy, J. M., and Wilson, C. (2006). Properties of therapeutic aptamers, in *The Aptamer Handbook* (Klussmann, S., ed.), pp. 363–416, WILEY-VCH Verlag GmBH & Co., Weinheim, Germany.

40. Conidi, A., van den Berghe, V., and Huylebroeck, D. (2013). Aptamers and their potential to selectively target aspects of EGF, Wnt/β-catenin and TGFβ-smad family signaling, *Int. J. Mol. Sci.*, **14**, pp. 6690–6719.

41. Da Pieve, C., Blackshaw, E. Missailidis, S., and Perkins, A. C. (2012). PEGylation and biodistribution of an anti-MUC1 aptamer in MCF-7 tumor-bearing mice, *Bioconj. Chem.*, **23**, pp. 1377–1381.

42. Dalziel, M., Crispin, M., and Dwek, R. A. (2013). Directing stem cell differentiation with antibodies, *Proc. Natl. Acad. Sci. U S A*, **110**, pp. 17608–17609.

43. Dougan, H., Lyster, D. M., Vo, C. V., Stafford, A., Weitz, J. I., and Hobbs, J. B. (2000). Extending the lifetime of anticoagulant oligodeoxynucleotide aptamers in blood, *Nucl. Med. Biol.*, **27**, pp. 289–297.

44. Douglas, S. M., Bachelet, I., and Church, G. M. (2012). A logic-gated nanorobot for targeted transport of molecular payloads, *Science,* **335**, pp. 831–834.

45. Eulberg, D., Jarosch, F., Vonhoff, S., and Klussmann, S. (2006). Spiegelmers for theapeutic applications: use of chiral principles in evolutionary selection techniques, in *The Aptamer Handbook* (Klussmann, S., ed.), pp. 417–442, WILEY-VCH Verlag GmbH & Co., Weinheim, Germany.

46. Fan, S., Wu, F., Martiniuk, F., Hale, M. L., Ellington, A. D., and Tchou-Wong, K. M. (2008). Protective effects of anti-ricin A-chain RNA aptamer against ricin toxicity, *World J. Gastroenterol.*, **14**, pp. 6360–6365.

47. Ferreira, C. S., Cheung, M. C., Missailidis, S., Bisland, S., and Gariépy, J. (2009). Phototoxic aptamers selectively enter and kill epithelial cancer cells, *Nucleic Acids Res.*, **37**, pp. 866–876.

48. Gannon, C. J., Cherukuri, P., Yakobson, B. I., Cognet, L., Kanzius, J. S., Kittrell, C., Weisman, R. B., Pasquali, M., Schmidt, H. K., Smalley, R. E., and Curley, S. A. (2007). Carbon nanotube-enhanced thermal destruction of cancer cells in a noninvasive radiofrequency field, *Cancer,* **110**, pp. 2654–2665.

49. Glazer, E. S., and Curley, S. A. (2011). Non-invasive radiofrequency ablation of malignancies mediated by quantum dots, gold nanoparticles and carbon nanotubes, *Ther. Delivery*, **2**, pp. 1325–1330.

50. Glazer, E. S., Zhu, C., Massey, K. L., Thompson, C. S., Kaluarachchi, W. D., Hamir, A. N., and Curley, S. A. (2010). Noninvasive radiofrequency field destruction of pancreatic adenocarcinoma grafts treated with targeted gold nanoparticles, *Clin. Cancer Res.*, **16**, pp. 5712–5721.

51. Glazer, E. S., and Curley, S. A. (2010). Radiofrequency field-induced thermal cytotoxicity in cancer cells treated with fluorescent nanoparticles, *Cancer*, **116**, pp. 3285–3293.

52. Gopinath, S. C., and Kumar, P. K. (2013). Aptamers that bind to the hemagglutinin of the recent pandemic influenza virus H1N1 and efficiently inhibit agglutination, *Acta Biomater.*, **9**, pp. 8932–8941.

53. Green, D., Ruparel, S., Roman, L., Henry, M. A., and Hargreaves, K. M. (2013). Role of endogenous TRPV1 agonists in a postburn pain model of partial-thickness injury, *Pain*, **154**, pp. 2512–2520.

54. Hampel, S., Kunze, D., Haase, D., Krämer, K., Rauschenbach, M., Ritschel, M., Leonhardt, A., Thomas, J., Oswald, S., Hoffmann, V., and Büchner, B. (2008). Carbon nanotubes filled with a chemotherapeutic agent: a nanocarrier mediates inhibition of tumor cell growth, *Nanomedicine (Lond.)*, **3**, pp. 175–182.

55. He, L., Deen, B., Pagel, A. H., Diez-Gonzalez, F., and Labuza, T. P. (2013). Concentration, detection and discrimination of *Bacillus anthracis* spores in orange juice using aptamers based surface enhanced Raman spectroscopy, *Analyst*, **138**, pp. 1657–1659.

56. He, W., Elizondo-Riojas, M. A., Li, X., Lokesh, G. L., Somasunderam, A., Thiviyanathan, V., Volk, D. E., Durland, R. H., Englehardt, J., Cavasotto, C. N., and Gorenstein, D. G. (2012). X-aptamers: a bead-based selection method for random incorporation of druglike moieties onto next-generation aptamers for enhanced binding, *Biochemistry*, **51**, pp. 8321–8323.

57. He, X., Hai, L., Su, J., Wang, K., and Wu, X. (2011). One-pot synthesis of sustained-released doxorubicin silica nanoparticles for aptamer targeted delivery to tumor cells, *Nanoscale*, **3**, pp. 2936–2942.

58. Healy, J. M., Lewis, S. D., Kurz, M., Boomer, R. M., Thompson, K. M., Wilson C., and McCauley, T. G. (2004). Pharmacokinetics and biodistribution of novel aptamer compositions, *Pharm. Res.*, **21**, pp. 2234–2246.

59. Hilger, I., and Kaiser, W. A. (2012). Iron oxide-based nanostructures for MRI and magnetic hyperthermia, *Nanomedicine (Lond.)*, **7**, pp. 1443–1459.

60. Hong, S. Y., Tobias, G., Al-Jamal, K. T., Ballesteros, B., Ali-Boucetta, H., Lozano-Perez, S., Nellist, P. D., Sim, R. B., Finucane, C., Mather, S. J., Green, M. L., Kostarelos, K., and Davis, B. G. (2010). Filled and glycosylated carbon nanotubes for in vivo radioemitter localization and imaging, *Nat. Mater.*, **9**, pp. 485–490.

61. Hu, H., Dai, A., Sun, J., Li, X., Gao, F., Wu, L., Fang, Y., Yang, H., An, L., Wu, H., and Yang, S. (2013). Aptamer-conjugated $Mn_3O_4@SiO_2$ core-shell nanoprobes for targeted magnetic resonance imaging, *Nanoscale*, **5**, pp. 10447–10454.

62. Huang, Y. F., Sefah, K., Bamrungsap, S., Chang, H. T., and Tan, W. (2008). Selective photothermal therapy for mixed cancer cells using aptamer-conjugated nanorods, *Langmuir*, **24**, pp. 11860–11865.

63. Hwang do, W., Ko, H. Y., Lee, J. H., Kang, H., Ryu, S. H., Song, I. C., Lee, D. S., and Kim, S. (2010). A nucleolin-targeted multimodal nanoparticle imaging probe for tracking cancer cells using an aptamer, *J. Nucl. Med.*, **51**, pp. 98–105.

64. Isogai, H., and Hirayama, N. (2013). In silico prediction of interactions between site II on human serum albumin and profen drugs, *ISRN Pharm.*, **2013**, Article ID 818364, 8 pages.

65. Jalalian, S. H., Taghdisi, S. M., Shahidi, Hamedani, N., Kalat, S. A., Lavaee, P., Zandkarimi, M., Ghows, N., Jaafari, M. R., Naghibi, S., Danesh, N. M., Ramezani, M., and Abnous, K. (2013). Epirubicin loaded super paramagnetic iron oxide nanoparticle-aptamer bioconjugate for combined colon cancer therapy and imaging in vivo, *Eur. J. Pharm. Sci.*, **50**, pp. 191–197.

66. Jayasena, S. D. (1999). Aptamers: an emerging class of molecules that rival antibodies in diagnostics, *Clin. Chem.*, **45**, pp. 1628–1650.

67. Jo, H., Her, J., and Ban, C. (2015). Dual aptamer-functionalized silica nanoparticles for the highly sensitive detection of breast cancer, *Biosens. Bioelectron.*, **71**, pp. 129–136.

68. Kanwar, J. R., Roy, K., Maremanda, N. G., Subramanian, K., Veedu, R. N., Bawa, R., Kanwar, R. K. (2015). Nucleic acid-based aptamers: applications, development and clinical trials, *Curr. Med. Chem.*, **22**, pp. 2539–2557.

69. Kavruk, M., Celikbicak, O., Ozalp, V. C., Borsa, B. A., Hernandez, F. J., Bayramoglu, G., Salih, B., and Arica, M. Y. (2015). Antibiotic loaded nanocapsules functionalized with aptamer gates for targeted destruction of pathogens, *Chem. Commun. (Camb.)*, **51**, pp. 8492–8495.

70. Keefe, A. D., Pai, S., and Ellington, A. (2010). Aptamers as therapeutics, *Nat. Rev. Drug Discovery*, **9**, pp. 537–550.

71. Kiel, J. L., and Kiel, A. L. (2013). Fundamental performance differences between antibodies and aptamers, in *Biomedical Applications of Aptamers* (Bruno, J. G., ed.), pp. 9–13, Nova Science Publishers, New York.

72. Kim, D., Jeong, Y. Y., and Jon, S. (2010). A drug-loaded aptamer-gold nanoparticle bioconjugate for combined CT imaging and therapy of prostate cancer, *ACS Nano.*, **4**, pp. 3689–3696.

73. King, D. J., Ventura, D. A., Brasier, A. R., and Gorenstein, D. G. (1998). Novel combinatorial selection of phosphorothioate oligonucleotide aptamers, *Biochemistry*, **37**, pp. 16489–16493.

74. Kobayashi, T. (2011). Cancer hyperthermia using magnetic nanoparticles, *Biotechnol. J.*, **6**, pp. 1342–1347.

75. Kober, A., and Sjöholm, I. (1980). The binding sites on human serum albumin for some nonsteroidal anti-inflammatory drugs, *Molec. Pharmacol.*, **18**, pp. 421–426.

76. Kuwahara, M., and Obika, S. (2013). In vitro selection of BNA (LNA) aptamers, *Artif. DNA PNA XNA*, **4**, pp. 39–48.

77. Kwon, H. M., Lee, K. H., Han, B. W., Han, M. R., Kim, D. H., and Kim, D. E. (2014). An RNA aptamer that specifically binds to the glycosylated hemagglutinin of avian influenza virus and suppresses viral infection in cells, *PLoS One*, **9**, p. e97574.

78. Lao, Y. H., Phua, K. K., and Leong, K. W. (2015). Aptamer nanomedicine for cancer therapeutics: barriers and potential for translation, *ACS Nano*, **9**, pp. 2235–2254.

79. Lauridsen, L. H. and Veedu, R. N. (2012). Nucleic acid aptamers against biotoxins: a new paradigm toward the treatment and diagnostic approach, *Nucleic Acid Ther.*, **22**, pp. 371–399.

80. Lee, J. F., Stovall, G. M., and Ellington, A. D. (2006). Aptamer therapeutics advance, *Curr. Opin. Chem. Biol.*, **10**, pp. 282–289.

81. Li, L. L., Yin, Q., Cheng, J., and Lu, Y. (2012). Polyvalent mesoporous silica nanoparticle-aptamer bioconjugates target breast cancer cells, *Adv. Health Mater.*, **1**, pp. 567–572.

82. Lian, S., Zhang, P., Gong, P., Hu, D., Shi, B., Zeng, C., and Cai, L. (2012). A universal quantum dots-aptamer probe for efficient cancer detection and targeted imaging, *J. Nanosci. Nanotechnol.*, **12**, pp. 7703–7708.

83. Lu, W., Singh, A. K., Khan, S. A., Senapati, D., Yu, H., and Ray, P. C. (2010). Gold nano-popcorn-based targeted diagnosis, nanotherapy treatment, and in situ monitoring of photothermal therapy response of prostate cancer cells using surface-enhanced Raman spectroscopy, *J. Am. Chem. Soc.*, **132**, pp. 18103–18114.

84. Mallikaratchy, P. R., Ruggiero, A., Gardner, J. R., Kuryavyi, V., Maguire, W. F., Heaney, M. L., McDevitt, M. R., Patel, D. J., and Scheinberg, D. A.

(2011). A multivalent DNA aptamer specific for the B-cell receptor on human lymphoma and leukemia, *Nucleic Acids Res.*, **39**, pp. 2458–2469.

85. Mann, A. P., Bhavane, R. C., Somasunderam, A., Liz Montalvo-Ortiz, B., Ghaghada, K. B., Volk, D., Nieves-Alicea, R., Suh, K. S., Ferrari. M., Annapragada, A., Gorenstein, D. G., and Tanaka, T. (2011). Thioaptamer conjugated liposomes for tumor vasculature targeting, *Oncotarget*, **2**, pp. 298–304.
86. Manoharan, M., Inamati, G. B., Lesnik, E. A., Sioufi, N. B., and Freier, S. M. (2002). Improving antisense oligonucleotide binding to human serum albumin: dramatic effect of ibuprofen conjugation, *ChemBioChem*, **12**, pp. 1257–1260.
87. Martincic, M., and Tobias, G. (2015). Filled carbon nanotubes in biomedical imaging and drug delivery, *Expert Opin. Drug Delivery*, **12**, pp. 563–581.
88. Mullis, K. B., Vivekananda J., Kiel, J. L., and Cook, R. M. (2008). Chemically Programmable Immunity. US Patent no. 7,422,746.
89. Murray, E., Gregg, D. A. Norton, M. L., Swick, J. T., and Towler, W. I. (2012). Method for a continuous rapid thermal cycle system. US Patent no. 8,163,489.
90. Najafabadi, M. E., Khayamian, T., and Hashemian, Z. (2015). Aptamer-conjugated magnetic nanoparticles for extraction of adenosine from urine followed by electrospray ion mobility spectrometry, *J. Pharm. Biomed. Anal.*, **107**, pp. 244–250.
91. Nguyen, D. T., Barham, W., Zheng, L., Shillinglaw, B., Tzou, W. S., Neltner, B., Mestroni, L., Bosi, S., Ballerini, L., Prato, M., and Sauer, W. H. (2014). Carbon nanotube facilitation of myocardial ablation with radiofrequency energy, *J. Cardiovasc. Electrophysiol.*, **25**, pp. 1385–1390.
92. Ozalp, V. C., Eyidogan, F., and Oktem, H. A. (2011). Aptamer-gated nanoparticles for smart drug delivery, *Pharmaceuticals,* **4**, pp. 1137–1157.
93. Patwardhan, A. M., Akopian, A. N., Ruparel, N. B., Diogenes, A., Weintraub, S. T., Uhlson, C., Murphy, R. C., and Hargreaves, K. M. (2010). Heat generates oxidized linoleic acid metabolites that activate TRPV1 and produce pain in rodents, *J. Clin. Invest.*, **120**, pp. 1617–1626.
94. Pednekar, P. P., Jadhav, K. R., and Kadam, V. J. (2012). Aptamer-dendrimer bioconjugate: a nanotool for therapeutics, diagnosis, and imaging, *Expert Opin. Drug Delivery*, **9**, pp. 1273–1288.

95. Pinheiro, V. B., and Holliger, P. (2014). Towards XNA nanotechnology: new materials from synthetic genetic polymers, *Trends Biotechnol.*, **32**, pp. 321–328.

96. Raoof, M., Cisneros, B. T., Guven, A., Phounsavath, S., Corr, S. J., Wilson, L. J., and Curley, S. A. (2013). Remotely triggered cisplatin release from carbon nanocapsules by radiofrequency fields, *Biomaterials*, **34**, pp. 1862–1869.

97. Samia, A. C., Dayal, S., and Burda, C. (2006). Quantum dot-based energy transfer: perspectives and potential for applications in photodynamic therapy, *Photochem. Photobiol.*, **82**, pp. 617–625.

98. Sapag, A., Salinas-Luypaert, C., and Constenla-Muñoz, C. (2014). First report of in vitro selection of RNA aptamers targeted to recombinant *Loxosceles laeta* spider toxins, *Biol. Res.*, **47**, p. 2.

99. Savla, R., Taratula, O., Garbuzenko, O., and Minko, T. (2011). Tumor targeted quantum dot-mucin 1 aptamer-doxorubicin conjugate for imaging and treatment of cancer, *J. Controlled Release,* **153**, pp. 16–22.

100. Sefah, K., Yang, Z., Bradley, K. M., Hoshika, S., Jiménez, E., Zhang, L., Zhu, G., Shanker, S., Yu, F., Turek, D., Tan, W., and Benner, S. A. (2014). In vitro selection with artificial expanded genetic information systems, *Proc. Natl. Acad. Sci. U S A*, **111**, pp. 1449–1454.

101. Sellins, K. S., and Cohen, J. J. (1991). Hyperthermia induces apoptosis in thymocytes, *Radiat. Res.*, **126**, pp. 88–95.

102. Shetty, S., and Ghosh, K. (2015). Novel therapeutic approaches to haemophilia, *Haemophlilia*, **21**, pp. 152–161.

103. Shi, L., Hernandez, B., and Selke, M. (2006). Singlet oxygen generation from water-soluble quantum dot-organic dye nanocomposites, *J. Am. Chem. Soc.*, **128**, pp. 6278–6279.

104. Shiratori, I., Akitomi, J., Boltz, D. A., Horii, K., Furuichi, M., and Waga, I. (2014). Selection of DNA aptamers that bind to influenza A viruses with high affinity and broad subtype specificity, *Biochem. Biophys. Res. Commun.*, **443**, pp. 37–41.

105. Silva, A. C., Oliveira, T. R., Mamani, J. B., Malheiros, S. M., Malavolta, L., Pavon, L. F., Sibov, T. T., Amaro, E., Tannús, A., Vidoto, E. L., Martins, M. J., Santos, R. S., and Gamarra, L. F. (2011). Application of hyperthermia induced by superparamagnetic iron oxide nanoparticles in glioma treatment, *Int. J. Nanomed.*, **6**, pp. 591–603.

106. Stecker, J. R., Savage, A., Bruno, J. G., Garcia, D. M., and Koke, J. R. (2012). Dynamics and visualization of MCF7 adenocarcinoma cell death by

aptamer-C1q-mediated membrane attack, *Nucleic Acid Ther.*, **22**, pp. 275–282.

107. Stovall, G. M., Bedenbaugh, R. S., Singh, S., Meyer, A. J., Hatala, P. J., Ellington, A. D., and Hall, B. (2014). In vitro selection using modified or unnatural nucleotides, *Curr. Protoc. Nucleic Acid Chem.*, **56**, pp. 1–9.
108. Suenaga, E., and Kumar, P. K. (2014). An aptamer that binds efficiently to the hemagglutinins of highly pathogenic avian influenza viruses (H5N1 and H7N7) and inhibits hemagglutinin-glycan interactions, *Acta Biomater.*, **10**, pp. 1314–1323.
109. Taghdisi, S. M., Lavaee, P., Ramezani, M., and Abnous, K. (2011). Reversible targeting and controlled release delivery of daunorubicin to cancer cells by aptamer-wrapped carbon nanotubes, *Eur. J. Pharm. Biopharm.*, **77**, pp. 200–206.
110. Tang, Y., Hu, H., Zhang, M. G., Song, J., Nie, L., Wang, S., Niu, G., Huang, P., Lu, G., and Chen, X. (2015). An aptamer-targeting photoresponsive drug delivery system using "off-on" graphene oxide wrapped mesoporous silica nanoparticles, *Nanoscale*, **7**, pp. 6304–6310.
111. Taylor, A. I., Arangundy-Franklin, S., and Holliger, P. (2014). Towards applications of synthetic genetic polymers in diagnosis and therapy, *Curr. Opin. Chem. Biol.*, **22**, pp. 79–84.
112. Tsay, J. M., Trzoss, M., Shi, L., Kong, X., Selke, M., Jung, M. E., and Weiss, S. (2007). Singlet oxygen production by peptide-coated quantum dot-photosensitizer conjugates, *J. Am. Chem. Soc.*, **129**, pp. 6865–6871.
113. Vance, S. A., and Sandros, M. G. (2014). Zeptomole detection of C-reactive protein in serum by a nanoparticle amplified surface plasmon resonance imaging aptasensor, *Sci. Rep.*, **4**, p. 5129.
114. Vivekananda, J., Salgado, C., and Millenbaugh, N. J. (2014). DNA aptamers as a novel approach to neutralize *Staphylococcus aureus* α-toxin, *Biochem. Biophys. Res. Commun.*, **444**, pp. 433–438.
115. Wahajuddin, S. A., and Arora, S. (2012). Superparamagnetic iron oxide nanoparticles: magnetic nanoplatforms as drug carriers, *Int. J. Nanomed.*, **7**, pp. 3445–3471.
116. Wang, K., Gan, L., Jiang, L., Zhang, X., Yang, X., Chen, M., and Lan, X. (2015). Neutralization of Staphylococcal enterotoxin B by aptamer antagonist, *Antimicrob. Agents Chemother.*, **59**, pp. 2072–2077.
117. Whitlam, J. B., Crooks, M. J., Brown, K. F., and Pedersen, P. V. (1979). Binding of non-steroidal anti-inflammatory agents to proteins- I.

Ibuprofen-serum albumin interaction, *Biochem. Pharmacol.*, **28**, pp. 675–678.

118. Xie, J., Zhang, H., Yea, K., and Lerner, R. A. (2013). Autocrine signaling based selection of combinatorial antibodies that transdifferentiate human stem cells, *Proc. Natl. Acad. Sci. U S A*, **110**, pp. 8099–8104.

119. Yaghini, E., Seifalian, A. M., and MacRobert, A. J. (2009). Quantum dots and their potential biomedical applications in photosensitization for photodynamic therapy, *Nanomedicine*, **4**, pp. 353–363.

120. Yang, X., Huang, J., Wang, K., Li, W., Cui, L., and Li, X. (2011). Angiogenin-mediated photosensitizer-aptamer conjugate for photodynamic therapy, *Chem. Med. Chem.*, **6**, pp. 1778–1780.

121. Yasun, E., Kang, H., Erdal, H., Cansiz, S., Ocsoy, I., Huang, Y. F., and Tan, W. (2013). Cancer cell sensing and therapy using affinity tag-conjugated nanorods, *Interface Focus*, **3**, p. 20130006.

122. Ye, F., Zheng, Y., Wang, X., Tan, X., Zhang, T., Xin, W., Wang, J., Huang, Y., Fan, Q., and Wang, J. (2014). Recognition of *Bungarus multicinctus* venom by a DNA aptamer against β-bungarotoxin, *PLoS One*, **9**, p. e105404.

123. Yea, K., Zhang, H., Xie, J., Jones, T. M., Yang, G., Song, B. D., and Lerner, R. A. (2013). Converting stem cells to dendritic cells by agonist antibodies from unbiased morphogenic selections, *Proc. Natl. Acad. Sci. U S A*, **110**, pp. 14966–14971.

124. Yoshida, Y., Waga, I., and Horii, K. (2012). Quantitative and sensitive protein detection strategies based on aptamers, *Proteomics Clin. Appl.*, **6**, pp. 574–580.

125. Zhang, H., Hou, L., Jiao, X., Yandan, J., Zhu, X., Hongji, L., Chen, X., Ren, J., Xia, Y., and Zhang, Z. (2014). In vitro and in vivo evaluation of antitumor drug-loaded aptamer targeted single-walled carbon nanotubes system, *Curr. Pharm. Biotechnol.*, **14**, pp. 1105–1117.

126. Zhao, X., Lis, J. T., and Shi, H. (2013). A systematic study of the features critical for designing a high avidity multivalent aptamer, *Nucleic Acid Ther.*, **23**, pp. 238–242.

127. Zhou, J., Soontornworajit, B., Martin, J., Sullenger, B. A., Gilboa, E., and Wang, Y. (2009). A hybrid DNA aptamer–nanomaterial for targeted cell labeling, *Macromol. Biosci.*, **9**, pp. 831–835.

128. Zsila, F., Bikadi, Z., Malik, D., Hari, P., Pechan, I., Berces, A., and Hazai, E. (2011). Evaluation of drug-human serum albumin binding interactions with support vector machine aided on line automated docking, *Bioinformatics*, **27**, pp. 1806–1813.

Index